INSCRIBING SCIENCE

WRITING SCIENCE

EDITORS Timothy Lenoir and Hans Ulrich Gumbrecht

CONTRIBUTORS

Gillian Beer

Lisa Bloom

Robert Brain

Lorraine Daston

Richard M. Doyle

David Gugerli

Hans Ulrich Gumbrecht

Friedrich Kittler

Timothy Lenoir

Alex Pang

Phillip Prodger

Hans-Jörg Rheinberger

Robin Rider

Brian Rotman

Simon Schaffer

Bernhard Siegert

INSCRIBING SCIENCE

Scientific Texts and the
Materiality of Communication

EDITED BY

Timothy Lenoir

STANFORD UNIVERSITY PRESS

STANFORD, CALIFORNIA 1998

Stanford University Press
Stanford, California
© 1998 by the Board of Trustees of the
Leland Stanford Junior University
Printed in the United States of America

CIP data are at the end of the book

Contents

❖❖

Contributors   xiii

1. Inscription Practices and Materialities of Communication   1
   Timothy Lenoir

2. The Language of Strange Facts in Early Modern Science   20
   Lorraine Daston

3. Shaping Information: Mathematics, Computing, and Typography   39
   Robin Rider

4. The Technology of Mathematical Persuasion   55
   Brian Rotman

5. On the Take-off of Operators   70
   Friedrich Kittler

6. Switchboards and Sex: The Nut(t) Case   78
   Bernhard Siegert

7. Politics on the Topographer's Table: The Helvetic Triangulation of Cartography, Politics, and Representation   91
   David Gugerli

8. Writing Darwin's Islands: England and the Insular Condition   119
   Gillian Beer

9. Illustration as Strategy in Charles Darwin's *The Expression of the Emotions in Man and Animals*   140
   Phillip Prodger

10. The Leviathan of Parsonstown: Literary Technology and
    Scientific Representation   182
    Simon Schaffer

11. Technology, Aesthetics, and the Development of
    Astrophotography at the Lick Observatory   223
    Alex Pang

12. Standards and Semiotics   249
    Robert Brain

13. Experimental Systems, Graphematic Spaces   285
    Hans-Jörg Rheinberger

14. Emergent Power: Vitality and Theology in Artificial Life   304
    Richard M. Doyle

15. Science and Writing: Two National Narratives of Failure   328
    Lisa Bloom

16. Perception Versus Experience: Moving Pictures and Their
    Resistance to Interpretation   351
    Hans Ulrich Gumbrecht

    Notes   367

    Index   443

Illustrations

FIGURES

2.1. German Broadside of Monster Born in Florence on
St. Jacob's Day, 1506   28

2.2. Celestial Light in the Constellations of Aries and Taurus,
Observed by Cassini, 1683   33

2.3. John Winthrop's *Piscis Echino-stellaris visciformis*   34

3.1. Gothic Bookhand   40

3.2. Dürer's Constructed Roman Alphabet   42

3.3. Dürer's Constructed Alphabet   44

3.4. Jammes's Alphabet: Capital Letters   46

3.5. Jammes's Alphabet: Lower-case Letters   47

3.6. Fournier's Scale of Point Sizes   49

3.7. Samples of Computer Modern Under Development   53

7.1. Federal Expenditures for Domestic Surveying, 1810–1865   100

7.2. Measurement of the Space Between Measuring Rods   103

7.3. Example of a Survey Protocol on the Baseline
Measurement in Aarburg   104

7.4. Stages of Transposition in the Cartographic Recording
System   107

7.5. *Triangulation primordiale de la Suisse*   112

7.6. Annually Printed Folios of the Dufour Map, 1845–1865   114

9.1. Plate 7 from Darwin's *Expression of the Emotions*   145

9.2. "Wonder and Astonishment" from Bell's *Anatomy and Philosophy of Expression*   153

9.3. "Cynopithecus niger, in a Placid Condition" and "The Same, when Pleased by Being Caressed"   159

9.4. "Discontent, Bad Humor"   167

9.5. "Horror and Agony"   168

9.6. Darwin's Interpretation of "Horror and Agony"   169

9.7. Oscar Rejlander's "Two Ways of Life," 1856   172

9.8. "Mental Distress"   174

9.9. The Original Photograph from Which "Mental Distress" Was Drawn   175

9.10. "Sneering"   176

9.11. "South Australian Aboriginal Female," ca. 1870   178

9.12. "Disgust"   179

10.1. The Leviathan of Parsonstown   188

10.2. William Herschel's 40-foot Reflecting Telescope at Slough   191

10.3. John Herschel's Drawing of the Orion Nebula   202

10.4. The Parsonstown Picture of a Region of the Orion Nebula, 1867   205

10.5. Rosse's Machine for Polishing Telescope Specula   210

12.1. Marey-Rosapelly Vocal Polygraph   262

12.2. Vocal Polygraph Inscriptions of Sanskrit *yamas*   263

12.3. Scott Phonautograph   265

12.4. Koenig Manometric Flame Apparatus   265

12.5. Photograph of Manometric Flames   266

12.6. Demeny Photophone   268

12.7. Edison Business Phonograph   269

12.8. Graphic Recordings of Patois   272

12.9. Rousselot Speech Inscriptor   273

12.10. Livi Anthropometric Cartography   275

12.11. Musée Phonographique Survey Form   278

13.1. Fractionation Diagram   293

13.2. Electron Micrographs of Microsome Fractions   299

13.3. Ultracentrifugal Analysis of Microsomes   300

13.4. Effect of Sodium Deoxycholate Concentration on the
       Recovery of RNA and Protein   301

14.1. Conway's Life   312

14.2. Glider   315

TABLE

9.1 Selected Treatises on Expression, and Techniques
     Used to Illustrate Them   152

Contributors

Gillian Beer is King Edward VII Professor of English Literature and President, Clare Hall, the University of Cambridge. She is the author of *Open Fields: Science in Cultural Encounter* (Oxford: Oxford University Press, 1996) and *Darwin's Plots: Evolutionary Narrative in Darwin, George Eliot, and Nineteenth-Century Fiction* (London and Boston: Ark Paperbacks, 1985).

Lisa Bloom is Assistant Professor at the San Francisco State University in the Inter-Arts Program. She is the author of *Gender on Ice* (Minneapolis: University of Minnesota Press, 1993).

Robert Brain is Assistant Professor of the History of Science at Harvard University. He is the author of *The Graphic Method: Inscription, Visualization, and Measurement in Nineteenth-Century Science and Culture*, forthcoming from the University of Chicago Press.

Lorraine Daston is Professor of History and Philosophy of Science and Co-Director of the Max-Planck-Institute for the History and Philosophy of Science, Berlin. She is the author of *The Vertigo of Scientific Progress* (Berlin: Max-Planck-Institut für Wissenschaftsgeschichte, 1995), *Ravening Curiosity and Gawking Wonder in the Early Modern Study of Nature* (Berlin: Max-Planck-Institut für Wissenschaftsgeschichte, 1994), and *Classical Probability in the Enlightenment* (Princeton, N.J.: Princeton University Press, 1988).

Richard M. Doyle is Assistant Professor in the Department of English at Pennsylvania State University. He is the author of *On Beyond Living* (Stanford, Calif.: Stanford University Press, 1997).

David Gugerli is Professor of the History of Technology at the Eidgenossisches Technische Hochschule (ETH) in Zurich. He is the author of *Redeströme: Zur Elektrifizierung der Schweiz 1880–1914* (Zurich: Chronos Verlag, 1996), and with Rudolf Braun coauthor of *Macht des Tanzes, Tanz der Mächtigen* (Munich: C. H. Beck, 1993).

Hans Ulrich Gumbrecht is Albert Guerard Professor of Comparative Literature at Stanford University. He is the author of the two-volume *Eine Geschichte der Spanischen Literatur* (Frankfurt am Main: Suhrkamp, 1990) and *In 1926* (Cambridge, Mass.: Harvard University Press, 1996).

Friedrich Kittler is Professor of Comparative Literature and Communications at the Humboldt University Berlin. He is the author of *Discourse Networks, 1800/1900* (Stanford, Calif.: Stanford University Press, 1990) and *Gramophone, Film, Typewriter* (Stanford, Calif.: Stanford University Press, forthcoming).

Timothy Lenoir is Professor of History of Science and Technology and Chair of the Program in History and Philosophy of Science at Stanford University. He is the author of *Politik im Tempel der Wissenschaft* (Frankfurt/Main: Campus Verlag, 1992) and *Instituting Science: The Cultural Production of Scientific Disciplines* (Stanford, Calif.: Stanford University Press, 1997).

Alex Pang is Editor for Science and Technology of the Encyclopaedia Britannica Online, Chicago, Illinois.

Phillip Prodger is a graduate student in Art History at the University of Cambridge completing his dissertation on the history of photography.

Hans-Jörg Rheinberger is Professor of Molecular Biology and of the History and Philosophy of Science and Co-Director of the Max Planck Institute for the History and Philosophy of Science, Berlin. He is the author of *Toward a History of Epistemic Things: Synthesizing Proteins in the Test Tube* (Stanford, Calif.: Stanford University Press, 1997).

Robin Rider is Curator of Special Collections, University of Wisconsin, Madison. She is the author of *The Quantifying Spirit in the Eighteenth Century* (Berkeley: University of California Press, 1990). With Henry Lowood, she is currently coauthoring *Publishing the Book of Nature*, to be published by the University of California Press.

Brian Rotman is Professor of Interdisciplinary Humanities at Louisiana State University. He is the author of *Signifying Nothing: The Semiotics of Zero* (Stanford, Calif.: Stanford University Press, 1993) and *Ad infinitum . . . The Ghost in Turing's Machine: Taking God Out of Mathematics and Putting the Body Back In: An Essay in Corporeal Semiotics* (Stanford, Calif.: Stanford University Press, 1993). He is currently exploring the material semiotics of virtual worlds.

Simon Schaffer is Lecturer in History and Philosophy of Science at the University of Cambridge. With Steven Shapin he is the coauthor of *Leviathan and the Air Pump* (Princeton, N.J.: Princeton University Press, 1987).

Bernhard Siegert is Assistant Professor of Comparative Literature and Communication at the Humboldt University Berlin. He is the author of *Relays: Literature as an Epoch in the Postal System* (Stanford, Calif.: Stanford University Press, forthcoming).

INSCRIBING SCIENCE

Inscription Practices and Materialities of Communication

Metaphors of inscription and writing figure prominently in all levels of discourse in and about science. The description of nature as a book written in the language of mathematics has been a common trope since at least the time of Galileo, a metaphor supplemented in our own day by the characterization of DNA sequences as code for the book of life, decipherable in terms of protein semantic units. An important recent direction in the fields of science and literature studies is to consider such descriptions as more than metaphoric, as revelatory of the processes of signification in science more generally. Two icons of this direction in science studies are Shapin and Schaffer's *Leviathan and the Air Pump*, which identifies a constitutive moment of early modern science in Boyle's efforts to construct a literary technology to facilitate the virtual witnessing of scientific facts, and Woolgar and Latour's *Laboratory Life*, which describes the modern scientific laboratory as an organized site for persuasion through literary inscription.[1] Similar concerns are evident within the field of literature studies. For instance, a recent bibliography spans some 768 titles.[2] Nearly everywhere we look, the "semiotic turn" is upon us.

In a certain sense, the semiotic turn is not new. Considerations about language, whether Kuhn-inspired interest in quantitative linkages between scientific publications or concerns about Wittgensteinian language games and forms of life, have always been part of science studies in one form or another. These are not the sources of the recent semiotic turn, but they point us in the right direction. If there are any "origins" of the recent turn, I would trace them to Paul Feyerabend's seminal work *Against Method*.[3] Those of us who were moved by Feyerabend read his work as an injunction to move away from theory-dominated accounts of knowledge production in science and toward an account sensitive to actual scientific practice, in which theory was simply one of the many important games in town, with experimenters and crafters of instruments and techniques being crucial but silenced laborers in the production of knowledge. The rehabilitation of skill and craft knowledge (even in the domains of theory and mathematical and computational practice), concerns about tacit knowledge and unarticulable skill, experimenter's regress, interpretive flexibility, and negotiated closure of debate all contributed to newer accounts of science as a disunified, heterogeneous congeries of activities.

The emphasis on practice and on the local context of investigation initiated by the first generation of lab studies prompted a new wave of inquiries into the ways these different domains of practice mesh with one another locally and translate globally to other sites. These inquiries have included exploration of the "articulation work" necessary in linking up different social worlds, as well as examination of how networks of heterogenous actors, practices, and different social worlds, including industry and markets, are knit together in usable, effective packages. Other lines of work have led directly from considerations of science as practice to the view of science as culture studies. Some of these studies have verged on the semiotic approach that interests me. It is evident, I think, that the farther we have moved down this path, the more we have come to embrace the view that science should be viewed within the framework of cultural production.

Most recently the field has been polarized by criticism from the scientific community, beginning with Paul Gross and Norman Levitt's 1994 book *Higher Superstition* and reaching a sort of zenith (or nadir, depending on how you view it) with an article published in the journal *Social Text* in May 1996 by Alan Sokal, a New York University physicist fed up with what he saw as the excesses of the academic left.[4] Sokal hoodwinked *Social Text* into publishing a parody thick with gibberish as though it were serious scholarly work. The article's title: "Transgressing the Boundaries: Toward a Transformative Hermeneutics of Quantum Gravity."

The dispute over the article—which was read by several editors at the journal before it was published—goes to the heart of the public debate over left-wing scholarship, and particularly over the belief that social, cultural, and political conditions influence and may even determine knowledge and ideas about truth. Sokal's critique has become a cause célèbre in what have been dubbed the "culture wars," the battles over multiculturalism, college curricula, and whether there is a single objective truth. Sokal, Gross, Levitt, and others have argued that there is truth or at least an approach to truth, and that scholars have a responsibility to pursue it. They have accused the academic left of debasing scholarship for political ends. "While my method was satirical, my motivation is utterly serious," Professor Sokal wrote in a separate article in the magazine *Lingua Franca*, in which he revealed the hoax and detailed his "intellectual and political" motivations.[5] "What concerns me," he explained, "is the proliferation, not just of nonsense and sloppy thinking per se, but of a particular kind of nonsense and sloppy thinking: one that denies the existence of objective realities." Professor Sokal, who describes himself as "a leftist in the old-fashioned sense," says he worries that trendy disciplines and obscure jargon could end up hurting the leftist cause. "By losing contact with the real world, you undermine the prospect for progressive social critique." Apparently putting his career as a physicist on hold while cleansing the universities of cultural studies, Professor Sokal recently announced he is completing a book aimed at exposing the errors that have led to our relativist malaise by tracing them to their source in the bastion of postmodern philosophy itself, namely, in the works of Derrida, Baudrillard, Virilio, Lyotard, and Foucault.

The essays in this volume provide concrete examples that counter the views of Sokal and others who find it ludicrous to think of analyzing science as a form of cultural production. The issues raised in these debates are quite important, and my purpose here is to defend the role of postmodern philosophies in science studies. A number of scholars with diverse interests—historical laboratory studies of protein biosynthesis, literary studies on science fiction and artificial life, history and philosophy of mathematics, historical work on psychophysics and linguistics—have all begun to find inspiration in the theoretical writings of Derrida and other "postmodern" philosophers. At the same time, scholars from the side of literature studies have begun to focus on the role of rhetorical practice and techniques of persuasion in scientific texts, on narrative structures and metaphor in the internal structure of scientific work, and on the semiosis among scientific narratives and grand cultural narratives, represented in literature, museum exhibits, and popular culture, as means for the construction and stabilization of scientific artifacts.[6] Finally, relevant to our concerns here have been studies of literature and media motivated by Derrida, Lacan, and Foucault but extending their work in interesting ways by emphasizing the materiality of literary and scientific inscriptions—graphic traces as well as the media for producing signs such as standardized paint pigments, photographic equipment, and phonographs—as a precondition for and constraint upon other forms of literal and literary sense-making. A number of the essays in this volume are interested—and it is important in the current debates—in articulating the rationale that these diverse fields have located in the writings of Derrida and in offering reasons for seeing a common thread linking Derrida's philosophical concerns with the interests of scholars of scientific practice and cultural studies of science. To put our thesis in its simplest terms: Derrida and other "postmodern" philosophers offer a way to reframe and move beyond the issues of realism and relativism that have so completely paralyzed recent debates and that led Feyerabend, more than two decades ago, to throw his hands up in disgust and suggest we pursue another course. Taking up Feyerabend's challenge, my proposal is that while these different approaches from poststructuralist semiotics and literature studies may seem strange companions for science studies, when brought together under what I am calling studies of the materialities of communication, this new synthesis offers a fruitful orientation for moving beyond the impasse currently represented by the "culture wars." In the remainder of this introduction, I analyze the relationship of these approaches to our purposes.

Despite its often forbidding language, Derrida's project in *Grammatology* poses important questions relevant to science studies. A central theme of Derrida's writings is his critique of Western logocentrism, the idea, which Derrida associates with Plato in particular but which is present nearly everywhere in Western philosophy, of the possibility of unmediated presence of the truth/logos, and of an originary, unitary source of truth, a "transcendental signified." Within this tradition speech has always been privileged as a more primary (direct) form of communication than writing, speech being considered to directly symbolize "ideas," while the written sign since Plato has been regarded as a second-order sign, the sign that stands in for speech. Writing is thus the sign of a sign. This leads Derrida to formulate the deconstructionist project:

The *signatum* always referred, as to its referent, to a *res*, to an entity created or at any rate first thought and spoken, thinkable and speakable in the eternal present of the divine logos. . . . If it came to relate to the speech of a finite being (created or not) through the intermediary of a *signans*, the *signatum* had an immediate relationship with the divine logos which thought it within presence and for which it was not a trace. And for modern linguistics, if the signifier is a trace, the signified is a meaning thinkable in principle within the full presence of an intuitive consciousness. The signified face, to the extent that it is still originarily distinguished from the signifying face, is not considered a trace; by rights it has no need of the signifier to be what it is.[7]

For philosophers such as Plato, Aristotle, Hegel, Husserl, and Saussure, the written sign was considered a mere supplement, a storage medium to assist memory inessential to the meaning of language. But Derrida challenges the claim to priority of speech over writing and reverses the role of the written sign/supplement. In the passage above, for instance, he calls into question the notion that the signified "face" has an existence independent of the signifier, questions indeed whether the signified is anything other than the signifier, the trace, and suggests that the signified may itself be a trace of the signifier. Writing, embedded within an entire economy of signs, is thus constitutive of meaning rather than a passive medium for restoring the presence of language to thought. This notion launches Derrida's deconstructive enterprise:

This reference to the meaning of a signified thinkable and possible outside of all signifiers remains dependent upon the onto-theo-teleology that I have just evoked. It is thus the idea of the sign that must be deconstructed through a meditation upon writing which would merge, as it must, with the undoing [*sollicitation*] of onto-theology, faithfully repeating it in its totality and making it insecure in its most assured evidences.[8]

Derrida concludes that once it becomes plain that the signified is indistinguishable from the trace, that, as he phrases it, "the trace affects the totality of the sign in both its faces," once, that is, it is realized that the signified is originally and essentially trace, that it is always already in the position of the signifier, the metaphysics of presence must recognize writing as its source.

Derrida's concern with essentialism and the metaphysics of presence parallels recent directions in science studies, particularly discussions of scientific practice and the role of instruments in scientific work, and discussions surrounding social constructivism that insist that truth not be considered as an objective, socially independent reality. Similar to Derrida's concerns about characterizations of writing as a second-order signification and the metaphysics of presence, a central plank of science studies has been the critique of all characterizations of the relation of theory to its object that regard the scientific instrument and experimental system as a passive and transparent medium through which the truth or presence of the object is to be achieved. The instrument is no longer regarded as simply an extension of theory, a mere supplement, useful for exteriorizing an ideal meaning contained within theory. When we treat the experimental system as a model of the theory, we no longer tend to regard it simply as an expression, an unproblematic translation of the ideal relations and entities of the theory into the representative hardware-language of the experimental system. Furthermore, if we consider the web of instrumentalities that mediate and stabilize our interactions with nature,[9] then rather than treating knowledge as stabilized by reference to an independent, objective reality antecedent to scientific work, we are free to develop a pragmatic realism based on the representations of nature as articulated through the technologies of experiment and intervention. From this perspective, it is through our machines that practices and simultaneously a nature capable of being theorized are stabilized.

The relevance to our project of Derrida's philosophical critique goes beyond the coincidence of his concern to reject any claim to knowledge based on an originary, unmediated communication. No less relevant is Derrida's concern with foregrounding the constitutive power of inscription, a move that parallels our interest in mediating machines. Applied to the domain of science studies, Derrida's manner of proceeding suggests that we should attend to the empirical, material character of the experimental system as bound up with the production of a graphic trace, a *grapheme* as he terms it.[10] To appreciate the relevance of this strategy to our consideration of scientific inscription, we should consider Derrida's discussion of writing machines.

Derrida begins by calling attention to Saussure's frustrations with the graphic image and written signs, which have "always been considered by Western tradition as the body and matter external to the spirit, to breath, to speech, and to the logos."[11] In a chapter on the prestige of writing and the reasons for its ascendency over the spoken word, Saussure complained that while the object of study in linguistics should be the spoken word alone,

the written word is so intimately connected with the spoken word it represents that it manages to usurp the principal role. As much or even more importance is given to this representation of the vocal sign as to the vocal sign itself. It is rather as if people believed that in order to find out what a person looks like it is better to study his photograph than his face.[12]

Saussure cites a number of reasons explaining how the written sign could usurp the power of the vocal sign. "The written form of a word strikes us as a permanent, solid object and hence more fitting than its sound to act as a linguistic unit persisting through time."[13] He also speculates that visual images are more lasting than auditory impressions, allowing writing to take over in the end. His third reason, however, brings Saussure to contemplate directly a social construction of the power of inscription. A literary language, he notes, can enhance the power of writing to control speech, for it has its dictionaries, grammar books, and rules for spelling taught from books in schools: "It is a language which appears to be governed by a code, and this code is itself a written rule, itself conforming to strict norms—those of orthography. That is what confers on writing its primordial importance. In the end, the fact that we learn to speak before learning to write is forgotten, and the natural relation between the two is reversed."[14]

Here we see the problem. The written sign is in fact an institution, backed up by other texts and embedded in a network of enforceable codes. The problem Saussure was left to ponder was that this artificial sign, this supplement of a natural connection between word and sound, could exercise a form of tyranny over nature: "But the tyranny of the written form extends further yet. Its influence on the linguistic community may be strong enough to affect and modify the language itself. That happens only in highly literate communities, where written documents are of considerable importance. In these cases the written form may give rise to erroneous pronunciations."[15] Whereas Saussure sees this usurpation by the written sign as unnatural, a literal sin against the primacy of meaning, Derrida celebrates it as crucial to his grammatological project, which seeks to erase discussions of meanings given prior to the act of inscription.

Derrida focuses on the exteriority of meaning, the materiality of the signifier. Accordingly, this exterior materiality is not molded to the demands of either pre-given objective reality or already constituted meaning; it resists and imposes its own constraints on the production of meaning. As Derrida notes, "the outside of indication/indexicality does not come to affect in a merely accidental manner the inside of expression. Their interlacing . . . is originary."[16] In Derrida's terms, there is never a "mere" supplement; the supplement—the specific characteristics of material media (in the case of technoscience, the materiality of the experimental system)—is a necessary constituent of the representation. Like Derrida, laboratory studies observe the striking congruence between literary inscriptions and "facts": discussions about facts are inseparable from their inscriptions; the acceptance of a scientific fact is tied to the strength of its links to layers of texts; the ostensibly factual nature of a statement can be undermined by drawing attention to the process of its inscription.[17] Equally central to recent laboratory studies is the dispersed character of inscription devices: what was once a piece of theory in the literature of one field is now reified as a technique or inscription device in another.[18] This too reflects a further dimension of Derrida's grammatological project: his insistence on the deferred/differed character of reference, the constitution of meaning through an unending process of slippage in a web of intertextuality, a process reminiscent of literary inscription in science.

At first glance, except for articulating an interesting philosophical conundrum, Derrida's insistence that we consider the materiality of the signifier and its implications for constructing meaning as an endless process of textual difference may not seem to translate easily into our interests with scientific inscriptions. Derrida seems more concerned with transforming linguistics into a generalized semiotics. Scientific practice, on the other hand, while bound up with the production of texts, claims to go beyond the "text." To grant Derrida's ideas wider applicability, we need to ask: is science a textual system? How, for instance, might the contents of thought be molded by the supplement? How would such an apparently speculative notion as the supplement look in concrete writing practice?

Derrida's discussion of Freud and the scene of writing provides interesting direction in answering these questions.[19] Derrida examines Freud's use of writing metaphors in the evolution of his theories of memory, dream work, and the psychical apparatus from the "Project" (1895) to the "Note on the Wunderblock (Mystic Writing Pad)" (1925). As he refines his model of the psychic apparatus in the different versions of his work, Freud draws upon analogies to optical instruments (e.g., the microscope and telescope), inscription devices (e.g., photographic apparatus), and systems of writing (e.g., hieroglyphics). Although it is not a theme of Derrida's essay, Freud retraces most of the issues about signification found in Saussure: the arbitrariness of the sign, the sign consisting of a union of signified (latent dream content) and signifier (manifest dream), the articulation of the sign as a process of spatial difference rather than one-to-one reference to an "essence." For Derrida, the most important feature of Freud's treatment of dream work is his insistence that the dream is not analogous to language, but rather to a system of writing, so that interpreting the dream resembles deciphering a pictographic script. Like Derrida himself, Freud treats the dream as a kind of writing in which speech is subordinated to graphic representation, including the treatment of logical connectives.[20] Freud, in short, was a grammatologist.

In "Freud and the Scene of Writing," Derrida points to the role of inscription devices and media technologies as metaphors and analogies useful to Freud's theorizing about psychic machinery rather than as constitutive of its content. At the end of the essay Derrida muses on where one might take this analysis if one were to pay closer attention to the implications of the materiality of the signifier. Among the directions he sees as fruitful are a history of writing and

a becoming-literary of the literal. Here, despite several attempts made by Freud and certain of his successors, a psychoanalysis of literature respectful of the originality of the literary signifier has not yet begun, and this is surely not an accident. Until now, only the analysis of literary *signifieds*, that is, nonliterary signified meanings, has been undertaken. But such questions refer to the entire history of literary forms themselves, and to the history of everything within them which was destined precisely to authorize this disdain of the signifier.[21]

A study of literature that focused on the materiality of the signifier, on the technologies of inscription, and on the practices of writing would be profoundly different from one focused on "meaning" and hermeneutics. Derridean grammatology sees the very possibilities of an "intentional consciousness" as emerging from a particular order of inscription:

Since "genetic inscription" and the "short programmatic chains" regulating the behavior of the amoeba or the annelid up to the passage beyond alphabetic writing to the orders of the logos and of a certain *homo sapiens*, the possibility of the *grammè* structures the movement of its history according to rigorously original levels, types, and rhythms. But one cannot think them without the most general concept of the *grammè*. . . . One could speak . . . of an exteriorization always already begun but always larger than the trace which, beginning from the elementary programs of so-called "instinctive" behavior up to the constitution of electronic card-indexes and reading machines, enlarges *différance* and the possibility of putting in reserve: it at once and in the same movement constitutes and effaces so-called conscious subjectivity, its logos, and its theological attributes.[22]

Derrida proposes that the conditions of writing make history.

The literary project envisioned by Derrida is taken up in large measure by Friedrich Kittler in *Discourse Networks, 1800/1900.* Kittler's premises are that literature is a form of data processing, storage, and transmission, and that writing is a channel of information transmitted through a discourse network of institutions, such as schools and universities, connecting books with people. In the discourse network of 1900, according to Kittler, writing in the alphabetic medium of the book was configured with gramophone, film, and typewriter. The juxtaposition of media technologies, psychophysics, and literature produced a transformation of the realms of the symbolic, the imaginary, and the real. In the discourse network of 1800 novelists and playwrights had created worlds by stimulating psychic processes of association, recollection, attention, and hallucination through streams of words. In 1900 these processes were technically implemented in film through techniques of projection and cutting, flashback and closeup.[23] Fantasy was converted into reality, and the figures of the film could be presented in such detail that the realistic was raised to the realm of the fantastic. Film became the imaginary. The new media technologies were not without effect on the content of literature. The writer became a media specialist, a technologist of the letter.[24] Literature after 1900 began to define itself against technological media. When it became possible to transpose texts to other media, for example, turning novels into filmscripts, the criterion for high literature became its inability to be filmed. Writers like Mallarmé renounced the visual imagination; Kafka rejected the idea that an illustrator for his *Metamorphosis* might draw Gregor as an insect. According to Kittler, "Literature thus occupies, with creatures or noncreatures that can only be found in words, the margin left to it by the other media. . . . The symbolic remained, autonomous and imageless as once only God had been."[25] Just as Freud mused on writing machines, writers in the discourse network of 1900 became the captives of media technologies: Henry James dictated to a secretary with a typewriter in order to present "free unanswered speech, diffusion or flight of ideas." Novelists like Joyce and Proust transposed the narrative techniques of film—the tracking shot, the zoom—into their own compositional strategies.

The approach to literature through the materialities of communication and media technologies exemplified by Kittler's work applies no less to the inscription technologies of science and mathematics. The relationship of this thesis to science was apparent to Derrida, although he did not pursue it. In a long footnote to his remark on linear writing and the end of the book, Derrida quotes Leroi-Gouhran's observation that linear writing constituted, during many millennia, independently of its role as conserver of the collective memory, by its unfolding in one dimension alone, the instrument of analysis out of which grew philosophic and scientific thought. The conservation of thought can now be conceived otherwise than in terms of books which will only for a short time keep the advantage of their rapid manageability. A vast 'tape library' with an electronic selection system will in the near future show pre-selected and instantaneously retrieved information. Reading will still retain its importance for some centuries to come in spite of its perceptible regression for most men, but writing [understood in the sense of linear inscription] seems likely to disappear rapidly, replaced by automatic dictaphones. . . . As to the long term consequences in terms of the forms of reasoning, and a return to diffuse and multidimensional thought, they cannot be now foreseen. Scientific thought is rather hampered by the necessity of drawing itself out in typographical channels and it is certain that if some procedure would permit the presentation of books in such a way that the materials of the different chapters are presented simulta-neously in all their aspects, authors and their users would find a considerable advantage. It is absolutely certain that if scientific reasoning has clearly nothing to lose with the disappearance of writing, philosophy and literature will definitely see their forms evolve. This is not particularly regrettable since printing will conserve the curiously archaic forms of thought that men will have used during the period of alphabetic graphism; as to the new forms, they will be to the old ones as steel to flint, not only a sharper but a more flexible instrument.[26]

Although written before computer-based writing and hypertext were familiar objects, Derrida's and Leroi-Gouhran's considerations of the power of electronic media to shape thought have become particularly salient in light of contemporary discussions about digital libraries and the future of the book. We hear frequently how on-screen reading, hypertext, and other dimensions of the new digital media force us to reconsider notions of authorship, text, and even writing itself, but in their reflections Derrida and Leroi-Gouhran suggest as well that media structure the con-tent of thought. Of course, this insight is not new. Pondering the power of alphabetic graphism, Condorcet observed that the material, technical bases of cognitive operations reside in the tables and charts made available by printing, enabling the reader to grasp the relations and combinations that link facts, objects, numbers, and formulas. In Condorcet's view the limitless perfectibility of humankind was thus closely bound up with the technical invention that alone could bring the possibilities opened by alphabetical writing to their fulfillment.[27] It is thus wrong to reduce texts to their semantic content. Such considerations lead the historian Roger

Chartier to muse that "when it passes from the *codex* to the monitor screen the 'same' text is no longer truly the same, because the new formal devices that offer it to its reader modify the conditions of its reception and its comprehension."[28] Chartier draws the further conclusion that if texts are emancipated from the form that has conveyed them since the first centuries of the Christian era—the codex, the book composed of quires from which all printed objects with which we are familiar derive—by the same token all intellectual technologies and all the operations at work in the production of meaning will be modified. Similarly, in the essay "Material Matters" in the volume *The Future of the Book*, Paul Duguid insists it is inappropriate to think of the contents of media as separate from the media themselves, like wine and bottles. "Rather than to think of wine in bottles, each of which has a separate identity, it is more useful to consider information and technology as mutually constitutive and ultimately indissoluble."[29] In this light, books are machines to think with, and as a machine the book is more than a useful conduit of ideas produced elsewhere. It is itself the means of production.[30]

In science studies reactions to Derrida's work have always focused on his notorious claim that "There is nothing outside the text" (*Il n'y a pas de hors-texte*),[31] a position that has conjured fears about a flight from the real world of nature and a neglect of nonhuman actors.[32] But if we consider that science constructs its object through a process of differential marking, and it makes that object stable through public forms for the construction and dissemination of meaning, then consideration of communication technologies and technologies of representation becomes fundamental. They are "machines" that mediate and stabilize our representations. Furthermore, as extensions of the senses simultaneously affecting persons dispersed over numerous sites, they are powerful sources of mediation, multiplication, and stabilization of technoscientific practice. In their material form media do not provide mere "representatives" of an object described by theory, they create the space within which the scientific object exists in a material form. Media are not a mere supplement enabling the extension of research into areas where theory is insufficient to tread. Rather, the stronger thesis explored by the essays in this volume is that attention to the materiality of inscriptions themselves will demonstrate the extent to which inscription devices actually constitute the signifying scene in technoscience.

A central thesis of several essays in this volume is that serious consideration of the implications of Derridean deconstruction does not lead to a flight from "reality" into a discourse that is speaking all by itself, a play of signifiers without signifieds, a reduction of subatomic particles to text.[33] On the contrary, Brian Rotman sees in Derrida a resource for critiquing a disembodied Platonism in mathematics; and Hans-Jörg Rheinberger has provided an explicitly Derridean account of experimental practice in molecular biology.

Brian Rotman's proposal for a semiotics of mathematics draws upon Derridean notions consonant with recent directions in science studies.[34] In *Signifying Nothing: The Semiotics of Zero*, Rotman offers an archaeology—in Foucault's sense—of zero, the vanishing point, and imaginary money. Rotman studies transformations in different sign systems—arithmetical signs, picture signs, and money signs—showing in each the same pattern of signification, its destabilization and deconstruction through the introduction of a new sign, and finally the "naturalization" of the new sign. Operative in each of these systems is the assumption of an independent reality of objects providing a pre-existing field of referents for signs assigned to them. Zero, the vanishing point, and imaginary money function in a dual fashion: they are signs within the system of signs (zero as a number, the vanishing point as an element in a picture, imaginary money—a bank note—as a currency capable of being exchanged for goods), but they are simultaneously *meta*-signs external to the system (zero as the origin of the numbers for the one who counts, the vanishing point as the organizing point for perspective viewing for the one who depicts a scene, the bank note as currency exchangeable for a certain amount of gold). Money originates the medium of exchange that allows money to become a commodity. Like Derrida's supplement, the introduction of a meta-sign deconstructs the anterior reality supposedly grounding the system of signs. Once the system is accepted as referring to some external reality, it will continue to claim this role however far removed its signs are from this putative reality—numerals can be written, for instance, that are impossible to achieve through any possible actual counting, impossible scenes can be depicted, transactions can be drawn up that bear no relation to achievable relationships between goods. The result, according to Rotman, is a reversal of the original relation and a subsequent naturalization of the meta-sign:

The signs of the system become creative and autonomous. The things that are ultimately 'real,' that is numbers, visual scenes, and goods, are precisely what the system allows to be presented as such. The system becomes both the source of reality, it articulates what is real, and provides the means of 'describing' this reality as if it were some domain external and prior to itself; as if, that is, there were a timeless, 'objective' difference, a transcendental opposition, between presentation and representation.[35]

The deconstructive move leads Rotman to examine more closely how these fictions of origin are maintained. He focuses especially on the operations in the new system of meta-signs. The "for-the-one-counting" (and its analogs in the sign systems for vision and money) of the system of numbers (namely, a person manipulating physical objects like an abacus or marks on a page) has a correlate in the meta-sign system, in what Rotman calls the meta-Subject. The meta-Subject, according to Rotman, is the addressee of all those mystifying commands such as "Integrate the function f," "Take the tangent to the curve $f(x)$." This subject is an idealized, truncated simulacrum, which Rotman calls the Agent, analogous to what Peirce in his discussion of thought experiments called a "skeleton diagram of the self," which is dispatched to perform these activities. For Rotman, mathematics is an immanently embodied form of semiotic practice.

Like Brian Rotman, Hans-Jörg Rheinberger's Derridean account of experiment breaks with the notion of a pre-existent referent grounding scientific representations. In fact, he finds the notion of representation insufficient to capture the interesting features of what goes on in the early molecular biology labs he examines for his essay in this volume. According to Rheinberger, the game of lab science is aimed at generating robust experimental systems for the production of what he calls "epistemic things," materialized interpretations that form the components of models. The referent of scientific work is the model according to Rheinberger, comparison never being made to nature but always to other models, a process Rheinberger analogizes to the operation of Derrida's supplement and Brian Rotman's xenotext. In Rheinberger's case study of the molecular biology of protein synthesis, for example, different cellular components are defined by centrifugation, sedimentation properties, radioactive tracers, and chromatograms using a DNA sequencing gel. The scientific object is gradually configured from the juxtaposition, displacement, and layering of these traces.

The experimental systems molecular biologists design are "future generating machines," configurations of experimental apparatus, techniques, and inscription devices for creating semistable environments— little pockets of controlled chaos—just sufficient to engender unprecedented, surprising events. When an experimental system is working, it operates as a difference generating system governed by an oscillatory movement—stabilization-destabilization-restabilization—what   Derrida calls the "play of possibilities" (*jeu des possibles*). At the heart of the laboratory/labyrinth are experimental arrangements for transforming one form of matter into another and inscription devices for transforming matter into written traces. The products of this complex of experimental arrangements and inscription devices are trace-articulations, which Rheinberger calls graphemes. They represent certain aspects of the scientific object in a form that is manipulable in the laboratory. Graphemes, in turn, are the elements for constructing models. They are the manipulable signs scientists use in "writing" their models.

In addition to the (misplaced) critique that Derrida's approach eliminates the real, the body, and reduces everything to "text,"[36] a second concern is that Derridean deconstruction harbors a kind of technology of the letter.[37] These concerns could be translated into the undesirable possibility that deconstruction, linked with the materialities of communication, reinstates a form of technological determinism. Thus, there are resistances in Rheinberger's world of experimental systems but, because science never has access to the real except through trace-articulations, they are the resistances of graphemes to being assembled in certain ways. If there are "agents" or "actants" in Rheinberger's lab, they come in the form of technologically embedded graphemes. A parallel concern is that Friedrich Kittler's approach to writing is sympathetic to a kind of technological determinism. Kittler encourages such interpretations through his positioning of his work with respect to other schools of literature studies and through provocative statements and titles such as "There Is No Software." Although Kittler's writing embraces rich historical detail, use of archival materials, and historical contextualization of literature within fields such as psychophysics, experimental physiology, and the engineering of technical media, Kittler explicitly rejects any characterization of his work as "new historicism" or sociology of literature—in other words, anything that would look outside media for the meaning and content of literature. Kittler's framing descriptions of his project frequently invoke McLuhan's deterministic media theories, for example, claims that the content of a medium is always another medium and that while we are engaged by media we are numbed to their transformation of our senses.[38] To be sure, Kittler, like McLuhan, considers man the shaper of technology, but once technology exists, it shapes man: "Man becomes as it were, the sex organs of the machine world . . . enabling it to fecundate and to evolve ever new forms."[39] In a similar vein, Kittler argues that "methodological distinctions of modern psychoanalysis and technical distinctions of the modern media landscape coalesce very clearly. Each theory has its historical a priori. And structuralism as a theory only spells out what has been coming over the information channels since the beginning of this century."[40] Thus for Kittler even Lacan's theories of the real, the imaginary, and the symbolic, which are so central to Kittler's own literary criticism, are historical effects of the surrogate sensory modalities offered by gramophone, film, and typewriter in the discourse network of 1900.

In *Inscribing Science*, an alternative to these deconstructivist-inspired approaches is represented by studies emphasizing the historically situated character of scientific representation, its multivalent and contested nature, and the investment of scientific argument in narrative structures, vocabularies, grammars, patterns of analogy, and metaphor both internal and external to the scientific text. In recent years, Gillian Beer has offered a number of pathbreaking studies on the cultural codes embedded in the scientific text, and her essay in this volume on Darwin's experience of "the insular condition" continues and expands this work. Her studies of Darwin's use of language in constructing the argument of *The Origin of Species* have revealed the manner in which Darwin's familiarity with literary resources of Shakespeare, Milton, and other authors provided habits of imagination that affected, for example, his reading of Malthus and his reception of the implications of Malthus's ideas.[41] In *Darwin's Plots*, Beer shows how, for example, the history plays of Shakespeare, which emphasize stable succession through blood relationship, provided Darwin with one genetic pattern of succession and change with accommodation. At the same time, Darwin's use of language in *Origin* resonated with authors such as Milton and Shakespeare, enabling him to fashion a receptive audience for his ideas. In Beer's view, metaphor and narrative, by tapping into cultural presuppositions and activities of association just below the level of attention, were very much a part of Darwin's construction of theory. Darwin was writing for a general audience as well as for a group of scientific peers. Beer notes that in the first edition of *Origin* Darwin's choice of language was multivocal, using the tendency of words to dilate and contract across related senses, or to oscillate between significations. This served his rhetorical purposes and enabled him, while working within a tradition of natural history by and large subservient to natural theology, to write against the grain; indeed, it enabled him to include humankind in the interstices of his text through language expressing kinship, descent, and genealogy. As Beer shows, however, in later editions of *Origin* Darwin struggled to make his key terms mean one thing only.[42] Beer's point is that readers no less than authors constitute the meanings of the text: "Not what is said, but the agreement as to constraints on its reception, will stabilize scientific discourse . . . The enclosing within a community is a necessary condition for assuring stable signification."[43]

The example of Darwin's struggles to master language and rein in the play of signifiers speaks directly to the issue concerning Simon Schaffer, Alex Pang, and Phillip Prodger in this volume. In contrast to Latour and Woolgar's suggestion that scientific inscription devices and the laboratory create obligatory passage points constraining interpretation and assigning authority, Schaffer explores the multiple zones between the laboratory and popular representations. He argues that power and authority over the representation of the Orion nebula in the mid-nineteenth century did not reside with the authors of scientific texts, the astronomers, whose personal authority as observers was supposedly sufficient in early Victorian England to guarantee the truth of what they saw. Multiple authorities competed for rights to be the representatives of the nebulae, including journalists, natural philosophers, priests, critics, popular lecturers, and political radicals. This contest was not between high versus low science or different professional cultures, but rather between different audiences and the different aesthetic standards associated with them.[44] Schaffer emphasizes that astronomers and other commentators all deployed both literary and pictorial technologies with the same terms occurring in a variety of forums, but the different intended audiences exercised control over the most intimate details of the representations.

Alex Pang's essay amplifies this thesis. Schaffer notes that one crucial element enabling the play of interpretations of observations made at the Leviathan of Parsonstown was that the technologies of print and of picturing were highly malleable. In cases such as the Orion nebula, in which the issue is whether there is a condensing gas or a cluster of individual stars too small to be resolved optically, one might assume that the problems of interpretation would be resolved with astrophotography. But as Pang shows, matters of interpretation and artifactuality persisted even when photography was introduced into the observatory. Even when explicitly in pursuit of mechanically reproduced images untainted by human hand, the realities of the printing process forced astronomers to rely on skilled labor and required procedures for restoring lost information and correcting the content of pictures to match the "truth" observed by astronomers. Aesthetic standards had to be negotiated for the proper signaling of an "objective" photograph.[45] Once again we find matters of control and authority surfacing, this time between observatory and print shop, forcing astronomers and engravers to come to agreement about the criteria for judging a good picture and the definitions of "intervention" versus "alteration."

The notion that objectivity is an aesthetic standard constructed within a semiotic field of representations—not manufactured by an inscription device alone—highlights the essay by Phillip Prodger on Darwin's use of photography. Prodger shows how Darwin's evolutionary narrative informed the selection and placement of photographs as illustrations in his text *The Expression of Emotions in Man and Animals*. According to Prodger, Darwin was the first scientist to make use of the heliotype method for illustrating a text; his photographs were carefully situated in his text together with traditional engravings to take advantage of the striking appearance and novelty of these images in light of the tradition of scientific illustration dating back to the seventeenth century. With the contemporary popularity of photographs as documentary records, Darwin used them to add mechanical objectivity to his claims, even though the images were actually staged and Darwin himself participated in removing props and other evidence of human intervention.

Schaffer argues—and the essays by Pang and Prodger support the thesis—that every literary form of fact-making is linked to local complexes of technical and social practice and that stabilizing any representation is always at the same time a problem of political order and moral discipline. We cannot understand these processes by focusing attention on the character of inscription devices themselves or on the elusive chain of signifiers they produce; rather we must investigate the labor of controversy that ties signifiers to specific interpretations.

Are these two approaches—Derridean deconstruction and historical-constructivism—at all compatible? At the end of the opening section of *Grammatology* Derrida suggests that inscription devices and media technology be understood as linked to the contents of science, literature, and philosophy on the one hand, but simultaneously to particular social, economic, and political orders on the other:

The enigmatic model of the line is thus the very thing that philosophy could not see when it had its eyes open on the interior of its own history. This night begins to lighten a little at the moment when linearity—which is not loss or absence but the repression of pluri-dimensional symbolic thought—relaxes its oppression because it begins to sterilize the technical and scientific economy that it has long favored. In fact for a long time its possibility has been structurally bound up with that of economy, of technics and of ideology. This solidarity appears in the process of thesaurization, capitalization, sedentarization, hierarchization, of the formation of ideology by the class that writes or rather commands the scribes. . . .

The end of linear writing is indeed the end of the book.[46]

Derrida's project is focused on "writing in the broadest sense" rather than on specific historical episodes of the material practice of writing; and although he merely gestures toward history in the ordinary sense in the passage above, he nonetheless recognizes that the questions forming his grammatology are also relevant to a deconstructive history focused on the material practices of inscription and the social, political, and economic institutions that they sustain. That these two approaches can shed interesting light on scientific practice is the premise we explore in *Inscribing Science*.

The Language of Strange Facts in Early Modern Science

"Description of the Appearance of Three Suns Seen at the Same Time on the Horizon," "Observables upon a Monstrous Head," "A Narrative of Divers Odd Effects of a Dreadful Thunderclap," "On a Species of Wild Boar That Has a Hole in the Middle of Its Back, Which Foams When It [the boar] Is Pursued by Hunters"—these are titles taken more or less at random from the *Philosophical Transactions of the Royal Society of London* and the *Histoire et Mémoires de l'Académie Royale des Sciences de Paris* during the last half of the seventeenth century. This was a time when facts, conceived as chunks of pure experience detached from inference or conjecture, were new, and many of them were strange. The annals of the first, fledgling scientific societies swarmed with reports of strange phenomena, as did journals unattached to a learned academy, such as the *Journal des Savants*. Readers who gleaned their natural history and natural philosophy exclusively from these sources might well have drawn the conclusion that nature was out of joint, or echoed the *Histoire et Mémoires* of the Paris Academy of Sciences when it proclaimed "Our century, so full of marvels of all kinds."[1]

Wherein lay the affinity between the new science of the seventeenth century and strange facts? Why should the reformed natural philosophy of this period, remembered and revered for its allegiance to empirical regularity and mathematical law, have trafficked so heavily in anomalies and exceptions? No simple historiographic strategy of separating out serious sheep from giddy goats—the Galileos and Newtons from the Ulisse Aldrovandis and the John Evelyns—will answer to the case. Among the titles cited above were articles by the astronomer Gian Domenico Cassini[2] and the chemist Robert Boyle[3]; even Galileo was for a time fascinated with the wondrous Bologna Stone, which glowed in the dark with a cold light.[4] Descartes attempted a mechanical explanation of why (as was often alleged in medieval and early modern books of secrets) the corpse of a murder victim might begin to bleed anew in the presence of the assailant;[5] Leibniz proposed a scientific academy in 1675 that was to include an exhibition of, inter alia, "Mandragoras and other rare plants. Unusual and rare animals," as well as "Ballets of horses" and "the man from England who eats fire, etc., if he is still alive."[6] In short, seventeenth-century scientific luminaries of the first as well as the third magnitude avidly collected, observed, and pondered strange facts.

In what follows I shall argue that strange facts were the first scientific facts, for reasons both epistemological and cultural. Although these facts in infancy were by no means identical with our facts—they were, for example, in general neither robust nor replicable—they did nonetheless exemplify certain factual virtues that survive in the modern ideal of factuality. Like our facts, the facts of seventeenth-century science were notoriously inert—"angular," "stubborn," and even "nasty" in their resistance to interpretation and inference. Indeed, their very strangeness stiffened this resistance to theory and conjecture. Although the strange facts retained in the *Philosophical Transactions*, the *Histoire et Mémoires*, and other seventeenth-century scientific journals were at best incomplete and at worst useless from our point of view, they distilled certain features of factuality in purest form, namely, opacity in meaning and fragmentation in form. Seventeenth-century scientific facts were paragons of the sheer, brute thing-ness that has characterized facts ever since.

Sheer, brute thing-ness requires its own genre and language, and the seventeenth-century language of strange facts created both. However, before I examine that language, I must first argue for two anterior claims that make that language significant as opposed to merely curious: first, that facts were new to natural philosophy in the seventeenth century; and second, that these facts were in the first instance strange ones. I shall argue both claims in the next section. I then turn to an analysis of the language of strange facts—its tropes, its syntax, its characteristic genre. In conclusion, I briefly contrast this form of factuality and its characteristic language with more modern forms.

The Empiricism of Experience Versus
the Empiricism of Facts

For Aristotle and most of his followers in the Latin West, *scientia*, properly speaking, was the corpus of demonstrated, universal truths, "that which is always or that which is for the most part" (*Metaphysics*, 1027a, 20–27). Under this definition, anomalies, singularities, and other exceptions to the ordinary course of nature did not admit of scientific treatment. Neither Aristotle nor his medieval followers denied the existence of such oddities, abundantly catalogued in the pseudo-Aristotelian treatise *On Marvellous Things Heard* and Pliny's *Natural History*, but they did deny that anomalies resulting from chance and variability could form the subject matter of true science, for "there can be no demonstrative knowledge of the fortuitous (*Posterior Analytics*, 87b, 19–20). Nicole Oresme's *De causis mirabilium* (comp. ca. 1370) shows how it was possible for Scholastic philosophers to maintain that individual prodigies, such as monstrous human births in the shape of a pig or a cat, or a fast of several weeks, were wholly natural in their causes but nonetheless not susceptible of scientific explanation: "Therefore these things are not known point by point, except by God alone who knows unlimited things. And why does a black hair appear on the head next to a white one? Who can know so small a difference in cause?"[7] Well into the seventeenth century, natural philosophy continued largely to restrict its investigations to common experience, as Peter Dear has shown.[8]

Aristotelian natural philosophy shunned not only singular events, but all particulars, however commonplace, unless these led to or could be encompassed within generalizations, preferably causal generalizations. The proper domain of particulars, of facts, was history, not philosophy. Hobbes rehearses this scheme in *Leviathan*: "The register of Knowledge of Fact is called History. Whereof there be two sorts: one called *Natural History*, which is the History of such Facts, or Effects of Nature. . . . The other is *Civil History*, which is the History of the Voluntary Actions of men in Common-wealths."[9] Natural history could contribute raw materials and illustrations to natural philosophy—thus Aristotle's *History of Animals* was to prepare the way for a genuinely philosophical zoology—but by itself it was an inferior sort of knowledge, subordinated to the study of universals in philosophy or poetry (*Poetics*, 1451b, 1–7). Jurisprudence, like history, also relied predominantly on facts and inferences drawn from facts, rather than on universals and demonstrations about universals. However, this was simply proof of the inferiority of legal reasoning, even in the view of the jurists themselves.

This does not mean that Aristotelian natural philosophy was not empirical, only that its empiricism was not that of facts, in the sense of deracinated particulars untethered to any theory or explanation. Examples drawn from daily experience pepper the pages of Aristotelian treatises in natural philosophy, but they are just that—examples, and mundane ones at that. Examples illuminate or illustrate a general claim or support a hypothesis; counterexamples contradict these claims only when an alternative universal lies ready at hand. Examples do not float free of an argumentative context; they are, in our parlance, evidence rather than facts.[10] To have served up particulars, even prosaic ones, without an explanatory sauce would have thereby demoted natural philosophy to natural history. To have served up anomalous or strange particulars would have added insult to injury in the view not only of orthodox Aristotelians but also of early seventeenth-century innovators who, like Galileo and Descartes, still upheld the demonstrative ideal of science. Only a reformer intent on destroying this ideal as well as specific claims, or Aristotelian philosophy, would have been able to embrace strange particulars with open arms, and such was Francis Bacon. Impatient with Scholastic logic and scornful of the syllogism as an instrument for the investigation of nature, Bacon also challenged the validity of the axioms upon which Aristotelian demonstrations were grounded. Human nature being what it is, we rashly generalize our axioms from an experience too scanty to reveal the true rules and species of nature.[11] Bacon prescribed a cautionary dose of natural history to correct these prematurely formed axioms. Nor would ordinary natural history of what happens always or most of the time ("nature in course") suffice, for common experience does not probe nature deeply enough. Natural philosophers must also collect "Deviating Instances, that is, errors, vagaries, and prodigies in nature, wherein nature deviates and turns aside from her ordinary course."[12] Bacon's message was that natural philosophy would have to take not only particulars, but also strange particulars seriously. A due attention to strange phenomena would act as an epistemological brake to over-hasty generalizations and axioms, "help[ing] to cure the understanding depraved by custom and the common course of things."[13]

Bacon's grounds for studying strange phenomena were metaphysical

as well as epistemological. Although he still spoke the apparently volun-
tarist language of nature in course and nature erring, he initiated a unified
and thoroughgoing determinism in natural philosophy. By including
oddities and anomalies within the purview of natural philosophy, Bacon
insisted upon causal explanations of *all* phenomena. In particular, marvels
and prodigies were no longer exempted from scientific explanation: "For
we are not to give up the investigation until the properties and qualities in
such things as may be taken for miracles of nature be reduced and comprehended under some form or fixed law, so that all the irregularity or
singularity shall be found on some common form."[14] Thus Baconian
collections of everything "that is in nature new, rare, and unusual" were
ultimately meant to strengthen, not subvert a natural philosophy of regularities and laws. It was just because such strange phenomena posed the
greatest challenge to a watertight determinism of causes that they became
objects of special attention in the reformed natural philosophy.

Baconian facts were new not because they were particulars, not even
because they were strange. Particulars had long been the stuff of history,
natural and civil, and strange particulars had been a staple of both sorts of
history since Herodotus. They were new because they now belonged to
natural philosophy, expanding its domain beyond the universal and the
commonplace. Within natural philosophy they supplemented the empiricism of examples used to confirm and instruct with a collection of
counterexamples that were a standing reproach to all extant theories.
Indeed, Baconian facts were handpicked for their recalcitrance, anomalies
that undermined superficial classifications and exceptions that broke glib
rules.[15] This is why the first scientific facts retailed in the proceedings of
the Royal Society of London and the Paris Academy of Sciences were
such strange ones, for natural philosophy required the shock of repeated
contact with the bizarre, the heteroclite, and the singular in order to
sunder the age-old link between "a datum of experience" and "the conclusions that may be based on it"; in other words, to sunder facts from
evidence.

Bacon's blueprint for a reformed natural philosophy was not, however, the sole source of the early modern preoccupation with strange
phenomena. Although these strange facts were intended on the one hand
to destroy old systems and theories with a barrage of counterexamples,
and on the other to offer privileged insights into the essential but hidden
workings of nature, they exercised a fascination for the seventeenth-
century mind that went beyond these destructive and constructive func-

tions in natural philosophy. A conjunction of circumstances conspired to thrust strange phenomena to the forefront of both learned and popular consciousness in the late sixteenth and early seventeenth centuries. Reports of voyages to newfound lands both east and west retailed marvels that rivaled those of Pliny, some of which—rhinoceros horns and ostrich eggs, crocodiles and coconuts—found their way into the cabinets of curiosities that were the vogue among gentlemen of means and leisure, as well as among some universities and, later, scientific societies.[16] Works on natural magic as well as collections of *histoires prodigieuses* (marvelous stories) created an enormously popular genre that dwelt primarily on strange effects like apparitions, sympathies and antipathies between plants and animals, astonishing feats such as those of a man who washed his face and hands in molten lead or another who could sleep three weeks at a stretch.[17] Finally, the battles of both pen and sword waged over the Reformation sharpened the eye for portents, as the avalanche of both vernacular broadsides and Latin treatises devoted to the interpretation of comets, apparitions in the clouds, two-headed calves, and the like attest.[18] All of these circumstances converged to make strange phenomena culturally salient in the first decades of the seventeenth century at all levels of society from learned humanist to illiterate milkmaid. Who had not at least heard a broadside ballad sung aloud at the local inn about the latest ominous comet over Augsburg or the horrifying monster born in Ravenna? Bacon's appropriation of these strange phenomena to reform natural philosophy was highly original, but his heightened awareness and interest in them was not.

This is a signal irony in the transformation of strange phenomena like the ominous comet or portentous monster into Baconian facts. Amidst the apocalyptic anxieties that sparked and in turn were fueled by the early modern wars of religion, strange events—rains of iron or blood, the birth of Siamese twins, swarms of insects, bleeding grapevines—were seen by learned and unlettered alike as "signs and prodigies, which are most often the heralds, trumpets, and advance couriers of [divine] justice."[19] Hence strange phenomena were saturated with significance; all were grist for the interpreter's mill, and were as often as not pressed into service as propaganda on one or another side of the raging religious controversies of the day. Yet within the context of late seventeenth-century natural philosophy, strange phenomena became the very prototype of the inert datum, defying not only theological but also theoretical exegesis. They began as signs par excellence and ended as stubbornly insignificant. The causes of this startling transformation are complex, having as much to do with demonology as with a belief in the autonomy and inviolability of natural laws, and I cannot go into them here.[20] What is important for the next part of my argument is that an aura of the numinous still clung to at least some of the strange phenomena described by the natural philosophers, and that one key function of the language of strange phenomena was to dissipate that aura without at the same time dissipating the singularity of the fact.

Describing Strange Facts

Three major problems of language confronted the natural philosophers who submitted accounts of strange facts to the journals of learned societies. First, they had to unburden their descriptions from the portentous associations that enveloped monstrous births, celestial apparitions, bizarre weather, and other strange phenomena common to both the religious and natural philosophical literature. I shall argue that here naturalization proceeded chiefly by a strategy of cultural separation of the "curious" or "ingenious" from the "vulgar," rather than by explanation by means of natural causes. Second, the granular texture of strange facts, which were, as Bernard de Fontenelle, Perpetual Secretary of the Paris Academy of Sciences, put it, "detached pieces" of knowledge wrenched apart from both theories and one another by a "kind of violence,"[21] required a genre and a narrative form that would preserve this detachment and the free-floating independence it implied. Here, the innovation of the short article (versus the long treatise) and the frequent recourse to lists and tables made it possible to talk about strange facts without smoothing their cherished angularity. Third, strange facts by definition strained the boundaries of everyday experience and therefore of everyday language. In order to describe what most readers had never seen before and to render the extraordinary both vivid and plausible, the chroniclers of strange facts created a language of multiple analogies, fine-grained and circumstantial detail, and frank aesthetic pleasure. Let me discuss each of these three points—dissociation, narrative, and description—by means of concrete illustrations drawn mostly from the *Philosophical Transactions* and *Histoire et Mémoires*.

It is no accident that the canon of strange facts of late seventeenth-century natural philosophy strongly resembled that of religious portents of the sixteenth and seventeenth centuries. Impressed into the learned and popular imagination alike by Reformation and Counter-Reformation polemics and hallowed by several apocalyptic biblical passages, these were the prototypical strange facts of the period. Although not all of the oddities purveyed in the pages of the *Philosophical Transactions* and the *Histoire et Mémoires* fit into this canon—rotting fish that glowed in the dark and stone-eating worms, for example, had never qualified as portents—the triple suns, Siamese twins, earthquakes, rains of ashes and blood, misshapen plants and animals, floods, and spectacular thunderstorms reported by members and correspondents of the early scientific societies could equally well have been hawked in the latest gloom-and-doom broadside. Given this overlap in subject matter, it is not surprising to find outright conflations of strange facts and ominous portents. For example, the New England divine Increase Mather published in 1684 *An Essay for Recording of Illustrious Providences*, in which he reported remarkable deliverances from Indians, cases of demonic possession, divine judgments visited by lightning, flood, and earthquake, ossified fetuses carried for years in the womb, and "what ever else shall happen that is Prodigious." Although Mather's intent was expressly religious, he had numerous occasions to cite the *Philosophical Transactions* for cases comparable to those in his collection and, moreover, promised a sequel of *Miscellaneous Observations, Concerning Things Rare and Wonderful* and a *Natural History of New England*, to be compiled according to "the Rules and Method described by that Learned and excellent person *Robert Boyle* Esq."[22] In order to safeguard the neutrality of strange facts with respect to religious (and therefore political) interpretations, as well as with respect to rival scientific theories, natural philosophers had to create a language that would not only forswear but also discourage interpretation.

A plain, unadorned account of a strange phenomenon, shorn of all explicit interpretation, would not alone suffice. Such reserve did not break sharply enough with the broadside literature, which did occasionally announce an astonishing object or event without comment, but also without diminishing its fearful connotations. For example, a 1506 German broadside of a hermaphroditic monster sporting two sets of wings, a horn, and two crosses above its breast bore only the simple caption that the creature had been born on St. Jacob's Day in Florence, and that "our holy father the Pope" had ordered that it be given no food and allowed to die (Fig. 2.1). Although the broadside writer ventured no account of the monster's significance, *that* it was significant was unmistakable from the iconography. Certain strange phenomena were also so freighted with

Figure 2.1. German broadside of monster born in Florence on St. Jacob's Day, 1506. The legend reports that "our holy father the Pope has in his holiness ordered that it be given no food and that it be left to die." Reproduced by courtesy of the Bayrische Staatsbibliotek, Handschriften Abteilung (Einblatt VIII, 18), Munich. On the images and propaganda surrounding this particular "monster of Ravenna," see Ottavia Niccoli, *Prophecy and People in Renaissance Italy* (Princeton, N.J.: Princeton University Press, 1990), pp. 37–46.

meaning that they invited a portentous reading even when the reporter scrupulously withheld comment. Indeed, silence could heighten the ominous effect, as when a correspondent of the Royal Society reported how a lightning bolt had hit a Pomeranian church in the midst of Sunday service on June 19, 1670: "the whole Town, and particularly the Congregation in St. *Nichlas* Church (when the Minister was Preaching) was strangely surprised with a most terrifying flash of Lighting and a fearful Thunder-clap. . . . The Candle on the South-side of the Altar was put out by the blow, the other remain'd burning. Two of the Chalices there, were overthrown, and the Wine Spilt, and the Wafers scatter'd about; but the empty Chalice stood firm."[23] Taken by itself, this account is severely noncommittal, and was no doubt intended to be so. But it was inevitably situated within a tradition in which lightning bolts struck down blasphemers in midoath, and in which the eucharist (only "the empty chalice stood firm") was the object of intense theological debate.

Many natural philosophers blocked the interpretive impulse with a language of lofty condescension, that of the "curious and ingenious" elite withdrawing from vulgar errors. However, what allegedly distinguished the learned from the lay observer in the case of strange facts was neither focus (both were riveted by the same astonishing phenomena) nor deeper knowledge of causes (the natural philosopher was seldom in a position to explain an oddity, and rarely hazarded even a hypothesis) but rather sensibility. Writing in 1716 of a "late surprizing Appearance of the *Lights* seen in the *Air*" (apparently an aurora borealis), astronomer Edmund Halley confessed himself "heartily sorry" to have missed the onset of the celestial display, "which however frightful and amazing it might seem to the vulgar Beholder, would have been a most agreeable and wish'd for Spectacle." He further hinted at how said "Vulgar Beholder" might well have mistaken streaks of light for "the Conflicts of Men in Battle," armies in the sky being a staple of the portent literature.[24] In a similar vein, the chemist Robert Boyle noted that while he himself observed with "wonder and delight" that the veal shank intended for Sunday's dinner was glowing in several places, the servant who brought him the news was terrified.[25]

Halley's and Boyle's stance toward phenomena that verged on the uncanny and the ominous was certainly a naturalizing one, but taken by itself this label is misleading. Naturalization is here nothing more than a promissory note, an a priori conviction that strange phenomena like celestial lights and shining veal would eventually be explained by secondary causes. As we have seen, much of the attraction of strange facts lay in their recalcitrance vis-à-vis all and sundry theories, and hypotheses about their causes were advanced hesitantly, if at all. It was not so much a gap in knowledge as an alleged gap in social standing that imparted force to such remarks about the vulgar understanding. Although it was not the case that only the lowly and the unlettered read providences and portents into strange phenomena—Increase Mather was, after all, an educated man— the guilt-by-association argument was cultivated by natural philosophers hoping to strip strange facts of their ominous overtones. The oft-repeated honorifics that designated and addressed the "curious" and "ingenious" of Europe (and later the citizens of the Republic of Letters) were badges of a certain cultural status, sometimes straightforwardly linked to a social rank, but sometimes also linked to other qualities—esoteric learning, a cultivated sensibility, connoisseurship—that could increasingly substitute for rank.

This curious sensibility, which allowed the cognoscenti to relish what the vulgar found "frightful and amazing," depended on the distinctly early modern affinity between inquisitiveness and acquisitiveness. This is a sense of intellectual property different from that of patents and copyrights. It is the property acquired by intense and minute scrutiny of an object; the sort the English essayist Joseph Addison conferred upon his "Man of Polite Imagination," whose preternatural attentiveness gives him "a Kind of Property in every thing he sees, and makes the most rude and unculti-vated Parts of Nature administer to his Pleasures."[26] Wonder and curiosity working in tandem created this proprietary stare: wonder caught the attention and curiosity riveted it. Even Bacon, uneasily ambivalent about wonderstruck curiosity in many respects, acknowledged its essential role as bait and motivation: "by the rare and extraordinary works of nature the understanding is excited and raised to the investigation and discovery of forms capable of including them."[27] This power to awaken, hold, and even deepen attention was, of course, most obvious in strange phe-nomena. But even those natural philosophers and natural historians who wearied of the exotic and the anomalous subscribed to a psychology that forced them to treat the prosaic and common as if it were foreign and extraordinary. The botanist Nehemiah Grew tried without success to include "not only Things strange and rare, but the known and the com-mon" amongst the items in the Royal Society Repository (as its cabinet was called), certain that these ordinary things would "yield a great abun-dance of things for any Man's reason to work upon."[28] His Royal Society colleague Robert Hooke realized, however, that reason unprovoked by curiosity would never bestir itself, and tried the opposite tack of making the common rare and the domestic exotic, in the express hope of sharpening attention: "In the making of all kinds of Observations or Experiments there ought to be a huge deal of Circumspection, to take notice of every least perceivable Circumstance. . . . And an Observer should endeavour to look upon such Experiments and Observations that are more common, and to which he has been more accustom'd, as if they were the greatest Rarity, and to imagine himself a Person of some other Country or Calling, that he had never heard of, or seen the like before."[29]

This strategy of stimulating attention by estrangement with respect to both ordinary and extraordinary phenomena left its marks on the language used to describe strange phenomena. The strange facts of seventeenth-century natural philosophy subverted generalizations and splintered natural kinds. The observation reports that record them were notoriously prolix, but they had to be, for who could tell which detail would turn out to be significant? For example, when Robert Boyle published an account of "a Diamond, that shines in the Dark," he faithfully recorded which side of the diamond he rubbed, and with what color cloth.[30] Fearful of excluding anything, observers strained every nerve to catch everything. The feats of concentrated attention required were herculean, demanding a state of exaggerated attention that attempted to focus simultaneously and with equal sharpness on every aspect of the sensory field. The resulting descriptions are the literary equivalent of the impossible accuracy of some of Albrecht Dürer's animal paintings, in which the rabbit's or squirrel's every tuft of fur is as crisply outlined as every other: there is no background in these pictures, only foreground. Here curiosity was displayed in its root sense (Latin *cura*), namely painstaking, even excessive care lavished on each and every detail.

The details most frequently and emphatically reported were visual, particularly colors: the thorax of a young woman of Suffolk anatomized after death by "jolking of the breast" was found to be filled with a fluid "like *Cream*, or rather like a size of *Spanish White*, having a cast of yellow";[31] the fire-damp at a coalworks in Flintshire turned the rocks "fire-red";[32] stone-eating worms in Normandy are "blackish," with heads "the colour of a Tortoise-shell, braunish, with some small white hair";[33] the sea surrounding Barbados "shew'd itself to be Azure, the rest of the wave being dark-coulor'd . . . yet did the top of the wave break and appear to be green";[34] a mock sun is "quite reddish."[35] Although the mechanical philosophy may have dismissed secondary qualities like colors, textures, and tastes as illusory, it was just these sensory appearances that the descriptions of strange facts strived to fix in words.

Not only do the descriptions dwell lovingly on the sensory surface, on "phenomena" in the strict sense of the word; they also register an aesthetic pleasure at the sensualism of those surfaces. It is just this sort of sensual pleasure that Robert Hooke promised investigators in his *Micrographia* (1665): "And I do not only propose this kind of *Experimental Philosophy* as a matter of high *rapture* and *delight* of the mind, but even as a *material* and *sensible Pleasure*."[36] The pursuit of this "sensible pleasure" sometimes imparted a certain levity to the reports of strange phenomena: Antony von Leeuwenhoek was enchanted by "the very pretty motion" of microscopic creatures he discerned in melted snow water, "often tumbling about and sideways";[37] Robert Boyle repeated an air pump experiment with shining wood "partly for greater certainty, and partly to enjoy so delightful a spectacle."[38] What is striking about the exclamations of admiration and delight that punctuate the observation reports of strange phenomena is that the emotions have been deliberately displaced to objects that aroused mostly terror or disgust in so-called "vulgar" spectators. We have already heard Halley and Boyle rhapsodizing over fearsome phenomena; here is Robert Hooke on the appearance of a blue fly under the microscope: "All the hinder part of its body is cover'd with a most curious blue shining armour, looking exactly like a polish'd piece of steel brought to that blue colour by annealing. . . . Nor was the inside of this creature less beautifull than its outside, for cutting off a part of the belly, . . . I found, much beyond my expectation, that there were abundance of branchings of Milk-white vessels."[39] In this passage the commonplace is estranged by means of minute inspection and at the same time transformed into a thing of beauty, worthy of a still life.

The aesthetic of the seventeenth-century still life, with its fondness for the rare and exotic and its shimmering rendition of surfaces, was closely akin to the natural philosophical aesthetic of strange phenomena. Moreover, there was an element of both literal and figurative connoisseurship in the natural philosophical reporting of strange phenomena: literal in that many of the strange phenomena—insects in amber, striped tulips, a doe with antlers—were coveted by collectors for cabinets that often promiscuously displayed the works of art and nature side by side; figurative in that the recognition of the extraordinary depended on an extensive knowledge of the ordinary. Thus Cassini, announcing his 1683 discovery of a "rare and singular" light in the constellations Aries and Taurus (Fig. 2.2), noted that his attention was first drawn to the new phenomenon because these stars were "much more luminous than usual" at twilight, an anomaly only a practiced astronomer could have detected.[40] Similarly, the

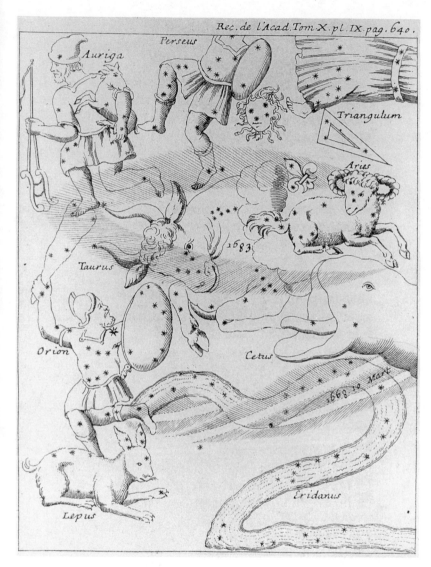

Figure 2.2. Celestial light in the constellations of Aries and Taurus, observed by astronomer Gian Domenico Cassini in spring 1683. Reproduced from Giovanni Domenico Cassini, *Oeuvres diverses de M.I.D. Cassini de l'Académie Royale des Sciences* (Paris: Compagnie des Libraires, 1730).

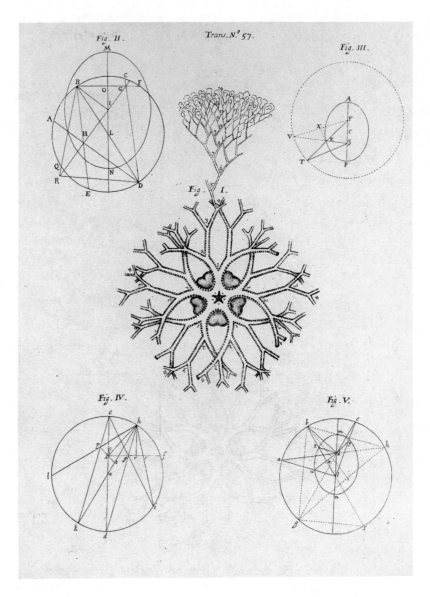

Figure 2.3. John Winthrop's *Piscis Echino-stellaris visciformis* (center), fished out of Massachusetts Bay. The surrounding geometrical figures relate to other articles; it was the custom of the *Philosophical Transactions* to cluster all figures for a given issue on a single page. Reproduced from *Philosophical Transactions of the Royal Society of London* 57 (Mar. 25, 1670).

chemist Wilhelm Homberg had to preface his "Curious Observation on an Infusion of Antimony" with an explanation of the empirical rule about metal salts to which antimony was an exception.[41] Conversely, those unsure of the commonplaces of natural philosophy and natural history were ill equipped to recognize the truly strange: John Winthrop of Connecticut could offer his observations of a "strange kind of *Fish*" (Fig. 2.3) and of dwarf oaks only with diffidence, for "whether it be a novelty to see such kind of dwarf-trees bearing acorns, I know not."[42]

Whether or not the strangeness of the phenomenon was evident or esoteric, its very strangeness drove reporters to analogy, and often to multiple analogies, heaped one upon the other. These multiple analogies decomposed the oddity into a mosaic of features, each to be mapped piecemeal onto a more familiar element of experience. Thus Martin Lister compared various parts of "star-stones" discovered in Bugthorp and Leppington to a "five-sided column," to the "blown Flower of the *Pentaphyllum*," to "the stalk of the *Asperula*," and to "the *antennae* of Lobsters."[43] Halley attempted a kind of subjective triangulation of the appearance of an aurora by juxtaposing the analogies of several spectators: "Some likened it to that Representation of *Glory* wherewith our Painters in Churches surround the Holy *Name* of God. Others to those radiating *Starrs* wherewith the Breasts of the Knights of the most Noble Order of the *Garter* are adorned. Many compared it to the *Concave* of the great *Cupola* of St. Paul's Church."[44] These composite analogies created a kind of chimera of the imagination, each part commonplace enough, but the combination of commonplaces as bizarre as the original. Paradoxically, the very analogies intended singly to connect the strange to the familiar achieved the opposite effect in the aggregate.

The tried-and-true means of dissolving variety into uniformity and of transforming the singular into the shared are numbers, and these were not lacking in the descriptions of strange phenomena. However, with the exception of astronomical observations and monetary amounts (for example, how much rubies and topazes cost in Ceylon), few of these numbers were actual measurements. Rather, they were quantitative approximations, which aimed more at precision than at accuracy. Precision refers here primarily to the clarity, distinctness, and intelligibility of concepts; accuracy refers to the fit of numbers to some part of the world, to be ascertained by measurement. When Leeuwenhoek claimed that there "were no less than 8 or 10,000" living creatures teeming in a drop of water,[45] or when a Florentine physician claimed that a lemon branch grafted onto an orange tree produced some fruit that "are half Citron-limon, half Orenge" and others "two-thirds of Citron-limon and one of Orenge; others the contrary,"[46] they made no pretense to exactitude. Hooke spoke to this penchant for precision without accuracy when he recommended mathematics because it "instructs and accustoms the Mind to a more strict way of Reasoning, to a more nice and exact way of examining, and to a much more accurate way of inquiring into the Nature of things," but warned not to expect too much reliability and accuracy in actual measurements, "for we find that Nature it self does not so exactly determine its operations, but allows a Latitude almost to all its Workings."[47] Instances of quantification that achieved precision rather than accuracy are fairly common in the early modern period, but such guesstimates served a special function in accounts of strange phenomena. First, they conveyed an idea of orders of magnitude and, at least potentially, formed the basis of comparisons with other phenomena. Second, and probably more important, they sharpened the mental outlines and thus simultaneously heightened the vivacity and plausibility of things and events barely imaginable and scarcely credible.

The descriptive texture created by accounts both dense with particulars and crisp with numbers was pointillist, but without the synthesis achieved by the right perceptual distance. The many, sharp-edged elements that made up such descriptions seldom converged into an impression of the whole phenomenon, unless supplemented by the occasional illustration. But even when they existed, the illustrations tended to supplant rather than to coalesce the descriptions, for the black-and-white, regularized (and often geometricized) woodcut forms filtered out most of the salient verbal details. Strange facts were thus detached in their descriptive elements both from natural philosophical theories on the one hand and from religious interpretations on the other.

The narrative form that best captured this detachment was the list. By no means all strange facts were listed; some were couched in narratives. But the connective tissue of these narratives derived less from the descriptive unfolding of the thing or event than from the actions and responses of the observers. There is a momentum of surprise, personal circumstance, and sometimes misfortune that motivates and impels these narratives: Halley rushes outdoors to catch the last glimpse of an aurora;[48] Boyle peers at a piece of shining wood "a while before I went to sleep";[49] George Garden of Aberdeen holds both the hands of "a Man of strange imitating nature";[50] a hurricane in Northamptonshire "storm'd the yard of one *Sprigge* . . . where it blew a Wagon-body off of the axle-trees, breaking the wheels and axle-trees in pieces, and blowing three of the wheels so shatter'd over a wall";[51] the investigation of a monstrous birth begins with the mother's labor pains beginning at midnight and ends with the "tumultuous concourse of people" and "the Fathers importunity to hasten the Birth to the Grave" interrupting the autopsy.[52] But the actual descriptions of the strange phenomena imbedded within these narratives of persons and deeds remain listlike. The sentences shrink, sometimes to fragments; the bridges between them crumble. Here, for example, is a description of a monstrous birth in Sussex: "It had two heads. Both the Faces very well shap'd. The left Face looked Swarthy; and never breathed. And the left Head was the bigger; and stayed longer in the Bearing. . . . The Breast (and Clavicles) very large; about seven Inches broad. But two hands. And but two feet."[53] Far-flung correspondents often listed and numbered their observations from Jamaica or Japan, as did authors of collections like the *Ephemeridum medico-physicarum germanicarum* (Leipzig, 1673–74), which contained 210 observations, including: "A Prince that lived a great while with great and dangerous diseases," and "A Girle of eight years old, greedily eating Mortar in great store, without any other harm than paleness of looks."[54]

These numbered lists were among the first tables of scientific data, direct descendants of Bacon's "Table of Essence and Presence" in Book Two of the *New Organon*. Their avowed purpose was to aid weak memory and to minister to the perplexed understanding, but their implications transcended the psychological. Lists carried with them a view of the structure of scientific experience, an experience that came in small chunks rather than smooth continua, an experience better suited to short journal articles than to long treatises. Strange facts highlighted this brevity and detachment, for they could not be explained by extant theory, subsumed under extant categories, or in any other way pressed into the service of a sustained argument that might mute their shrill peculiarity. Even the new genre of the letter-turned-journal-article demanded too much narrative integrity and flow to accommodate these cornery fragments of experience comfortably. The natural form of the strange fact was not the observation report but the list, in which one item was linked to the next only by an arbitrary enumeration that could be permuted at will. The nature of strange facts resembled nothing so much as the inventory of a museum or, better, of a *Wunderkammer* (wonder-cabinet).

Strange facts no longer play a central role in scientific empiricism. They have been replaced by the facts of the replicable, the homogeneous, and the countable. Facts retain their reputation for intractability, but that reputation now rests on different grounds: the first scientific facts were stubborn not because they were robust, resisting all attempts to sweep them under the rug, but rather because they were outlandish, resisting all attempts to subsume them under theory. The literary forms required differed accordingly: descriptions that emphasized differences at the expense of resemblances, that prized rather than ignored peculiarities and variability, that captured surface appearances rather than probed underlying causes. Numbers belong to such descriptions but statistics do not, for statistics assume identical, repeating units. Qualities and quantities can coexist peacefully in these descriptions, but not generalizations, for strange facts are immiscible. Analogies to the commonplace and the familiar combine with one another to accentuate, not weaken, the strangeness of the phenomenon. The language of strange facts aimed neither at transparency nor probability but rather at density and distinctness. It is a deliberately epidermal language, a language of surfaces rather than essences, of secondary rather than primary qualities—a true language of phenomena.

Shaping Information: Mathematics, Computing, and Typography

In the mid-1970s, the computer scientist Donald Knuth "regretfully" stopped submitting papers to the American Mathematical Society because, as he put it, "the finished product was just too painful for me to look at." He was referring to the result of applying relatively new technologies for typesetting to the particular challenges of mathematical text—what typesetters kindly label as "penalty copy."[1] Knuth's first response to this technical challenge was a system for mathematical typesetting called TEX (Tau Epsilon Chi), now adopted by the American Mathematical Society and used by many authors of "penalty copy."[2] Knuth's objective in developing TEX was to create "a method for unambiguously specifying mathematical manuscripts in such a way that they can easily be manipulated by machines." But he also envisioned a second step in the solution of the problem: using classical mathematics (and computers) "to design the shapes of the letters and the symbols themselves."[3] In fact, deploying techniques of science in the design of letterforms was scarcely new, as Knuth would soon discover: scholars and artists in the Renaissance applied themselves to the same tasks. The story of shaping the face of the printed page by scientific means is thus a long one.

A Matter of Proportion

We pick up the story around 1450—the era of humanism, dedication to the glories of ancient Greece and Rome, and the first successful use of movable type for printing in the Latin West. Italian humanists, sur-

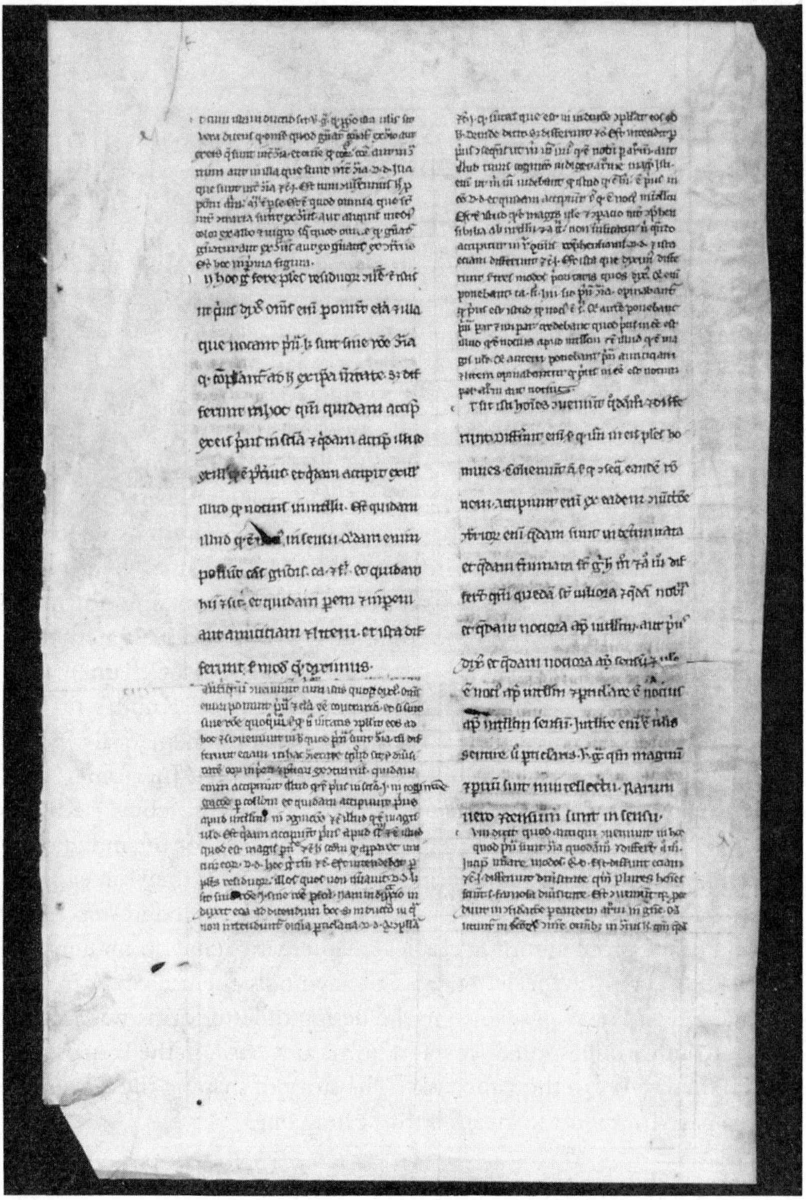

Figure 3.1. Gothic bookhand, in a commentary on Aristotle's *Physics* (England, latter half of the thirteenth century). From the Philip Bliss collection of medieval manuscript fragments. Courtesy of Stanford University Libraries, Department of Special Collections.

rounded by examples of classical Roman architecture, began at that time to take a fresh look at Roman architectural inscriptions. Their concern was with not merely the textual content of such inscriptions, but also, quite literally, the form. The classical elegance of Roman capital letters preserved in stone stood for them in sharp contrast to the gothic obscurity of medieval letter forms (see Fig. 3.1) and, true to their passion for ancient learning, these humanist scholars were convinced that Euclidean geometry, that paragon of ancient learning, lay behind the beauty of Roman capitals. Thus, they set out to discover how to draw the letters properly using only ruler and compass—that is, how to construct them geometrically.

One of the first to undertake the task was Felice Feliciano of Verona (1433–?)—humanist, calligrapher, sometime printer, and so committed to the recovery of ancient learning that he signed his work Felix the Antiquarian. Feliciano trekked through Italy, carefully measuring the letters in Roman inscriptions and tabulating their proportions. "And this is what I, Felice Feliciano, found in old letters by taking measurements from many marble slabs, both in the noble city of Rome and in other places."[4] He noticed that Roman stonecutters favored the Vitruvian ratio of 1:10 to relate the thickness of the main stroke of the letter to its height, and he made much of the fact that Roman letters could always be framed by a square and its inscribed circle. The letters also had as ornaments tails (or serifs) at the head and foot of the vertical strokes. In the late 1450s Feliciano designed an alphabet based on his findings and composed a manuscript treatise by way of explanation.[5]

He occasionally strayed from the geometrical straight and narrow in designing his alphabet, confessing that the tail of the letter R "cannot be perfectly described with the compass; let yourself be guided by your eye."[6] In other words, when aesthetics demanded, geometry gave way.

There is, by the way, solid evidence that Feliciano's advice was followed. On the earliest Renaissance building in Verona is an inscription dated 1468, with letter forms that correspond exactly with the designs in Feliciano's manuscript.[7] The manuscript itself, however, passed fairly rapidly into obscurity, and was only rediscovered in the Vatican Library in the nineteenth century.[8]

After Feliciano's manuscript came a series of published works by other authors, all of them describing and illustrating constructed Roman alphabets on the classical model. The main difference among them is the key ratio, which varies from author to author: some prescribe 1:9, some 1:10,

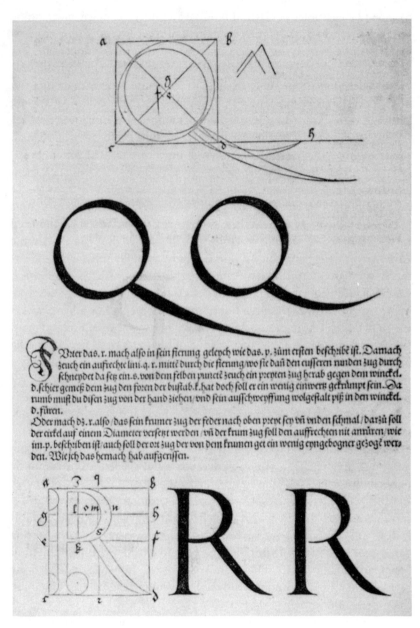

Vter das. r. mach also in sein sierung geleych wie das. p. zům ersten beschribe ist. Damach
zeuch ein auffrechte lini. q. r. mittē durch die sierung/wo sie dañ den aussern runden zug durch
schneydet da setz ein. s. von dem selben punctē zeuch ein preyten zug herab gegen dem winckel.
d. schier gemeß dem zug den foren der bustab. f. hat doch soll er ein wenig einwertz gekrümpt sein. Da
rumb must du disen zug von der hand ziehen/vnd sein aufschweyffung wolgestalt piß in den winckel.
d. füren.
Oder mach dz. r. also /das sein krumer zug der feder nach oben prеyt sey vñ vnden schmal / darzů soll
der cirkel auff einem Diameter versetzt werden /vñ der krum zug soll den auffrechten nit annüren/ wie
im. p. beschriben ist/ auch soll der ort zug der von dem krumen get ein wenig eyngebogner gezogē werə
den. Wie ich das hernach hab auffgerissen.

Figure 3.2. Dürer's constructed Roman alphabet. From Albrecht Dürer, *Under-weysung der Messung* (Nuremberg, 1525). Courtesy of Stanford University Librar-ies, Department of Special Collections.

some 1:12.[9] Among the most ambitious is the treatise *On Divine Proportions* (published in 1519) by Luca Pacioli, a professor of mathematics in Italy who was also responsible for the first printed account of algebra as well as an edition of Euclidean geometry. (His name is honored today by an accounting software package.) Pacioli saw evidence of divine proportion in the human body (à la Leonardo) as well as in the work of human hands, in beautiful architectural forms as well as in the inscriptions ornamenting them. He therefore thought it fitting to include in his treatise on proportions not only diagrams of geometric solids and a traditional account of arithmetic and geometric proportions but also a geometrical analysis of Roman letters—a task perfectly consistent with the Renaissance passion for uncovering and reconstructing the knowledge of the Ancients.[10]

Other authors followed suit, notably Albrecht Dürer, who added a constructed alphabet to his guide to proportions and perspective for artists.[11] (See Figs. 3.2 and 3.3.) All these authors, whether they addressed their lessons to stonecutters, calligraphers, or typefounders, were harnessing both classical geometry and the new technology of printing in the service of a humanist ideology that made a virtue of any association with antiquity.

The Academy Gets into the Act

Though other, similar examples could intervene, the next major stage in the story opens in the late seventeenth century, when the Paris Academy of Sciences threw the weight of institutional prestige and authority into making typography a science. To serve the interests of the state, the Academy was assigned by Minister Colbert the task of producing descriptions of practical arts and trades.[12] They chose first to describe "the Art which preserves all others—namely printing," and appointed a small committee to study the issue.[13] Gilles Filleau Des Billettes was a well-connected bibliophile; Jacques Jaugeon was a mathematical entrepreneur best known for his educational table games.[14]

They were joined by the learned Father Jean Truchet (known as Père Sébastien), a great fan of mechanics and measuring. Over the course of his career Truchet would, for example, make a series of barometric altitude measurements in mountainous regions; study the design of measuring devices used in commerce and experimental physics, carefully arraying his data in elaborate tables; and, inspired by patterns of floor tiles, take up the

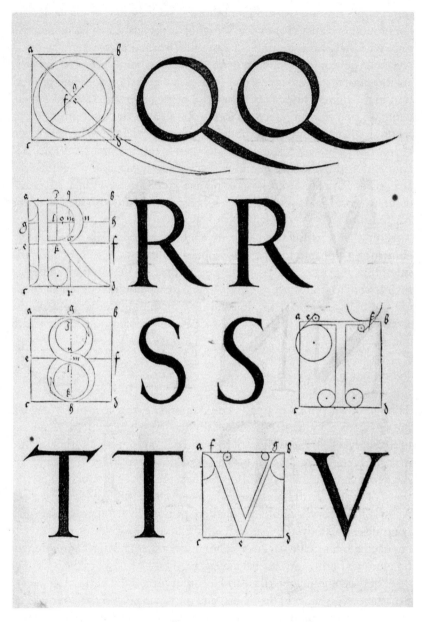

Figure 3.3. Dürer's constructed alphabet. From Albrecht Dürer, *Underweysung der Messung* (Nuremberg, 1525). Courtesy of Stanford University Libraries, Department of Special Collections.

study of geometrical patterns in general.[15] Because neither Truchet nor the other two committee members had any practical experience in printing, two craftsmen were called in to assist them: an engraver and a letter-cutter for the royal printing house.

The committee, which set to work in January 1693, interpreted its charge to include the scientific study of type design and, eventually, the design of altogether new roman and italic typefaces. Ample manuscript records of the committee's deliberations on printing survive in Paris and elsewhere, but—ironically—their report on printing never saw print.[16]

In its study of typefaces, the committee systematically combined the empirical and geometrical with the aesthetic and practical. The royal library, put at Truchet's disposal, was filled with examples of fine, ordinary, and wretched typography, and Truchet systematically browsed its holdings. In the process, the books of Pacioli, Dürer, and Tory came to his attention. Like Feliciano, whose work he did not know, Truchet measured the proportions of more than 50 different varieties of letter design and carefully tabulated the results. The committee then identified the letter designs it found most agreeable, compared them with the Renaissance treatises on geometrical construction of alphabets, and decided what to add to or remove from strict geometrical constructions in order to create truly beautiful letters. Here, too, the geometry provided basic scaffolding, subject always to artistic embellishment.

Truchet then turned the results of these discussions into drawings of new type designs for the committee to consider (and tweak). Eventually the committee reached a consensus on this "matter," as they put it, "of taste and . . . extreme delicacy."[17] To record their design decisions and eventually guide the craftsmen who would cut the actual type, they imposed a rectilinear grid (of more than 2,000 tiny squares) on each new letter design.[18] The grid, square for roman letters and rhomboid (with a well-defined slope) for italics, was clearly borrowed from the Cartesian coordinates of analytic geometry (Figs. 3.4 and 3.5). One can also superimpose connections with Truchet's geometry of floor tiles (and even hints of modern rasters).[19] The combination of the grid plus the rule and compass of traditional geometry—a system nearly as complicated and elegant as the geometry of Ptolemaic astronomy—saved the appearances of the letterforms devised by the committee. The production of more than 165 large-scale copperplate engravings of the letters, following the committee's design specifications, began in 1695 and continued for two decades thereafter.[20]

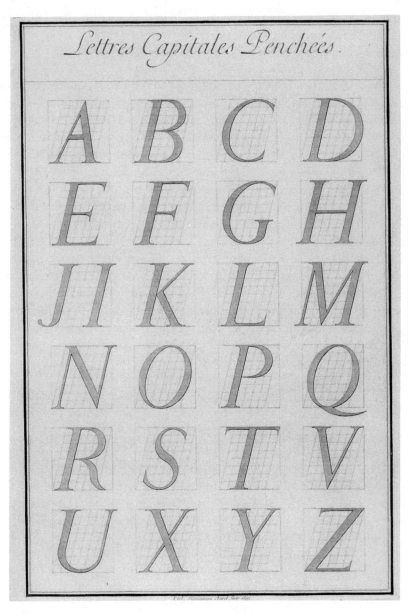

Figure 3.4. Jammes's alphabet: capital letters. From André Jammes, *La réforme de la typographie royale sous Louis XIV* (Paris: P. Jammes, 1961). Courtesy of Stanford University Libraries, Department of Special Collections.

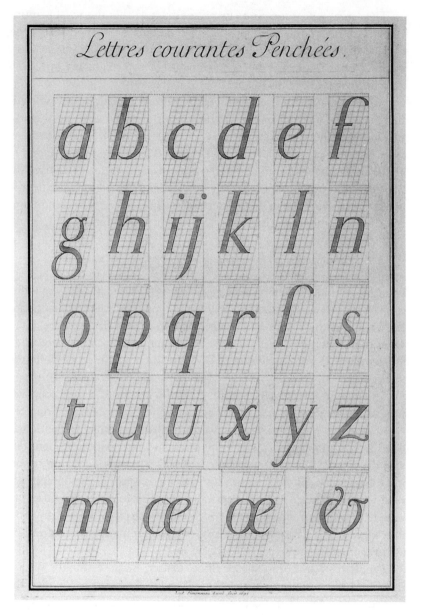

Figure 3.5. Jammes's alphabet: lower-case letters. From André Jammes, *La réforme de la typographie royale sous Louis XIV* (Paris: P. Jammes, 1961). Courtesy of Stanford University Libraries, Department of Special Collections.

The large-scale engravings were so attractive that the royal printing house asked that the committee's type designs be turned into type, whence the name *romains du roi*.[21] The first use of this new type was a lavish volume honoring the achievements of the reign of Louis XIV, including the work of his Academy of Sciences.[22]

Readers who have used eighteenth-century French books will find the *romains du roi* familiar; the circumstances of their birth speak to the relevance of science, broadly defined, for typography. The impulse toward order and regularity, empirical investigations classified and reduced to tabular form, the appeal to geometry as an instrument for describing the shape of letters—all of these characteristics are noteworthy and predictable effects of involving the premier scientific institution of the ancien régime in the improvement of printing.

The Point System: Fournier and Typography

Truchet, in filling out his table of proportions for well-crafted type styles, came up against a long-standing problem is typography—the lack of standardization in the names and dimensions of various sizes of type. Even when the French government tried to set standards for type sizes, their choice of a basis for measurement was so arbitrary as to prove useless in practice.[23]

The matter of type sizes was still fraught with disorder and complication when an erudite and ambitious printer, Pierre Simon Fournier, began in 1742 to publish a series of works on printing and typefounding.[24] (See Fig. 3.6.) Fournier attacked the problem in a spirit often identified with the *Encyclopédie* of Diderot and d'Alembert: applying reason to improve arts and trades. Indeed, the *Encyclopédie*'s articles on typography are filled with Fournier's name, words, and ideas.[25] Moreover, his efforts to remedy imprecision and impose order on the chaos of type sizes show traits typical of the *encyclopédistes*: reason, measurement, standardization, and instrumentalism.[26]

Type size is an eminently practical matter, because straight typeset lines demand consistency in the size of type body. Fournier divided the type body into equal and definite parts called points, 72 in what is roughly an inch. With this measuring scale, a printer could finally "know exactly the degree of difference and the relation of type bodies to one another," Fournier explained. "One may combine [these signs] as numerals are combined; [as] two and two make four. . . . To combine the bodies it is

Figure 3.6. Fournier's scale of point sizes. From Pierre Simon Fournier, *Manuel typographique* (Paris, 1764–66), vol. 1. Courtesy of Stanford University Libraries, Department of Special Collections.

sufficient merely to know the number of typographical points in each. To do this," Fournier insisted, "these points or given units should be invariable, so that they will serve as standards in printing-offices [shops]."[27]

Fournier presented his proposed standard in a work he called his *Tables of Proportions*; even its title echoes some of Pacioli's and Dürer's concerns. There is also an obvious similarity between Fournier's standard points and the rigid regularity of the Academy of Sciences' grids, though Fournier insisted, at times vehemently, that he owed nothing to the Academy's committee members, whom he dismissed as amateurs meddling in the art and science of typography.[28]

Fournier recognized that a printed paper scale presented certain practical difficulties if it was to serve as a standard of measurement: presses of that era printed on wet paper, and wet paper shrinks as it dries. Fournier tried to compensate by allowing for this shrinkage, but shrinkage rates themselves varied widely.[29] In fact, one twentieth-century study suggests that the actual length of Fournier's scale of 144 points (approximately two inches) varies by as much as a full point from copy to copy of the book, as preserved in various rare book collections.[30] Fournier offered a second way around these difficulties. He "devised an instrument . . . called a prototype," made of iron and copper and designed to "make certain in every case of executing [casting] the body of a letter accurately."[31] This instrument, meant to serve "as a basis in calculating,"[32] shows clearly Fournier's debt to mathematical instrument makers of the period and lends credence to his claim that typography qualifies as an exact science. Fournier's work thus exemplifies the instrumentalism and quest for precision and standardization typical of the quantifying spirit of the eighteenth century.[33]

The point system promptly took hold in one national variant or another because it promised predictability, interchangeability, and accuracy, which translated into savings of time, spoiled pages, and type. The scheme thus offered printers ample economic justification for switching their thinking, and this economic factor did much to ensure general acceptance of the scheme.[34]

Printing from the Renaissance through the Enlightenment can thus be viewed as a durable alloy of art, technology, and science, emblematic of the efforts to make the mechanical arts rational. I have described here several stages whereby craftsmen and scholars tried to link the thinking and techniques of science to the processes of reproducing the written word.[35] All these developments should be viewed as stages toward the mastery of the medium, as attempts to control and shape printing and its products.

Constructing a Metafont

Fast forward two centuries, and we come again to Donald Knuth. During the late 1960s and 1970s he was publishing his multivolume work, *The Art of Computer Programming*, which, along with a prodigious number of papers in mathematics and computer science, would help earn him an endowed chair at Stanford, the National Medal of Science, and numerous other honors. Printing also caught his attention. Perhaps at first Knuth saw the problem of mathematically defined letterforms and their computer-based production as relatively straightforward. In any event, with characteristic speed, and in the unlikely setting of a camping trip, Knuth sketched out a program for creating "a mathematical definition of type fonts."[36] Recognizing that printing equipment may come and go, but assuming that "any new machines are almost certain to be based on a high precision raster,"[37] he aimed at a definition "that can be used on all machines both now and in the future." Whatever the precision, Knuth reasoned, "the letter shapes can stay the same forever, once they are defined in a machine-independent form."[38]

Knuth quickly recognized that the technical challenge of automatic generation of letterforms had to be coupled with an aesthetic imperative, so that the symbols thus defined might be viewed as beautiful according to traditional aesthetic norms. Knuth's adventures in type design thus led him back to the Renaissance, to the work of the Paris Academy of Sciences, and eventually to productive collaborations with renowned alphabet designers of the present. From this research, he concluded that the invocation of mathematics and reliance on computers and their peripherals could provide modern type designers "with an exciting new medium"[39] and equip them with new tools and the ability to experiment with ease.

Like his many predecessors in the geometric construction of letterforms, Knuth studied typefaces he admired, drawing on the technology at hand. He made 35mm slides of letterpress proofs of one of his books—one he liked!—and projected them onto his living room wall, then traced the fuzzy outlines of letterforms onto graph paper. So far this sounds like an updated version of the Paris Academy approach.[40]

But how to duplicate mathematically the crucial curves he saw in an attractive letterform? And how to generate equally attractive new letterforms? Here Knuth called up tools of analytic geometry: given a few specific points on a plane, he sought a "mathematical formula that defines a pleasant curve through these points." Thus armed, he could then define mathematically not only one typeface, in one size, but a whole family of similar fonts—in other words, a metafont.

The curves Knuth wanted turned out to be those defined by Bernshteyn polynomials,

$$z(t) = (1-t)3z^1 + 3(1-t)^2tz^2 + 3(1-t)^2z^3 + t^3 + t^3z^4,$$

which trace out curves called Bézier cubics. Varying the parameters yields a family of curves; the curve that suits the typographer's eye corresponds to particular values for the parameters, which can then be fixed for use in other letters in the same typeface.[41]

The program also borrows from the ancient art of the calligrapher and uses the language of pens and their widths. To construct the numeral 9, for example, a METAFONT program instructs the printer, "First draw a dot [at a given point]. . . . Then take a hairline pen and, starting at the left of the dot, draw the upward arc of an ellipse; after reaching the top, the pen begins to grow in width, and it proceeds downward in another ellipse in such a way that the maximum width occurs on the axis of the ellipse," and so on.[42]

Knuth found that the program, once completed, permitted him to "define a decent-looking complete font" of 128 characters in two months.[43] On the path to this "decent-looking" font, he haunted Stanford Special Collections, seeking in rare books guidance from his Renaissance predecessors, and experimented with different values for crucial parameters (see Fig. 3.7).

Knuth announced these results with pride in his Gibbs Lecture for the American Mathematical Society early in January 1978. METAFONT would facilitate visual experiments with type design by dramatically reducing the time between drawings and proofs, Knuth explained. It would also admit the possibility of a little randomness, to soften the rigid symmetry of mathematical constructions and add "warmth and charm" to typefaces.[44] He labeled the parameter "craziness" and provided a means for capturing its desirable results.

In recognition of practical realities, a single METAFONT typeface definition "can produce fonts of type for many different kinds of printing equipment, if the programmer has set things up so that the resolution can be varied."[45] Preliminary proofs can be pulled from lower-resolution, less expensive printers; the final stages of design tinkering require higher-resolution rasters.

Knuth continued to tinker with both METAFONT and the typefaces he had designed, settling eventually on a design he called Computer Modern. He used it for the second edition of his study of algorithms, published in 1981. The result was a midlife crisis. What had looked fine in intermediate proofs proved awful in the printed book. As Knuth recalls, "Ev-

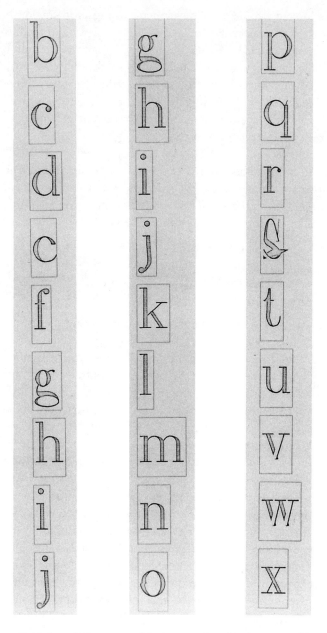

Figure 3.7. Samples of Computer Modern under development. Courtesy of Stanford University Archives, Donald Knuth papers.

erything looked wrong! . . . I developed a strong antipathy for the shapes of the numerals, especially the '2' and '6.' . . . I wanted to think about elegant mathematics, but it was impossible to ignore the ugly typography."[46] The experience sent Knuth back to the drawing board, literally and mathematically. He had learned a valuable, if surprising, lesson: "Type design can be hazardous to your other interests. Once you get hooked, you will develop intense feelings about letterforms; the medium will intrude on the messages that you read."[47]

He undertook to revise METAFONT completely, then used the new program to improve his letterform designs. He entered into fruitful collaborations with professional type designers and experts in computer-based typography, and tried out the new METAFONT on Stanford students in 1984. Like Jonathan Swift, he exhorted all to

> Blot out, correct, insert, refine,
> Enlarge, diminish, interline;
> Be mindful, when invention fails,
> To scratch your Head, and bite your Nails.[48]

Many of the exercises from the Stanford course found their way into *The METAFONTbook*, published in 1986, which introduced the program and its possibilities to Knuth's intended audience—type designers not afflicted with computerphobia. Like others of Knuth's works, *The META-FONTbook* took a breezy, friendly approach: whimsical illustrations, trenchant quotations, insider jokes. However user-friendly the book and the program, the problem Knuth had undertaken to solve was one of daunting complexity, in which centuries-old aesthetic considerations rubbed elbows with the latest in computer technology and some challenging mathematics. The late twentieth century seemed to him the right time for a new synthesis of letterform design, mathematics, and printing technology, one that mixed "more than 2000 years of accumulated knowledge about geometry and curves,"[49] computers capable of drawing "new fonts of characters in seconds,"[50] and friendly collaboration linking the typographic community and computer professionals. But Knuth does not underestimate the task, nor does he "ever envision the problem becoming simple."[51] In that his worthy predecessors would surely concur.

4. BRIAN ROTMAN

The Technology of
Mathematical Persuasion

My aim here is to outline a conception of mathematics-as-language and indicate, contrary to the received view, how mathematical reasoning is inseparable from persuasion and the exercise/creation of a certain mode of subjecthood. Let me frame my approach to the question of language and mathematics by means of two historically separated and very different citations:

Philosophy is written in this grand book—the universe, which stands continually open before our gaze. But the book cannot be understood unless one first learns to comprehend the language and to read the alphabet in which it is composed. It is written in the language of mathematics.   (Galileo)[1]

Against Galileo's confident invocation of language there is its contemporary problematization:

However the topic is considered, the *problem of language* has never been simply one problem among others. But never as much as at present has it invaded, *as such*, the global horizon of the most diverse researches and the most heterogeneous discourses, diverse in their intention, method, and ideology.   (Derrida)[2]

Where to begin? One could start from the outside, from mathematics' evident and inescapable immersion in the sociocultural matrix, and track its historical production as a form of instrumental reason and source of so-called objective, true, value-free knowledge. The last two decades have seen much work of this kind—heterogeneous but operating under the unifying premise that all forms of knowledge together with their legitimating claims to "truth" and the like are social constructions (mathe-matical no less than any other).[3] In relation to the present aim, the hope would be to close in on mathematical practice and, by focusing in on its communicational and symbolic functioning, produce an understanding of mathematics-as-language. Or one could start from the inside, which is what I shall do. This means assuming (and bracketing its complexities) the ultimate success of an external characterization of mathematics as a communicational activity, and attempting to complement this descrip-tion by giving a quasi-phenomenological, internal account that might re-flect what it means to *do* mathematics—write its symbols and think its thoughts—from within. Because there is virtually nothing in the way of a history for such a project, one starts from the present.

The cultural scene of the humanities is being shaped by an engage-ment with a certain linked set of contentiously theorized problems and intellectual moves that goes under the mark—simplifying but nonetheless useful for our purposes—of the *Post* (poststructuralist, postmodern, post-patriarchal, posthumanist, postindustrial, post-Enlightenment, postreal-ist). Though no single issue, discursive move, or program captures the boundaries of the Post, a key site of conflict and innovation is the nature and business of language. By this I mean "language" in the large sense, including natural speech, but also all forms of writing, recording, com-municating, representing, cyphering, and notating through the use of spoken words, written and gestured signs, iconographic devices, ideo-grams, symbol systems, and the like.

A discourse within the humanities that is resistant and unsympa-thetic to the destabilizing subversions of the Post is mainstream Anglo-American analytic philosophy, whose focus on logic and the nature of mathematical and scientific knowledge makes it naturally relevant to any examination of mathematics. Here the refusal of the Post (essentially, of poststructuralist, deconstructive theorizing) takes the form of a deep, entrenched, and perhaps irreconcilable conflict between an old and much worked through orthodox tradition regarding language and a newer philosophical outlook opposed and alien to it: a conflict between the so-called continental outlook, whose conception of language—dominated by Nietzsche, Husserl, Heidegger, Wittgenstein, and Derrida—assigns it a constitutive role and whose slogan might be, "Language speaks man into the world," and the current analytic mindset that understands language as an inert, transmissive medium, and its empiricist forebears, whose banner might read "Man speaks language about the world."

To say, then, of some human endeavor, in our case the practice of mathematics, that it is a "language," or symbol system, or mode of discourse, is to confront this conflict; something that seems inevitable if one is to place mathematics, at this late point in the century, in relation to the current of late twentieth-century thought. But such an engagement brings with it certain risks and difficulties because, as is generally acknowledged, the status of "language" (and not just within philosophy) is in flux—contentiously theorized, open-ended, problematic. Evidently, any description of the linguistic and signifying capacity of mathematics is likely to raise issues that will appear artificial or foreign to mathematics' previous unlanguaged or inadequately languaged image of itself.

Leaving such problematics aside, there is, on the face of it, little need to insist that mathematics is a language: who, after all, among those familiar with it would deny the proposition? Certainly not those users— accountants, engineers, economists, actuaries, statisticians, cliometricians, meteorologists, and the like—who have no choice but to translate in and out of mathematical expressions and terminology on their way to other interests and forms of engagement, and certainly not the professional practitioners of the physical and life sciences and technologies who have always appealed to and depended on mathematics as an essential linguistic-cognitive resource. Is not the contemporary technoscientific picture of physical reality literally unthinkable outside the apparatus of mathematical notations and terms used to articulate it? For Galileo, reading these terms was not so much articulating the language of science as deciphering the writing of God; and if in the four centuries since his proclamation science has progressively sought by a series of disclaimers to distance itself from the more blatant aspects of Galileo's theism, it has left his understanding of mathematics as the language and alphabet of the physical universe very much in place, adhered to, with remarkably little questioning, by innumerable scientists from the seventeenth century to the present.

Yet, despite the many much-repeated recognitions of mathematics' linguistic nature, there has been little sustained attempt (from either side of the continental/analytic philosophical divide) to develop the theoretical and conceptual consequences of saying what it means for mathematics to be a language or be practiced as a mode of discourse. A notable, important exception would be Wittgenstein's extended remarks on the foundations of mathematics, with their intent to characterize the doing of mathematics (at least in its elementary computational aspects) as a motley of certain kinds of language games. But, though canny, interesting, and provocative, Wittgenstein's fragmentary, deliberately unsystematic dicta do not address the *theoretical* question of mathematical language and discourse in any way. Nor do they indicate how one might do so. Notwithstanding this, however, the viewpoint here is sympathetic to his radical nominalism, which sees any mathematical "object" as an effect of the notation system that supposedly describes it.

Galileo's amalgamation of language, writing, books, and alphabetic letters raises certain crucial issues: What do languages—divine, natural, or manmade—have to do with alphabets? Is mathematics a "language" or a form of writing? What is the difference? And related to these, is there a distinction here between "depth" and "surface," between, for example, a project for a "grammar" of science and the apparently more subjective, impressionistic study suggested by "language" of science? And why the genitive here? Why *of* science? Why not the grammar, the language, the code, the symbolic system, or whatever, of mathematics itself—cut free from all questions of its instrumentality?

These are large-scale questions. In the present context we need to be more specific and focused. Thus, one might ask whether mathematical texts produce meanings that differ fundamentally in kind from those of speech and its alphabetic inscription: what is the relation, for example, of mathematical proofs to written narratives or to the arguments and dialogues of everyday discourse? And one wants to know what it means linguistically to *do* mathematics: what manner of symbolic activity, of imagining and scribbling, are mathematicians engaged in when they think/write mathematical signs?

Semiotics—professing itself to be the study of any kind of sign, language, or communicational system and with an ongoing finger in every kind of signifying pie imaginable—is the obvious place to look for a means of addressing the question of mathematics-as-language. Semiotics, stemming separately from the work of Saussure, a structural linguist, and Peirce, a metaphysically minded pragmatist, is a two-headed affair: both a kind of abstract formalism and a philosophico-taxonomic study of what it means to signify. From Saussure comes the definition of a sign as a coupling—said to be unmotivated and arbitrary—of signifier ("hundt," "chien") and signified (doghood); and the insistence on downgrading referential meaning in favor of a sign's structural relations to other signs. Peirce held that anything whatsoever that stands to somebody for something was a sign, and that any sign always created or involved an "interpretant"—variously another sign or interpreting agency—as an essential part of its action.

These definitions have been much discussed. This is not the place to elaborate them or to compare the formalist/structuralist and interpretive/discursive semiotics to which they give rise. Instead, I shall sketch two very different ways of approaching the question of mathematical signs that I have pursued over the past few years, which can be seen—very crudely and schematically—as instances of these two semiotics. Significantly, though employing divergent methods, aims, and starting places, each ends up foregrounding the role of the corporeal, sign-writing/reading subject.

The first, about which I shall be very brief, concerns the analysis of one particular and rather special mathematical sign, namely zero. As is well known, neither Greek nor Roman mathematics had a concept of zero. In the form we have, it is a Hindu invention transmitted into Europe via the Arab Mediterranean. Its introduction marked a discontinuity in western thought concerning the concept, use, theorization, metaphysics, and writing of numbers.

At the end of the sixteenth century, some three centuries after its introduction in northern Italy, the Dutch engineer and mathematician Simon Stevin, engaged in the project of extending the place notation from finite to infinite decimals in his treatise *The Dime*, was moved to a kind of wonder at the creative power of zero—"the true and natural beginning" of the numbers, he called it. As the point was to the geometrical line, so the nought—"poinct de nombre" he wanted to call it—was to the arithmetical progression. What are we to make of his perception of a *natural* beginning and the analogy it suggested to him? What is it about zero *as a sign* that might give it this singular generative quality for Stevin, if not for us?

My approach to this question in *Signifying Nothing* was to give a sufficiently abstract, generic characterization of zero and then construct a kind of archaeology, à la Foucault, of the zero sign in relation to other, similarly conceived nonmathematical signs.[4] Starting from a formulation of a particular kind of metasign—specifically a sign for the absence of certain other signs—I identified a series of semiotic relations between zero in the code of arithmetic, the vanishing point in the code of perspectival painting, and so-called imaginary money, or "Bank money" as Adam Smith called it, in the code of economic exchange. These relations were built on parallels between the generative, originating roles played by the metasigns in their respective codes: zero engendering an infinity of new number signs, the vanishing point delivering an endless supply of a new type of visual image, and imaginary money making possible an unlimited range of previously unavailable transactions.

On the basis of the resulting isomorphisms between the codes of arithmetic, painting, and money it becomes possible to articulate certain features shared by the semiotic actors who operate within these codes; between, that is, the subject of arithmetic (the one-who-counts) and the other subjects involved (the one-who-paints and the one-who-transacts). In particular, it becomes possible to apprehend the dual status of these subjects as certain kinds of discursive agencies, that is, semiotic subjects who both use and are created by their respective codes.

One consequence of this, radically antagonistic to prevailing Platonistic interpretation of mathematics (and to cognate realist or objectivist accounts of scientific knowledge), but familiar within Post (particularly poststructuralist) thinking about language, is that the "objects" concerned here—numbers, visual scenes, values—far from belonging to prior, pre-semiotic worlds waiting to be signified, are inseparable from, and in a radically uneliminable sense owe their being to, the very codes thought to be referring, notating, and describing them.

The comparative insights offered by such an archaeology, the homologies between mathematical counting, visual depicting, and economic transacting it makes available (which extend further than my brief summary indicates), are novel and suggestive. But its capacity to illuminate the overall nature of mathematical discourse is evidently limited. After all, there are mathematical signs other than zero and mathematical activities—giving definitions, executing operations, introducing notations, making assertions, following proofs and arguments—that are not reducible to acts of counting.

Another approach to mathematics, employing a discourse-based semiotic rather than an archaeological/structuralist one, is called for; one that would respect the fact that "the world of rigorous fantasy we call mathematics," as Gregory Bateson put it, is imagined and *thought* into existence.[5]

The thought experiment (*Gedankenexperiment*) has been a widely used element of scientific practice since Galileo. Only fairly recently, however, have these scenarios and narratives of imagined activities—actions conceivable in principle but difficult or impractical to realize—received any serious attention. Being used as devices of illumination, explanation, and persuasion, they were seen in relation to proper, theoretically grounded explanation and *real* experiments as merely rhetorical. The putting into question of this "merely," as part of a growing recognition of the role of persuasion in the constitution of scientific practice, is a historiographical achievement of the last two decades. The historian Thomas Kuhn, for example, has posed a series of provocative questions about thought experiments and their constitutive function, observing how they have "more than once," in the mid-seventeenth and early twentieth centuries, "played a critically important role in the development of physical science."[6]

Proposed here is a semiotic model of mathematical activity fabricated around the idea of a thought experiment. Such a model would identify all mathematical reasoning—proof, justification, validation, demonstration, verification—with chains of imagined actions that detail the step-by-step realization of a certain kind of symbolically instituted, mentally experienced narrative. Thus, unlike physical science, in which thought experiments are contrasted with real ones as one ratiocinative and persuasional device among others, mathematics, as presented here, will be exclusively founded on them.

Peirce seems to have been the first to suggest that mathematical reasoning was akin to the making of thought experiments, "abstractive observations" he called them. Peirce's contemporary, the physicist Ernst Mach, not only saw in them essential tools of physics but also, like Peirce, understood them as the basis for any kind of planning and forward-directed rational thought. The relation between real experiments and thought experiments is, I would argue, less obvious than it looks. To see this it will be helpful, before introducing Peirce's characterization, to go back historically to the point when the category "experimental" was itself being legitimated as the foundation of empirical science.

In their study of the origins of English experimental science as instituted by Robert Boyle in the late seventeenth century, the historians Steven Shapin and Simon Schaffer set out to demonstrate that "the foundational item of experimental knowledge [that is, the very category of scientific "fact"] and of what counted as properly grounded knowledge generally, was an artifact of communication and whatever social forms were deemed necessary to sustain and enhance communication."[7] To this end they identify several technologies—"knowledge-producing tools"—employed by Boyle to establish the categorial identity and legitimacy of scientific "facts" and their separability from purely theoretical observations. Chief among these was what they call the "technology of virtual witnessing." This amounted to a rhetorical and iconographical apparatus, a carefully controlled manner of writing and appealing to pictures and diagrams which, by causing "the production in the reader's mind of such an image of an experimental scene as obviates the necessity for either direct witness or replication,"[8] allowed the "facts" to speak for themselves and be disseminated over the widest possible arena of belief. Behind the employment of this apparatus was Boyle's desire to distance his method of empirical verification, and with it the claim that such a method produced and dealt in the natural category of "facts," from the kind of experiments involving only pure "ratiocination," unconnected to the rigors of actual experimentation, that he attributed to others, notably Pascal. Pascal was, of course, a mathematician, and making experiments out of a process of pure ratiocination—thought experiments—is, I contend, precisely how mathematicians manufacture mathematics.

The proposal, then, is that Boyle's rhetorically accomplished replacement of the actual by the virtual, the shift from doing/watching to the imagining of doing/watching, from real, witnessed and executed experiments to their virtual, reproduced-in-the-mind versions, charted by Shapin and Schaffer, points, contrary to Boyle's efforts to separate them, to a certain duplication of persuasive technique between the establishment of empirical *factuality* and mathematical *certainty*. Each comes about by the use of an elaborately designed apparatus able to mask its rhetorical features under the guise of a neutral method for discovering preexisting "facts" of nature or, in the case of classically conceived arithmetic, objective "truths" about the so-called natural numbers. Identifying the thought-experimental mechanism for mathematics might, then, provide a way of explicating the persuasional and rhetorical basis for mathematical reasoning and logic. Of course, there are differences between real experiments—however their meanings and outcomes are rhetorically processed, negotiated, and disseminated—and imagined ones. In particular, what corrresponds in mathematics to an empirical world studied by science—mathematics' external "reality," as it were—is already a symbolic domain, a vast field of ideal, semiotic objects. As a result, the distinction between direct manipulation of such symbolic objects and their virtual or thought-experimental manipulation is less obvious and identifiable. To avoid all sorts of confusion, we need to pay attention to the language of mathematical texts. More specifically, because the issue concerns the imagining of actions rather than their direct execution, we need to focus on the way texts invoke agencies and instruct them to manipulate mathematical signs. We can start from a division of mathematical discourse into two distinct modes: one valorized by the mathematical community as a proper, formal mode and the other an improper, supplemental, and extramathematical one.

The *formal* or rigorous mode—mathematics "proper"—consists of all those written texts (on paper, blackboards, screens) presented according to very precise, unambiguous rules, conventions, and protocols regulating what is a permitted and acceptable sign-use with respect to making and proving assertions, giving definitions, manipulating symbols, specifying notations, etc. I call the sum total of all such rules, conventions, and associated linguistic devices accepted and sanctioned as formally correct by the mathematical community the Code. The *informal* or unrigorous mode consists of the mass of signifying activities which, in all but the most austere and artificial instances, accompany the first mode: drawing diagrams, providing examples, underlying motivations, narratives, intuitions, applications, and intended interpretations, and generally using natural language (figuratively, ostensively, descriptively) together with iconographical devices to convey all manner of mathematical sense. I call this heterogeneous collection of semiotic means the metaCode.

As indicated, the standard account—enshrined, for example, in the writing of purely internalist histories of the subject—would be that the metaCode is epiphenomenal, a matter of mere affect, psychology, heuristics, and handwaving, subordinate to the "real" business of doing mathematics taking place in the Code, and in principle completely eliminable. I shall indicate that such a view of the metaCode as a practically necessary but theoretically dispensable supplement, a tool or prop to be discarded once the passage into pure reason and formal truth has been accomplished, is entirely misconceived.

Mathematical signs, formal and informal, are communal and intersubjective. They are transmitted by, rely on, and are addressed to subjects. In the case of the Code, I call the common, idealized sign-using agency— the universal reader/writer addressee of the Code—the mathematical Subject. An examination of the signs addressed to such a Subject reveals two principal features of formal mathematical discourse. The first is the total absence of any expression connecting the Subject to the inhabited world: no mention of his/her immersion in culture, history, or society nor any reference to psychological, physical, temporal, or spatial characteristics. Crucially, the user of the Code is never asked to interpret a message that makes reference to the Subject's *embodied* presence. This elimination of the corporeal from what is allowed as correct mathematical language has a many-sided history related to effects of idealization and demands of reductive abstraction to deliberately prosecuted programs of rigor and the need to deliver to the physical sciences a suitably "objective" formalism. But the complexities of the history of mathematical rigor aside, the result is that a group of fundamental devices, present, it seems, in all natural languages—what Peirce called "indexical" signs, Jakobson "shifters," Quine "token-dependent" signs, and others "deictic" elements—are absent from the mathematical Code. Such terms as *I, you, here, now, this, was,* and *will be,* which depend for their interpretation on the located or locatable presence of their utterers, are simply unavailable to the mathematical Subject.

According to the linguist Emile Benveniste, the indexicals, chiefly the first person *I,* are the vehicle through which subjectivity itself—via discourse—gets produced:

Language is . . . the possibility of subjectivity because it always contains the linguistic forms appropriate to the expression of subjectivity, and discourse provokes the emergence of subjectivity because it consists of discrete instances. In some way, language puts forth "empty" forms which each speaker, in the exercise of discourse, appropriates to himself and which relates to his "person" at the same time defining himself as I and a partner as you.[9]

Evidently, the mathematical Subject—though an embodied sign-writing and -reading user of the Code—is an I-less subject, unable to articulate her or his embodied subjectivity within the Code. It is rather in the metaCode, permeated by ostensive gesture and the indexicals of natural language, that this articulation can take place. In contrast to the Subject, let us call the sign user of the meta Code, the one who has access to the "I" of the metaCode and the subjectivity it enjoys and makes available, the Person.

The second principal feature of the language of mathematical texts is the giving of commands. Mathematical discourse is permeated with injunctions: define *A,* compute *B,* consider *C,* prove *D,* integrate *E,* construct *F,* iterate *G,* and so on. In fact, so dense is the network of injunctions that mathematics can appear to be entirely an operative discourse, its communications little more than ensembles of orders to be carried out. Who issues these exhortations and to whom are they variously addressed? And, given that mathematicians do no more than hang around thinking and scribbling, what sort of *actions* are demanded of them?

Structural linguists distinguish two sorts of imperative: the speaker-inclusive ("Let's go!") and the speaker-exclusive ("Go!"). In mathematics this division corresponds to a radical split between two types of mathematical command: *inclusive* ones, such as "let us consider a Hausdorff space," "define a mapping *m*," and "show that alpha is the case," ask that speaker and hearer institute and inhabit a common imagined world; they are issued by and addressed to the mathematical Subject. *Exclusive* ones, such as "invert the matrix *M*," "iterate *S*," and "integrate the function *g*," require that certain actions meaningful in some already imagined world be executed. But who carries out such actions? Clearly, it is not the Subject who can be asked to iterate indefinitely or well-order the continuum or invert an arbitrary matrix. Rather, it is an idealized and truncated version, a model or simulacrum of the Subject, which I shall simply call the Agent, and Peirce, in his description of a thought experiment, calls a "skeleton diagram of the self," that is dispatched to perform these activities:

It is a familiar experience to every human being to wish for something beyond his present means, and to follow that wish by a question, "Should I wish for that thing just the same, if I had ample means to gratify it?" To answer that question, he searches his heart, and in so doing makes what I term an abstractive observation. He makes in his imagination a sort of skeleton diagram, or outline sketch of himself, considers what modification the hypothetical state of things would require to be made in that picture, and then examines it, that is, observes what he has imagined, to see whether the same ardent desire is there to be discerned. By such a process, which is at bottom very much like mathematical reasoning, we can reach conclusions as to what *would be true* of signs in all cases.[10]

I have introduced three agencies—Person, Subject, and Agent—which I claim operate in any mathematical thought experiment. The roles of Subject (imaginer) and Agent (imagined) are evident enough, but what of the Person? Pierce's description makes no mention of a third figure or agency. Ought he to have done so?

There is an analogy here between mathematics and waking dreams. The Agent is the imago, the figure dreamed about; the Subject is the imaginer, the dreamer dreaming the dream; and the Person is the dreamer awake in language consciously observing, articulating, and interpreting the dream. The necessity for this third level occurs because the dream code is restricted, making it impossible to articulate the meaning of the dream—*as* a dream—within the frame available to the dreamer. In just the same way, the language in which the mathematical Subject operates—

the Code—is restricted by its lack of indexicality. And what is unavailable to the Subject—precisely as a result of this lack—is, as we shall see, the *basis* with respect to which the whole technology of thought experiments achieves persuasion.

How *do* thought experiments persuade? They furnish the Subject with a scenario enacted by the Subject's proxy, the Agent, of what he/she would experience. This can only impinge and have persuasional relevance for the Subject by virtue of the similarity between Subject and proxy: only on the basis of the affirmation "it is like me" is the Subject persuaded that what happens to the Agent mimics what *would* happen to him or herself. And it is just this affirmation—which rests on a recognition of a sufficient likeness between imaginer and imago—that the Subject cannot articulate, because to do so would require access to an indexical self-description denied it. Only in the metaCode, the domain of the Person, can such a description be given. Of course, if, as Peirce seems to do, one takes the imaginer to be carrying out his reflections within natural language, then language and metalanguage coincide, and the need for a Person distinct from a Subject does not arise. That being so, it is Peirce—the author advocating the thought experiment as the means of self-persuasion—who is the missing person/Person.

The logician's picture of a mathematical proof is a step-by-step assent to a sequence of logical moves ending in the assertion being validated. But, as all mathematicians know, it is perfectly possible to agree with (fail to fault) every step of a proof without experiencing any *conviction*; and without such experience, a sequence of steps fails to *be* a proof. Presented with a new proof, mathematicians will seek the idea behind it, the principle or story that organizes the logical moves into a coherent whole: as soon as one understands *that*, the persuasional structure of the proof can emerge. Proofs embody arguments—discursive semiotic patterns—that work over and above, before, the individual steps and which are not reducible to these steps; indeed, it is by virtue of the underlying story or idea or argument that the sequence of steps is the sort of intentional thing called a proof and not merely an inert string of formally correct inferences. The point of this in relation to thought experiments is that such underlying stories are not available to the Subject confined to the Code; they can only be told by the Person from within the metaCode.

Implicit in this characterization of proof as thought experiment is the idea that mathematical assertions, statements of content, are to be seen as *predictions*; specifically, predictions about the Subject's future encounters with signs. They foretell what will happen if he/she does something. Thus, the assertion "$2 + 3 = 3 + 2$" predicts that if the Subject joins 11 to 111 the result will be the same as if he/she had joined 111 to 11. The assertion "2 is irrational" predicts that no matter what particular integers the Subject substitutes for p and q in the expression $p^2 - 2q^2$ the result will not be zero. Not all assertions yield so directly to a construal in terms of predictions as these examples, and one needs, more than this, some overall argument based on the recursive construction of mathematical assertions to secure the point in any general way.

Let me summarize the tripartite structure of the technology of mathematical persuasion sketched here. There are three semiotic figures: the Agent, an automaton with no capacity to imagine, who performs imaginary acts on ideal marks, on signifiers; the Subject, who manipulates not signifiers but signs interpreted in terms of the Agent's activities; and the Person, who uses metasigns to observe and interpret the Subject's ongoing engagement with signs. In terms of these agencies any piece of mathematical reasoning is organized into three simultaneous narratives. In the metaCode the underlying story organizing the proof-steps is related by the Person (the dream is told); in the Code the formal deductive correctness of these steps is worked through by the Subject (the dream is dreamed); and in what we might call the subCode the mathematical operations witnessing these steps are executed (the dream is enacted) by the Agent.

It is possible, as I've shown elsewhere,[11] to use this tripartite scheme to give a unified critique of the three standard accounts of mathematics— Hilbert's formalism, Brouwer's intuitionistic constructivism, and Fregean Platonism. Briefly, the move one makes is to consider the triad of signifier, signified, and Subject and show how each of the standard accounts systematically occludes one of the three elements. Thus, intuitionism, relying on an idealized mentalism, denies any but an epiphenomenal role to signifiers in the construction of mathematical objects; formalism, fixated on external marks, has no truck with meanings or signifieds of any kind; Platonism (the current orthodoxy), dedicated to discovering eternal, transhistorical truths, repudiates outright any conception of the (in fact, any humanly occupiable) Subject position in mathematics. Plainly, the valorization of a proper, formally sanctioned Code over an improper and merely supplemental metaCode deeply misperceives how mathematics traffics with signs. It is a misperception intrinsic to and formative of Platonism, because in order to deny the presence of persuasion within mathematical reasoning it has to understand the language of mathematics as a transparent, inert medium that manages (somehow) to express adequations between human description and heavenly truth. On the contrary, only by understanding language as *constitutive* of what it "describes"—only through such a post-realist reversal of mathematical "things" and signs in which, for example, numbers are as much the result of numeral systems as numerals are the names of numbers which antedate them—can one make sense of a historically produced apparatus of persuasion and a historically conditioned account of the (human) engenderment of the numbers. But this is in the future: the history of the Subject, Agent, and Person, no less than the history of mathematics as a sign practice of which these semiotic agencies would be a part, has yet to be written.

Such a reversal works to dissolve a split in "number," a hierarchical division characteristic of western thought and operative since the beginning of the classical period, which valorized *arithmetica* (numbers contemplated philosophically; ideal, perfect objects) over *logistica* (numbers as empirical objects calculated with by slaves). The hegemony of *arithmetica* now appears to be over. Computer technology, the contemporary manifestation of *logistica*, by insisting on the materiality of counting/calculating as a business of real time and memory cannot but reinstate the material (electronically) embodied slave upon the mathematical scene. Computer technology's effect on and interaction with mathematics is already so massive that it has become difficult to consider the fate of mathematics separate from that of *logistica*: certainly it seems strange and increasingly artificial (outside Platonism/realism) to talk about our relation to number without invoking some conception of *agency*, some suitably idealized version, that is, of an embodied mathematical sign-using subject. In the wake of this destabilizing of *arithmetica*, that is, the putting into question of a prior, theoretically given "number," one can rephrase Galileo's challenge to the Schoolmen. Instead of asking, "Why should the physical world conform to a prior, Aristotelian metaphysical theorization of it?" one asks, "Why should counting and hence numbers be imagined within a prior metaphysical scheme that knows nothing of their materiality and is silent about how they come to be?"

Why indeed? To illustrate what is at stake here let me, finally and very briefly, indicate a recent application I have made of the semiotic model outlined here,[12] namely, putting into question and making problematic the idea of the mathematical infinite—that simple and all-pervasive idea we are asked to invoke when we interpret the ideogram " . . . " in the expression "1, 2, 3, . . . " to mean "go on counting without end, *forever.*"

The persuasive force of a thought experiment works, as I have indicated, through the resemblance between ourself as imaginer and the imago we conjure as our proxy; between the mathematical Subject and his/her Agent. This relation between imaginer and imago lies between two limits: that of total identity (which, if reached, would mean that we perform real, not imagined actions) and total difference (which would destroy any sense of the idea of a resemblance being in play). Suppose now that the thought-experimental activity is that of counting, and we ask the question: Is the Agent imagined to be corporeal in any way whatsoever? Because the Subject, as a sign–reading/writing agency, is clearly and irremediably embodied, the question is manifestly one of resemblance between imaginer and imago. If we answer *yes*, and maintain that our Agent has some—however idealized, vestigial, or attenuated—immersion in the material world, then its action will not escape the regimes of space, time, energy use, decay, and so on that govern all physical process, and it will not, as a result, be able to go on endlessly. If we answer *no*, as we must if we are to cognize infinity and counting without end, then our Agent will be a disembodied ghost, and we are confronted by the question of persuasion. Why should a Subject whose embodiment is inseparable from its engagement with signs create an incorporeal phantom as its proxy? In what way can the imagined manipulation of signifiers by such a transcendental imago persuasively image the Subject's manipulation of signs? The negative answer would be that of current infinitistic mathematics were it to admit to an analysis in terms of Subjects and Agents; but short of so admitting, neither the answer nor the difficulty it leads to is likely to make a direct impact on such mathematics. The positive answer has radical, large-scale, and interesting consequences: it would eliminate all talk of infinity from mathematics and put in place of classical arithmetic with its endless progression of transhistorical, transcendental integers—the so-called *natural* numbers—a non-Euclidean arithmetic of realizable, this-universe constructible numbers; numbers that fade into indeterminacy and nonexistence as counting them into being is ever further prolonged.

On the Take-off of Operators

Let us say the function of literature and literary studies is to make transmittable the cohesiveness of the net in which everyday languages capture their subjects. Those to whom this designation sounds strange should, first, be reminded that without communications-technical designations, there could hardly be any talk about literature and literary studies. Second, somewhat more philosophically, Goethe's earth-spirit spoke of "weaving the living garment of divinity." But, as Faust's collapse at the theatrical entry of this spirit illustrated, such nets or reference totalities of an everyday language cannot themselves be captured. In order that everything be woven into a whole, a lid sits on the loom. With it, hopeful theoreticians of the post–World War II era created—with an eye toward the threat to the intellect and its sciences posed by the formal languages of this century—the beautiful theorem that everyday languages are their own metalanguages and consequently capable of being viewed from behind and below.

But there are certainly opportunities for bringing the margins or edges of such nets to actuality without at the same time switching over to the side of formalization and thus sacrificing the communicability of everyday language. This approach toward the marginal values of language becomes more necessary the more consistently modern communications technologies, in forming compound media systems, knot together their closed and locked nets. The assurance that everyday language is its own metalanguage could soon offer little consolation in a situation where conversation, which is what we are with each other according to Gadamer, says nothing about the factual way of the world or signal flow.

Of course, this situation appears new and disconsolate only under the humanistic premise that language is entirely subsumed by conversation, and conversation by the people who carry it out. Should one invert the premise for testing purposes, then the disappearance of the subject does not reside with language, as one might suppose, but instead has always proceeded via writing and media: an escape so boundless that the modern basis theorem of humanity as the master of language becomes questionable.

Everything that lies beyond the margin of everyday language can be related to the historical vectors of this escape. In each case, what takes place are processes of take-off (*Peenemünde*), because it has been experimentally proven that no return or landing must follow them. (Perhaps Patti Smith, when she sang "Landing," was for this reason so fervent.)

Concerning those take-offs that remain in, but also with, writing, Derrida is authoritative. According to Derrida's analyses, all margins of a text—from the title to the epigraph to the footnote—are operations that simply cannot be made to speak. Quotation marks have proven to be the general prerequisite of such instances of distancing, which must always be read in even when nothing is written. They are operators of writing to such an extent that it is customary, for instance, in reading a paper at a solemn professional meeting to frame all quotations with the words "quote" and "unquote," a practice that routinely provokes laughter at the corner pub.

Derrida's deconstructions themselves, of course, operate in a domain where all operators of writing are readily available as typographical options. Thus his analyses are able to withstand Nietzsche's philosophical critique simply because Nietzsche himself, as the only philologist among the philosophers, raised quotation marks to the rank of a category.[1] But with regard to old European texts, which did not even have the spacing of structuralists at their disposal, let alone metalingual operators, their inclusion, as anachronistic as it is systematic, threatens to transform analysis into over-interpretation.

Rather than making unconditional use of all the operators in current writing practice, a more systematic approach would begin with their archaeology and determine when and for what reason a particular operator was introduced, and thus also when and for what reasons it did not exist.

For the time being, let us confine our discussion to alphabetical space without numbers and focus on scholastical commentaries from the thirteenth century, best typified perhaps by Petrus Lombardus's *Sentence Commentaries*, in which he undertook the task of explaining which speech act

Jesus had actually executed with the sacramental words of the Last Supper. Whereas the Vulgate could simply and without commentary write, "Hoc est corpus meum," theological commentaries had to trace the reference structure of this speech and be able to indicate whether the deictic expression *hoc* referred to the bread as it had come from the baker of the Host, or rather to the same bread that had been transformed into the body of Christ by the speech act itself. This delicate question led to the hopelessly corrupt Latin of William of Auxerre's *Summa aurea*: "Sed queritur, cum dicitur hoc est corpus meum, quid demonstret ibi hoc pronomen hoc" ("It is now asked, however, to what, if it means that this is my body, the pronoun this refers").[2] Lacking any operators that could have set a distinction between use of and reference to words, the impossibility of not using, but simply referring to a distinct part of the quoted sentence could hardly be more dramatic. Without plumbing any of the philosophical niceties at stake in these matters, this move reduced the then-current dispute over nominalism to the mere necessity of presenting the missing operators to an essentially commentating culture. With the nominalist differentiation between *suppositio formalis* and *suppositio materialis*, that is, between reference to the subject matter of words and reference to words as terms, it became possible for the first time to comprehend the functional difference between the two sentences, "Angels have a nature," and "Angels have five letters."

But because it is equally true of thoughts and categories that nothing exists that is incapable of being switched, the nominalist knife needed a notational operator between the two ways of referring to words that had not existed in classical Latin and could not have existed. When Richard Fishacre returned to the problem of bread and wine, this and that, in his *Sentence Commentaries*, for example, the "hoc" that William of Auxerre had been able to indicate only as "hoc pronomen hoc" was suddenly preceded by a sequence of letters, as inconspicuous as they were nonsensical, but which nonetheless eliminated Wilhelm's entire problem of formulation: "Sicut hic diceretur, quod li hoc non est demonstrativum, sed stat materialiter."[3] ("Thus would be said here that the this is not demonstrative, but instead a *suppositio materialis* [reference to its own substance as a word] takes place.")

Landgraf's comments on these examples in his *History of Dogma* exhibit charming innocence: "Those who occupy themselves with scholastics encounter at a certain moment the small word *ly*, which assumes the place of the article unknown to classical Latin. . . . In the entire twelfth century, this *ly* is not yet encountered. The article *li* thus entered the theological lecture hall from the streets of Paris and was able to sustain itself so well that it won the day over the Lombard *lo*. Of course this is not surprising when one considers that Paris was of paramount importance during the period in which this transformation took place and had a major influence on the technology of school operations."[4]

In truth, nothing is more surprising than the admission of an operator from vulgar speech into the technical language of Middle-Latin theology, and nothing more banal than to explain it simply as an Old French article, which surviving texts of this vernacular scarcely exhibit in place of pronouns or prepositions. As a definite article, *ly* would only have restored a possibility, which was known to Greek philosophers but lost in Latin, of creating any number of categories by making substantives out of verbs, prepositions, and other classes of words: in Aristotle's case, for example, the where, the that-for-the-sake-of-which, and so on. Richard Fishacre or even Thomas Aquinas, whose summas probably did the most for propagating the use of *ly*, did not refer to a category of this sort but instead to the functioning of the word itself. Predicaments were not at issue, but rather predicables. While concepts in the Greek, and thus also categories in Aristotle, automatically conformed to the field of reference being discussed at the moment, that is, could deal with the world and the *logos* of that world equally well, reference to subject matter and reference to language were separated in the discourses of scholastics, if only for the reason that, according to a thesis of Johannes Lohmann, the texts were conceived in the vernacular but written in Latin.[5] That the operator *ly* originated in a vulgar language and had to be grafted onto Middle Latin is itself a symptom of the take-off that made languages at least conceptually manipulable if not yet technically. In other words, *ly* appeared in exactly the same place where our quotation marks would be inserted after the invention of printing, and thus the invention of titles, tables of contents, and word-addresses as well, but where, under the conditions of medieval hand-writing, there was a gaping typographical lacuna.

It is probably only the ineradicable familiarity with which readers look at books that prevents them from recognizing the invention of the ability to quote individual parts of a sentence in the thirteenth century as an historical rupture. "How this page," reads Enzensberger's Gutenberg poem, "resembles a thousand other pages, and how difficult it is to be amazed by that!"[6] It is therefore only by looking at operators that are not readily available in every type composition box, because they do not belong to the basic alphanumerical accoutrements of schoolchildren or even to word-processing programs, that one can make a plausible case that operators, far more than any battles or plagues, have made history. Those who tamper with the relationship of people to signifiers, according to Lacan, change the mooring of their being—even and indeed precisely when the newly introduced operators are legible only to an elite or, in an extreme case, to machines. Compared to the take-off of numerical or even algebraic symbols, that of the alphabet is in any case merely a prelude.

In Greek there was clearly no possibility of writing the sentence "two and two is four" differently than it would be spoken. The operator "plus" coincided with the "and" of everyday language, which was fine and good only so long as no one wished to substitute intricate correlation commands for addition commands that can usually be done "in one's head." Even when a symbol for subtraction of two numbers appeared in Diophantus's surviving works, it did not imply a corresponding sign for addition. Only in 1489, in Johann Widmann's *Nimble and Handsome Calculations for Businessmen*, were the two operators of cross and horizontal line explicitly used for functions of inversion. In doing this, however, Widmann apparently still found it necessary to provide his readers with a translation, just barely pronounceable, into their everyday business language: "the − that is minus and the + that is more."[7] Once this translation could be completely forgotten, however, number columns could be manipulated independently of speech. It became historically unimportant whether Widmann's plus sign originated from the Latin *et* and his yet unexplained minus sign perhaps from Diophantus after all, simply because the two operators could henceforth prove their silent efficiency. Novalis's wish that "numbers and figures rule world history no more" was already obsolete at the moment of its formulation.

The actual take-off of operators, however, only takes place when operators issue from other operators, as if an avalanche had begun. Just like Widmann's innovations, the importation of the Arabic zero in the thirteenth century (which witnessed more than merely the introduction of the ability to quote) probably did not take place according to plan. Without philosophers noticing any signs of trouble at all, small, innocent symbols revolutionized the business of bankers and trading agents. Nevertheless, or precisely for that reason, the algebraic operators of the early modern period constructed a consistent system, at the latest after Viète's cryptographic trick of inserting letters of the well-known alphabet for unknown numbers, which only needed feedback in order finally to allow operations via operators as well.

It was Leibniz who took this most important of all steps. Just as he had drawn the logical conclusion from Gutenberg's invention with his proposal for library catalogs,[8] he also recognized nearly all of the implications of the historical accident of symbols such as zero. His correspondence with every important mathematician of the time—from the two Bernoullis to Huygens and L'Hospital to Tschirnhausen—asked all of his colleagues not only to introduce new operators for new operations, but also, "in the interest of the republic of scholars," as Leibniz wrote, to reconcile these innovations among themselves.[9] When Tschirnhausen replied that the new terminology and new symbols made scholarship less comprehensible, Leibniz wrote back that one could have made the same objection even to the replacement of Roman letters with Arabic numerals or the introduction of zero.[10] In other words, the mathematics of a Leibniz learned from the contingency of its own operators how to discover their power. Never before had anyone begun the experiment of manipulating not words or people, but instead bare and mute symbols. From Leibniz come not only the extremely technical symbols for integration or congruency, for example, but also symbols of such familiarity that it has become hard to tell just by looking at them that they are inventions at all.

There was, for example, no specific sign for division before Leibniz, only the familiar horizontal line that separates numerator and denominator. That might have been good enough under the conditions of medieval handwriting, but Leibniz explicitly criticized the fact that two- or three-line expressions were an additional burden for the typesetter and no doubt for the reader's eyes as well. Thus Leibniz, and he was the first, replaced the fraction stroke with our colon, a success that was at least Europe-wide.[11] With the feedback between symbol and symbol, alphabet and algebra, mathematics reached the technical state of Gutenberg's print.

In the summer of 1891, Conrad Ferdinand Meyer made plans for a novella about an early medieval monk who begins his career as a copyist of pious parchments and ends it as a forger of juristically and economically relevant parchments. To wit, pseudo-Isodor discovers while copying (to quote him) "what wonderful power lies in these dashes and numbers! With a small dot, with a thin line I change this number, and in so doing I change the relations of property and force in the most distant regions."[12] Because of this fraudulence, which he gradually ceases to recognize, the hero of Meyer's novella was, in the end, to go insane. Only instead of the monk, who remained a fragment, the author himself unfortunately ended up in the insane asylum because of his novella.

And that was perhaps not without reason. At the historical end of the monopoly of the book, that is, in the age of telegraph[13] and telephone, the project of dating the wonderful power of mathematical symbols and dashes back to medieval chirography was a perfect misidentification of precisely the typography that had set this power for mathematicians as well as writers.

This misidentification resided in Europe's holiest concepts, however. Leibniz measured the operators, of which he invented more than anyone else, against a truth whose opposite could then oscillate between falsehood and fraud. Symbols, he wrote to Tschirnhausen, should represent the essence of a thing, indeed, should paint it, as it were, with just as much precision as conciseness.[14] But even Gauß was still alarmed by his own insight that "the character of the mathematics of recent times" was, in contrast to antiquity, "that in our language of symbols we possess a device through which the most complex argumentation is reduced to a certain mechanism." In the good society of Goethe's time, Gauß warned against using "that device only mechanistically" and called instead for a "consciousness" "of the original determination" "in all applications of concepts."[15]

In point of fact, all of this talk about essence or consciousness was simply an imprint of philosophy onto the operators, an imprint that only the mathematical contemporaries of Meyer were able to do away with altogether. In 1849, Augustus De Morgan wrote of Euler's symbol i, which commonly designates the (imaginary) value of the square root of -1, that the repeated complaints about its "impossibility" fell by the wayside "as soon as one simply gets used to accepting symbols and laws of combination without giving them any meaning." Simply because mechanical calculations, even with complex equations, led to verifiable results, mathematics was able and allowed to use its operators for every possible kind of experiment.[16] With this explicit departure from meanings, and therefore with the last remaining link to everyday language, a symbolic logic took off, which was able to leave behind experimentation in the technical sense even of De Morgan himself, which means it was able to move into silicon circuitry. The 1936 dissertation in which Alan Turing presented the principal circuit of all possible computers no longer made the slightest distinction between paper machines and calculating machines, where the term *paper machine* was Turing's euphemism for mathematicians and himself.[17]

In order to begin this final take-off, Turing and John von Neumann only had to remove a tiny but sacred distinction that had still enjoyed inviolable authority in the time of De Morgan or Babbage: the distinction between data and addresses, operands and operators. When Babbage designed the first universal calculating machine in 1830, he shuddered at the thought of entering operations or commands into his machine in the same punch-card format that he had provided for arbitrary number values.[18] Von Neumann's machines, on the other hand, write commands and data in the same format into the same, undifferentiated databank; that is their stupidity and their power.

The take-off of operators does not complete a world-historical course of instruction that brings to maturity abstractions of ever higher levels. On the contrary, the differentiation between use and reference, signification and quotation, as it was introduced by the *ly* of the thirteenth century, can and must once more implode in order to make operators so universal that they also operate on operators. To add a number with the binary value of the plus sign itself is no problem at all for von Neumann's machines, but instead an—at least by the standards of everyday language— always lurking address error of the programming. No one can say in everyday language, however, whether such mistakes beyond human beings do not after all start up programs in the world that keep on running effectively and without crashing the system. For which reason Alan Turing, no sooner than he had gotten the first computer running, delivered the oracle that we should already now prepare ourselves for the takeover of machines.[19]

TRANSLATED BY KEVIN REPP

Switchboards and Sex: The Nut(t) Case

The days in which media history is discussed in terms of the "public sphere" are numbered. Considering the captivating power discourse on the public sphere has in the area of computer technology, it is of interest to distinguish provisionally between strategies for writing a discourse-analytical history of the media as opposed to a history of the public sphere. It is well known that Habermas's inaugural dissertation, published 30 years ago, assumes nothing short of a complete reversal in the meaning of the concept "public," said to have occurred in the eighteenth century. The concept, which up until that time had referred to the sphere of government, began now to mean every platform upon which a "civil public" admired itself as the subject of history. Another media history, less infatuated with the ideal ego of the historical subject, opposes this view and insists instead upon the literal meaning of the Latin word *publicus*, according to which a medium such as the Roman *cursus publicus* (the postal service created by Augustus) was "public"/*publicus* in the sense that private persons were strictly excluded from its use. The central issue to be pursued by a genealogical study of the history of the term *public* is the suspicion that the opening of the postal service to private persons (since the end of the sixteenth century) did not involve a reversal in the meaning of the word *public*, but instead—in system theoretical terms—an increase in the complexity of the state system, which enabled it to integrate private persons as a subsystem. In this process they came to understand themselves as the public. Of course, if one understands this differentiation as the appropriation of the "public sphere" through an autonomous act of the people, the historically untenable thesis of Sombart follows, with which Habermas agrees: one can speak of a "postal service" only "when the regular opportunity to transport mail becomes available to the general public."[1] *Nihil est sine ratio reddenda*: nothing exists, if its origin cannot be attributed to the private citizen. In this view, media are the surfaces through which private citizens communicate with one another. This ontologization of the consumer standpoint transforms media analysis into a social science operating on behalf of what it intends to criticize—the transformations of the "reasoning public" into consumers—because it has adopted this consumer standpoint at the core of its own method. Consequently, applied to the current state of media technology, the Sombart-Habermas thesis on the public sphere completely coincides with the self-advertisement of the software industry, for example, which tries to convince us that one can only speak of the "computer" when its hardware is made available to the public through user interfaces. We are the masters of the media as long as they are domesticated by user-friendly interfaces in the system of everyday language and as long as social science and the historical writing it inspires perpetuates the discourse that hardware technologies and their history, for which engineers and military purposes are supposedly responsible, are nothing more than the environment of the system of everyday language.[2] Through such means the myth of The Person is kept alive, despite the fact that since the 1940s, when Claude Shannon and other Bell Labs engineers developed a mathematical theory of communication, program technicians have long pursued goals other than understanding people.

The subsumption of media history into a *Structural Transformation of the Public Sphere* has not only more or less successfully obstructed a science of the media that would deserve the name; it also excludes women, who as agents of the media cannot be directly repatriated to capitalist production as the holy origin of the public sphere. Thus Habermas's inaugural dissertation of 1962 remains true to the Marxist-Social-Democratic tradition, which—put bluntly—knows women only without media and media only without women.

Social Democrats such as August Bebel were so uninformed with regard to the first government-paid female "telephone assistants," who began to populate telephone offices from 1887, that they had to be enlightened in the Reichstag about the technological properties of women's voices: a message, however, that apparently fell upon deaf ears. The 50th edition of Bebel's classic *Woman and Socialism* (1909) is a blaring document of socialist media deafness. Infatuated by production in general and by the classical branches of production of the eighteenth and nineteenth centuries in particular, Bebel considered white-collar workers in service industries such as the "transport system" to be "more or less parasites."[3] Although Bebel celebrated traditional women's careers in agriculture, laundry services, and house-cleaning, despite relatively stagnant employment figures, he passed in silence over the women's boom in the area of "commerce and transportation"[4] attributable to the typewriter. The fact that women occupied 76.7 percent of the jobs among stenographers in America around 1900 is mentioned only in the context of a statistic,[5] with regard to "telegraphs, telephone systems, railroads, and postal services."[6] Furthermore, not a single word about women is to be found, even though women claimed 87 percent of the public service positions in telephone offices as early as 1907, not to mention the fact that "thanks to the typewriter and the telephone, the percentage of American women who worked in careers other than agriculture or 'domestic services' rose from 20 percent (according to the results of the first United States Census) to 42 percent."[7] From Bebel to Habermas a deafness has persisted with respect to the media or to women, as the case may be, according to which production is the only, and of course, the holy calling of Woman. Until the regulations of the Imperial Postal and Telegraph Administration were revised in the 1920s, this discourse of bureaucratic rights inaugurated by Fichte allowed female telephone operators to be public servants only as long as they did not marry.

If, on the other hand, taking up a thesis of Friederike Hassauer, gender difference is itself a media-historical variable of the distribution of positions between the sexes, it would be necessary to relate social historical phenomena such as the public sphere not to production and thus in the last resort to a romantic nature, but instead to a discourse and its media historical assumptions. Thus—to return to the inaugural dissertation of 1962 one last time—even Habermas's celebrated publicly oriented subjectivity and privacy of correspondence[8] has existed only since people like Gellert created programs that institutionalized gender differences in postal matters,[9] or since a printer like Samuel Richardson, under pressure from booksellers interested in developing a market for literature in the countryside, wrote a letter writer's guide for young ladies in rural locations, which resulted in *Pamela* and the birth of the genre of epistolary novel.[10]

The invention of the telephone was not an effect of the capitalist means of production, as even a cursory glance at its history reveals. Rather my point is that the new materiality of signifiers transformed gender distribution in the discourse network simultaneously with the technical restrictions it imposed on discourse. In doing this, the telephone offers a rare case where the scene of origin coincides with the reorganization of gender distribution. Telephone systems demand a third person, in whom all first and second persons are grounded—namely, the exchange office; historically the telephone also owes its invention to the mediating office of a third person. Her name was Mabel Hubbard. It was through her that the pedagogy of deaf-mutes and communications technology made a connection.

Scarlet fever caused Mabel Hubbard to lose her hearing at the age of five, whereupon her father, Gardiner Green Hubbard, a lawyer who had made a fortune from patent cases, set out to reform the system of care for deaf-mutes in America. In order to prevent his daughter's complete loss of speech, a fate considered unavoidable at the time, Hubbard imported the so-called "oral method" from Europe, which was predicated on the scandalous conviction that deaf persons could only speak if they had not been taught classical sign language to begin with. In the new theory, deaf persons speak by reading the lips of others. In other words, they speak by becoming *operators*.

Concern for his daughter was reason enough for Hubbard to make contact with Alexander Graham Bell, a teacher of deaf-mutes who had emigrated from London and whose oral methods of teaching according to his father's system were widely celebrated in Europe. Melville Bell, Alexander's father, had developed a system of phonetic transcription called Visible Speech, the characters of which were symbolic representations of the articulation apparatus and the types of articulation. In reading, deaf-mutes educated according to this system did not deal with signifiers at all, which according to Derrida imply an imaginary hearing of oneself, but instead decoded directional signals for the mouth and tongue. (In principle, Visible Speech should have been able to teach the art of kissing as easily as the articulation of speech.) Hubbard sent his daughter to Sarah Fuller's "School for Deaf-Mutes," where Alexander Graham Bell gave lessons. Soon after, Hubbard discovered that his daughter's teacher was tinkering with a machine able to increase the data transmission rate of expensive copper cables: a multiple telegraph. While Baudot and Edison were developing phase multiplex procedures, Alexander Graham Bell had been working since 1873 on what he called a "harmonic telegraph," later dubbed a frequency multiplex. The transmission of a harmonic chord and the use of resonators on the receiving end to resolve it into its component tones to carry frequencies for signals was an idea that Bell had been pursuing since the great phonetician Alexander J. Ellis had personally referred him to Helmholtz's treatise, *On the Sensations of Tone as a Physiological Basis for the Theory of Music.* Helmholtz had succeeded in using resonators tuned to the characteristic tone of an electric tuning fork and then opened and closed over a piano to synthesize the vowel sounds "u," "o," "ö," and "a."[11] At almost the same time, the young Bell had become acquainted with Charles Wheatstone, no less famous than Helmholtz, who demonstrated for him an improved model of Kempelen's speech machine.[12] It has been suggested that Bell misunderstood Ellis, thinking that he had implied that Helmholtz had succeeded in transmitting vowel sounds electrically with his resonator piano: a momentous mistake.

Imitating Helmholtz may have led Bell to invent the telephone, but it nevertheless hindered him at the same time. Bell occasionally experimented with Léon Scott's phonautograph (which he improved by using the ear of a human cadaver) in order to produce feedback between written and spoken oscillation curves for his students.[13] But in terms of communications technology, Bell, the reader of Helmholtz who returned again and again to his principle of intermittent currents—even when the possibility of speech transmission through "undulating" currents had been accidentally proven—remained (in the words of his biographer) "incapable of thinking in terms of a membrane transmitter and receiver."[14] Hubbard, on the other hand, recognized the potential capital in the telephone experiments of this impoverished teacher of deaf-mutes. The only problem was that Bell wanted nothing other than to teach deaf-mutes to speak. Well, almost nothing: Bell wanted Mabel Hubbard as well. His desire led to the deal that produced the largest media conglomerate of all times: Mabel wrote to Bell, whom she feared more than loved, to say that she would not marry him until he developed a patentable telephone.[15] Hubbard subsequently financed Bell's telephone experiments and in 1875 founded the Bell Patent Association. In 1877 Bell married a woman who, like his mother, was deaf and mute. On the same day, his father-in-law founded the Bell Telephone Company. No wonder Bell wrote to Mabel: "The telephone is mixed up in a most curious way with my thoughts of you."[16] Telephones have since their first hour been intimately connected with women. Since the invention of the telephone, women have always been "especially good at making it work."[17] The technical media are haunted by their ghosts.

In Ma Bell, the mother of Alexander Graham, a shift in the materiality of discourse can be demonstrated that implies the silencing of speech in the first person as a feminine career possibility and comprises the immediate prehistory of the telephone. That this demonstration occurred within the function of male authorship is characteristic of that muting itself. My case in point is George Bernard Shaw's *Pygmalion*.

The parallels between the cockney speech of the flower girl and the deaf-mute condition of Bell's students is no coincidence. The empirical Eliza Doolittle was very probably Eliza Grace Symonds, Alexander Graham Bell's mother, who was deaf and mute from birth.[18] And the connections extend further: Shaw owes the plot of *Pygmalion* to the Bell family. A. G. Bell had a grandfather named Alexander Bell, Senior, who taught speech etiquette classes for young ladies at Cavendish College for the Instruction of Ladies in Languages, Arts, and Sciences, located on the same Wimpole Street as the fictional laboratory of Professor Higgins. This same Bell published a five-act play in 1847 called *The Bride*, which begins with the following remark by a valet named Allplace: "How much I have improved the manners of this family. . . . Polishing a prosy lawyer into a tolerable baronet is a task to break a man's back."[19] A copy of this rare text was sent to Dublin to Grandfather Bell's first son, David Charles, who left it to his own son, Chichester, from whom it eventually found its way into the hands of his friend, George Bernard Shaw. The empirical Henry Higgins, the eccentric phonetician Henry Sweet, was actually acquainted with the empirical Eliza, because he was a frequent visitor in the home of his colleague, Melville Bell. Shaw not only bequeaths Higgins Sweet's first name; he also has him mention in the same breath Alexander Melville Bell's *Visible Speech* and Henry Sweet's *Broad Romic*,[20] an adaptation of *Visible Speech* and the direct precursor of the International Phonetic Alphabet.[21] In the preface to the first edition of *Pygmalion*, Shaw himself refers explicitly to the "illustrious Alexander Melville Bell." Finally, a striking line uttered by Higgins in Act Five, when he doubts whether Eliza "has a single idea that has not been funneled into her head," recalls almost literally Eliza Bell's situation: her ears were in fact accessible only with a megaphone. When A. G. Bell, en route to the telephone, connected Koenig's manometric flame to a speaking tube, just as Professor Higgins does in *Pygmalion*, he commented upon the construction with these words: "just like Mama's tube."[22]

At the very center of *Pygmalion* lies this question: Is it possible for Eliza, a flower girl—or speaking machine—to establish an identity using language of which she is not the subject? Shaw dramatizes nothing other than the discursive role of deaf-mutes whose acquisition of language is based on the "oral method." At the same time, Higgins judges Eliza's discourse with the same argument as the opponents of the "oral method." Sign language supporters complained: "We can impart to the deaf a certain deficient degree of vocalization, we can teach a few of them like parrots to pronounce [words] somewhat like we do, but we can never bring them to understand the use of vocal language in its articulation, its emphasis, its meaning."[23] Higgins argues along precisely the same lines that if Eliza had to "speak for herself," one would "find out very quickly [whether she] knows even one word that I did not put into her mouth."[24] Since the end of the eighteenth century, pedagogy had urged the importance of students finding their "own words" if they were to become humans rather than parrots[25]—but this emphasis evidently left female students of phonetics cold at the end of the nineteenth century: Eliza professionalizes her parrot status at the end of the play, at the cost of the classical comic ending, in order to become an assistant to a phonetics specialist.[26]

Because the first person of Eliza Doolittle is spoken by Higgins, as the first person of Eliza Bell was spoken by her husband and the first person of Mabel Hubbard by her father, nothing remains for them but speaking, and speaking on, extraneous lines—no differently than the tens of thousands of operators in the switching offices after them, who were in principle neither transmitters nor receivers of a discourse, but rather third persons, in whom all first and second persons ground themselves. The office of the *operator*, the female telephonist, implements Heidegger's insight that "only because in everyday speech language itself does *not* come up for discussion, but instead keeps itself in check, are we able to speak of a language."[27] In the words of Brenda Maddox: "The operator was not allowed to use her own words."[28]

This keeping oneself in check offered uncounted women the chance to occupy roles in the discourse that had previously been reserved exclusively for men. For technical reasons alone, the telephone made unavoidable the creation of a governmental post for the third person, in whom the capacity to establish an individual discourse was lost. Because the renunciation of all claim to the discursive authority of the ego, with which, according to Benveniste, "every speaker takes possession of the entire language at his own risk,"[29] could not be expected of male subjects, who since the Prussian educational reform were supposed to be educated by the state to become persons, and thus to be capable of appropriating language, women as third persons became indispensable. Women alone could commit the act of speech dictated since 1881, which combines a third person with a deixis and is, according to Benveniste,[30] completely impossible: "Switchboard here, what is required?"[31] Because civil servants had, as a result of their training, developed a responsible ego that could hardly be subjected to a prohibition on the use of one's own words,[32] the Department overturned Fichte's verdict against women in governmental posts[33] and created a position for women, the impossibility of which yet corresponded to the impossibility of the sentences spoken by them: the position of an untenured civil servant—that is, of an *operator*. As telephone operators, women professionalized a discursive authority that is the definition of insanity itself. The beginning and the end of auctorial speech was accomplished in 1878 by Miss Emma Nutt, the first telephone operator, hired by the Boston Telephone Dispatch Company, the company founded under the license of Hubbard and Bell. Noting that the operator's discourse coincides with the discourse of insanity—that is, not being the one who speaks oneself—one could observe that telephone operators are nut(t) cases. Soon, thousands from the army of job-seeking women left behind by the American Civil War followed Nutt. And it was all the easier because men as *operators* at the switchboard were (in the words of one of the first female telephone operators) "complete and consistent failures."[34]

In Germany, state secretary Heinrich von Stephan at first stood in the way of such an elimination of men from the communications service. As early as 1872, even before the era of the telephone, he explained to the Reichstag that "no institution is less suitable for providing women with employment than the Imperial Offices of Communications."[35] Although Stephan had been able after only two years to mend the breach opened by the hiring of 250 female telephone operators in 1874, this precedent was finally overturned in 1881. In the same year the Remington marketing department discovered that the armies of unemployed women seemed destined to make a commercial hit out of the first mass-producible typewriter,[36] the first telephone switchboard was set up on Berlin's Französische Strasse.[37] After the multiple changeover switch (which allowed several thousand connections to be made from a single seat) was added in 1885, Stephan finally had to surrender, under pressure from the Crown Princess and later Empress. In 1887, the first "assistant workers for the telephone service" were hired as nonbudgeted civil servants.[38] The massive entry of women into the rooms of the unyielding postal administration ensued only after the death of Stephan in the year 1897. His successor, General Major Victor von Podbielski, opened up all areas of the postal administration to women with an edict in February 1898.

With Lilian Sholes, daughter of the inventor of the Remington and the first female typist,[39] and with Emma Nutt, the first female telephone operator, the beginning of the end had come for asymmetrical distribution of genders in discourse. While the typewriter delivered "the expediency of a printing press" into the hands not only of poets but also of stenotypists,[40] the switchboard led to the perforation of gender boundaries in German civil service rights. These developments, incidentally, mutually augmented each other, just as new media in general are in fundamental solidarity among themselves and mutually interchangeable. The telephone "helped to expand the stenotypist's area of responsibility immensely. . . . In no time, the telephone made the amount of work to be taken care of grow to enormous proportions. Pyramids of paperwork piled up because of a small network of in-house connections in a single firm."[41] The acceleration of business and command routes by the telephone, which only typists with their record strokes per minute were capable of handling, left chancery clerks, who wrote calligraphically but slowly, with their elite consciousness but without jobs. Permanent—by which is meant male—civil servants survived only in ecological niches they owed solely to the prudery of directors. In 1912 the "greater dexterity (of women) at the machines"[42] trumped the higher education of the men once again when the adding machine allowed them entry into the postal check offices opened in 1909.[43]

The First World War finally ended the exclusion of women from communications institutions. In 1917, fully 75,000 women civil servants and wartime auxiliaries stood in the service of the RPTV, and in the period after the war they were not dismissed: a regulation issued in view of the strategic situation on December 15, 1917, assured that.[44]

Of course, this regulation, like all other social safeguards for women, applied only under a restriction: their single status. Until 1929, "wedlock dissolved the contract of employment on the day of the marriage ceremony."[45] Should their highest purpose, producing and raising new civil servants for the Fatherland, catch up with the irreducibly numerous women at the switches, then they would be excluded from service to the machine of state, as they always had been in Fichte's interpretation. The truth in the transcendental proviso of the bureaucratic state demands that mothers be withdrawn from service. Marriage becomes the solitary and state-sustaining criterion of gender: "The difference between man and woman in the civil service, however, that cannot be cleared away without shaking the foundations of a disciplined civil service lies in the exclusion of the married woman."[46] Women are either married to machines or to men.

At the same time, however, telephone technology newly ordained the gender traditionally excluded from the sacrament by the priestly bureaucracy. Analogous channels of communication had destroyed the monopoly on transmitting discrete signals, which had existed from time immemorial. The transcendental origin of a self-presence of the mind (identified as The Individual or The Civil Servant) had previously justified the exclusion of women from state institutions. Now there was reason enough for their inclusion: the voice. Because the frequency range of a woman's voice was more completely encompassed by the frequency band transmitted by the telephone (originally 1,000 to 1,500 hz, after the introduction of the first intermediate amplifier to 2,000 hz, since 1929 to 2,400 hz),[47] it was simply more suitable than a male voice for telephone transmission. In 1898, after Stephan's death, the Reichstag was able to hear it from the mouth of the State Secretary of the Imperial Postal Service himself: "With regard to the ladies' question I must repeat that women in particular are especially well suited for the telephone service, because their vocal range is three tones higher and as a result they are more easily understood than men."[48]

That the voice includes rather than excludes women from the discourse implies a fundamental shift of the linguistic *episteme*. As long as the voice was a presence that exists in that it disappears, the designation of its true essence fell to the fields of philosophy, classical-romantic poetry, psychology, and anthropology. All that changed abruptly after 1877. The media of sound recording and transmission not only revealed to physicists and physiologists that the existence of the voice is grounded in the parameters of frequency, amplitude, and phase; more dramatically, such media allowed these parameters to be individually manipulated, making possible an analysis of the very thing that, during the romantic era, had been thought to be the indivisible foundation of all syntheses. Of course, there had been voice physiologists even before the invention of the telephone, just as there had been tonal research in physics. But their sound-emitters never really advanced beyond the technical standard of romantic machine-magic or carnival sensations. Johannes Müller acquired the larynx of a cadaver in 1835 and blew upon it "in order to study the production requirements of specific vocal sounds in concreto"; it sounded "like a carnival pipe with a rubber membrane."[49] Helmholtz, as already mentioned, experimented with tuning forks, the sound energy of which was regulated by resonators; others, like Toepler and Boltzmann in 1870 or Rayleigh in 1877, were still tinkering with the organ pipes Kempelen had used long before.[50] The voice itself was ruled out as an object of research or as standard output for lack of the necessary equipment. It was no wonder, then, that after 1877—four years before Stephan set out to spread news of the telephone to the rest of the world—a stampede of physiologists, medical scientists, and physicists to the new media of sound transmission and recording began. Because the signal sent by a telephone transmitter reaches the receiver through induction at a phase delay of $P/2$, Du Bois-Reymond was able to confirm Helmholtz's theory that timbre was independent of phase delay as early as 1877.[51] In 1878, on the other hand, Ludimar Hermann presented the bold thesis that, because of the telephone, either the general law of induction or Helmholtz's theory of the intensity of timbres dependent upon partial tones had to be discarded. Even if induction by itself did not bring about a change in timbre, this still had to occur as a result of the increase in amplitude of higher partial tones in relation to the amplitude of lower ones caused by induction.[52] Picking up on this, H. F. Weber countered Helmholtz only two weeks later by pointing out that timbre was in fact completely altered through telephone transmission and that Hermann, in preserving the laws of induction, had only forgotten to take the self-induction of electrical circuits into consideration in his calculations.[53] Around 1900, Max Wien finally demonstrated that the condition necessary for recognizing voices on the telephone had to be located in the sensitivity of the ear, because the intensity of partial tones, and consequently the timbre, was severely altered according to the properties of the characteristic sounds of the telephone membrane.[54] In addition, investigators made telephone calls using frog muscles and nerves as components of the circuit; while in St. Petersburg Tarcharow used the telephone as a receiver for detecting electrical currents in frog muscles (and thus as a galvanoscope)—and heard "a distinct musical tone on the telephone."[55] Hermann used it conversely as a transmitter: "Of the vowel sounds, 'a' affects the frog preparation most intensely, next come 'o,' 'ou,' 'e,' and the weakest is 'i.' "[56]

In 1892, Hermann Gutzmann, the first private instructor of phoniatry, began to use the restricted telephone frequency band of 1,000 and 1,500 hz for his "Investigations of the Limits of Linguistic Perception," after the consistent perceptive distortion of his name into "Butzmann" or "Dutzmann" on the telephone made him aware that "sounds with similar acoustic character [in this case *b*, *d*, *g*, phonetic stops] are consistently confused" in the process.[57] Finally, in 1921 Carl Stumpf seized upon the telephone in order to confirm his formant theory of consonants.[58] Whereupon the president of the Imperial Office of Telegraph Technology, Karl Willy Wagner, seized upon the formant theory: damper series connected to telephone lines made the formant regions predicted by Stumpf unmistakably audible and allowed Wagner to use them as a basis for redefining the frequency range necessary for sufficient comprehension on the telephone.[59]

Such conflations of measurement standards with the objects they measure had become commonplace since Boston otologist Clarence John Blake provided Alexander Graham Bell with the cadaver's ear that revealed the possibility of membrane-directed currents, and since Blake for his part used the result of this revelation, Bell's telephone, to develop a supra-individual method for testing hearing.[60] Thus the hearing norm owes its existence to the gift of a cadaver's ear.[61] Collaboration in Germany was similarly pursued by intertwined but different interests: on the personal advice of Helmholtz, Werner Siemens made the membrane of his telephone in the form of a drum head.[62] But such connections between engineers and physiologists only marked the beginning of a symbiotic coalescence of two entire research disciplines. Specifically, to the same extent that technical laboratories were set up at the physiological and anatomical institutes of the universities, departments for research on speaking and hearing came into being at media conglomerates and governmental institutions. In 1905 the British General Post Office began to administer articulation tests. A physiology of the voice and hearing that actually suited the specific needs of the media was finally elaborated in 1913, however, when physics and chemistry professor Irving B. Crandall joined the research staff of Western Electric (an AT&T daughter company, of course) with the declared goal of "arriving at an accurate physical description and a standard of measurement for the mechanical operation of the human ear in terms that we can directly apply to our electrical and acoustical instruments."[63] The voice and ear became elements of a technical communication system: "It was apparent that great advantages would come from similarly analyzing speech and hearing; for an accurate knowledge of every part of a system, from the voice through the telephone instruments to and including the ear, would permit more intelligent design of the parts under control."[64] And at the very latest since Harvey Fletcher, Crandall's successor who joined the Department of Physical Research at Western Electric in 1916, initiated "the first major phase in the investigation of language, speech, and comprehension,"[65] "the nature of speech" has no longer been the soul or "self-manifesting subjectivity," but instead relations of frequency and amplitude. Because of this, the "interpretation" of speech is no longer the business of philosophers and linguists, but instead of engineers and physicists for whom the voice is nothing more than a part of the telephone system. As a result, the study of speech includes rather than excludes women. Their vocal range is not only better situated in the telephone frequency band; it is also significantly more conducive to the articulation of language. Specifically, while the elimination of all frequencies below 500 hz reduces speech energy by 60 percent without any loss of articulation, the elimination of frequencies above 1,500 hz reduces articulation by 35 percent, but speech energy only by 10 percent.[66] Conclusion: since the invention of the amplifying valve, nothing speaks in favor of men's voices on the telephone. Women's voices, on the other hand, are not the obscure, prearticulate origin of the clear and distinct articulation of men; they are instead *claires et distinctes* in a manner that has nothing to do with the philosophy of language.

Since psychogenetic voices began to count less as system compatible ones counted more, writing thus lost its power over women. Kafka's observation that writers, discourse clerks in anonymous offices, "through all their labors still do not acquire any right to be treated with love,"[67] expresses this insight, so extraordinarily painful for authors. Precisely this right of auctorial speech to love, which sent the hearts of women readers flying to the poets since Goethe's *Werther*, had founded an entire classical tradition along with its cult of authors' names. Engineers and telephone operators, on the other hand, waited for the day when a happy distortion of the partial tone intensity of their voices would produce a siren's song on the other end of the line.

TRANSLATED BY KEVIN REPP

Politics on the Topographer's Table: The Helvetic Triangulation of Cartography, Politics, and Representation

Maps are crucial elements in the construction of states. They are essential, of course, in delimiting the physical site, the territorial boundaries, the natural resources, and even political constituencies that make up the state. But maps are also representations saturated with national ideologies. Maps are mythopoetic structures, formative iconographical elements in the physical embodiment of the myth of the nation, and as such participate in constructing and stabilizing the state's defining powers.[1] Maps are made to appear as disinterested scale models, inert records providing a transparent window on the world analogous to the photographic image—a message without a code.[2] Yet maps, Bruno Latour reminds us, should be understood as "immutable and combinable mobiles" constructed from networks that accelerate the passage and control of information from periphery to centers of control and calculation through the establishment of obligatory points of passage. Mapmakers achieve their purposes by subjecting the networks of information gathering to a metrology, a grid, systems of measurement and accounting; and at the same time the map's ability to carry out the purposes of those centers of calculation depends upon inscribing the system of signs constituting the map onto the outside world. "The outside world is fit for an application of the map only when all its relevant features have themselves been written and marked by beacons, landmarks, boards, arrows, street names, and so on."[3] This feature of mapmaking alerts us that maps are interested, and that in order for their interests to be achieved the system of signs must be accepted as a basis for interpretation.[4] Accepting the map as a natural description of the world is acknowledging the sovereignty asserted by the state or center of calculation in whose name the map is produced as its instrument. When we consider that maps are part of a semiological system, a system of values and signs constituting and sustaining the myth of the state, then it makes sense to consider that achieving consensus on the formal properties of representation embodied in the map should be taken as a crucial step in nation building.

The central thesis of this essay is that mapmaking played a central role in the principal stages of state building usually identified by political historians as crucial to the formation of modern Switzerland, from the first stirring of efforts to conceptualize a Helvetic Republic in the 1750s through the intensive period of state building between 1830 and 1874. Producing a map of the Swiss Federal State entailed more than providing a reliable system of cartography. Agreements on the map were intimately bound up with ideological debates at the heart of the emerging new nation. To accept the map of the Federal State as a legitimate description of the geographical region between the Lake of Constance and Geneva was to accept the defining power of the Swiss state. In the next section I will discuss the "patriotic dreams" of the proponents of the Enlightenment for the transparency of spaces, whether cartographical, literary, or political, which formed the point of departure for both mapmakers and state builders in Switzerland. These early enlightened efforts in cartography were, however, without issue until the 1830s, when the political mood shifted suddenly in favor of liberalism and the accompanying intensification of the national cartography project in Switzerland. For this renewal of state crafting, science, as I will argue in the section "Political Constellation and Federal Cartography," proved to be both instrumental as well as politically expedient. The following two sections concentrate on the technical cartographic aspects of the national topography project led by Guillaume-Henri Dufour. In these sections, while showing that the path from landscape to map is a sequence of transpositions of the organization and labor involved in the processes of recording,[5] I examine the legitimizing function that came to be attributed to precision survey techniques, which served the purpose of both increased domination over nature and the construction of the bourgeois state as the nation. In the final section I discuss how "cartography," a system of recording that appeared to subject nature to the rule of measure, reached its full incarnation in photogrammetry, a supposedly uncoded, objective mechanical reproduction of the landscape. This extremely popular system of recording promoted the "assemblage" of Switzerland, but it did so only by eliminating certain images of nature through rigorously disciplining the senses.

The Transparency of Spaces

Designed in accordance with the scientific methods of the French cartographic school around Jacques and César-François Cassini[6] and geared toward the creation of a central "bureau" for topographical surveying, the first plan for a national survey that was to cover Switzerland with a trigonometric net originated in 1754 in the prison of Aarburg. Jacques Barthélémi Micheli du Crest (1690–1766), the author of the project, had been at Aarburg as a state prisoner of the Canton of Bern since 1749.[7] The cartographer an enemy of the state? The case is much more complicated. Micheli du Crest belonged to the family of Geneva magistrates, became captain of a regiment in the French service in 1713, was elected to the Great Council of Geneva in 1721, and maintained contacts with renowned scientists of his day both at home and abroad.[8] Published in 1730, his "Carte des environs de Genève" garnered him international respect as a cartographer, just as his other publications did in the areas of astronomy and experimental physics.[9] An additional field of scientific engagement, devising plans for the construction of fortresses, brought him less success. On account of the criticism of a government project to restore the city's fortifications,[10] which he published in 1728, Micheli du Crest was expelled from the Council in 1730, and his property was confiscated; a defense written in Paris[11] led to his decapitation "in effigy" in 1735. Micheli du Crest's political activity in Paris brought him just as little fortune, and for that reason he also had to leave Paris in 1742, traveling to further stations of exile in Zurich, Bern, and Neuchâtel. This, at least in the eyes of the Bern magistrates, must have brought him close to the Henzi Conspiracy,[12] the only serious attempt to depose the patrician ancien régime in Bern through violence. On August 18, 1749, he was sentenced to life in prison.[13]

Although it was thus not his scientific work that put Micheli du Crest behind bars, it would nevertheless be naive to assume that his political struggle against the aristocratic, exclusionary tendencies of the Bern patriciate had nothing to do with the *scholar* Micheli du Crest at the ripe age of 59. The question is, however, what constitutes the political dimension of a thermometer or a topographical map? An answer can be reconstructed from a letter to Albrecht von Haller of May 1755, in which Micheli du Crest informed the doyen of the Swiss Enlightenment about his previous cartographic work in Geneva:[14]

I was surveying the terrain for a detailed map of Geneva and its surroundings, in which every house, every hedge, every path, every different variety of planting, every stream, every escarpment, slope, and hillock would have to be precisely measured. I had done this with the chain in French territory, and it had given me no little difficulty; I saw that the difficulty would be greater still in Savoy, where I had twice the terrain to survey. In my room I had a table seven feet long and five feet wide on which I drew a provisional map with great precision and consequently saw the empty space that I would have to fill in Savoy. From elevated sites all over this district, first from the steeple of St. Peter's and then from a number of such points throughout the region, I sketched large maps of various landmarks and their alignments (trees, houses, towers, rocks, crosses, woodpiles—in a word, everything that I could use as an indicator), so that I obtained, on different maps, fully a dozen stations of alignment in every direction. Then, on my table, using these maps, I caused the aligned objects drawn from these different stations to intersect, and when I could see that all my alignments crossed at the same point, I could be sure of the perfect accuracy of an object's placement.[15]

Micheli du Crest's method of work was precise, unerring, self-critical, and did not stop short of even the smallest topographical details. There was practically nothing, his letter to Haller stated, that had escaped his view. And this view was of a new quality, because unlike the absolutist view—a view predicated on a centralized perspective—it did not proceed from a particular point in a prescribed direction, but instead allowed a change in perspective as often as one wished. In a manner similar to the school of Dutch painters, which as early as the seventeenth century had begun to replace the (Italian) central perspective with a perspective of several vanishing points, the enlightened cartography of the eighteenth century introduced a recording technique that no longer permitted a single, exclusively valid standpoint of the observer. The depicted space was transformed into a "sequence of rooms or vistas successively viewed"—"additive works that could not be taken in from a single viewing point."[16] In Micheli du Crest's terrain recording, the observer can direct his view from different stations "en tout sens" onto the landscape, until he is sure of having correctly recognized its form. In doing this, he makes use of a technical system of recording that allows him to verify the surveyed lines and points of intersection on the survey table from virtually any other position, "so that, with the finished work divided up into smaller sections, each section thus had seven or eight landmarks, designated by numbers and small red circles marking their positions for my records, and serving as focal points in that section."[17] In this way, the topographer could emancipate himself from the constraints that the landscape imposed on him and move freely within it, or rather, within its depiction.[18] The statement is clear: Once a perspective that is quasi-independent from the observer is introduced, anyone in the possession of his rational faculties can be an observer, reenact and verify the work of the cartographer. Cartography was to cease to be an opaque science of decree.[19]

Haller, the founder of experimental physiology, would have understood very precisely what was at issue in Micheli du Crest's letter. Experimentation and empirical knowledge were the only reliable foundations of scientific research for him as well. Moreover, Haller had himself taken a first step toward the "emancipation of perspective" in his poem. "The Alps," a paradigm of Enlightenment. The poem presents the reader with a breathtaking motion by describing from an elevated position—perhaps the Chasseral—[20]the alps, the midlands, the lakes of the countryside surrounding Bern, the mountains of Freiburg, and—*turned around backwards*—the hilly outlines of the Jura as a "streak of green valleys, which, twisted here and there, vanishes in the distance."[21]

Micheli du Crest had taken the emancipation of perspective prefigured by Haller a step further when he consistently applied vertical projection in his Geneva map.[22] He did this, perhaps, at the cost of relinquishing the "ever novel delight" of which Haller had spoken, but certainly out of obligation to the utilitarian primacy of measurability—"[everything] would have to be precisely measured." The landscape was no longer simply represented as a panorama by Micheli du Crest, but rather described according to a perspective that was only conceivable in exclusively geometrical terms. Unfortunately, Micheli du Crest the political prisoner had to be satisfied with panoramas;[23] but as a free cartographer he wanted to use "vertical projection from beginning to end."[24] In order to do this, of course, he needed to emancipate himself from the hands of the state. In Aarburg, however, his view was held captive—"[he was] a man held captive in a stronghold, and one who could only work from a single position"[25]—and his topographical work remained to the end of his life a "patriotic dream."[26]

Micheli du Crest's enlightened intention to represent space in a geo-metrically transparent or critically rational manner, and thus to configure it appropriately for the emerging bourgeois public sphere, becomes ap-parent the moment one takes a closer look at the "Nova Helvetiae Tabula Geographica." Johann Jakob Scheuchzer (1672–1733), a doctor of medi-cine and professor of mathematics, presented it to the Zurich government in 1713, after eighteen years of work. The "Scheuchzer Map" appealed to a large market, both domestic and foreign, was reprinted in Amsterdam and Paris as early as 1715, and attained an extraordinary degree of popu-larity in Switzerland in numerous reprints and copies.[27] It was not merely Scheuchzer's submissive cover letter to the authorities that made his a map of the ancien régime, but precisely his understanding of nature, which differed completely from that of Micheli du Crest and Haller. Rivers and lake shores were rendered in a thoroughly schematic fashion, mountains "stylized in a formal wavy line and placed somewhat closer together for high mountains, without any characteristic stamp."[28] The bourgeois crit-icism of the nineteenth century would accuse Scheuchzer of having had no "sense for true natural forms."[29] His map was not—through change of perspective and of the observer—intersubjectively verifiable ("true"), but instead, despite barometric elevation measurements, resembled a mag-isterial decree: officially decreed perception of natural space. Even the villages, towns, and cities were distinguished as Catholic or Protestant according to the preeminent political-confessional criteria, recently re-affirmed by the Second Villmerger War of 1712.[30]

The difference from Enlightenment topography emerges particularly clearly in the decorative border of the Scheuchzer Map. The border contains, in addition to the allegorical representation of the most im-portant Swiss rivers, a frightening depiction of the Devil's Bridge, and beyond this the "Climb up to Gemmi," engraved by Melchior Füssli, including waterfall and rainbow. Landslides, avalanches, the Lucerne Dragon, and ball lightning keep them company. Whereas Haller emphati-cally called for "Reason's light [to] brighten the vault of the Earth,"[31] Scheuchzer drew threatening natural phenomena and forced them to the margins of his map, but even so he did not achieve the radical new dimension of the "Carte des environs de Genève" by Micheli du Crest.[32]

Micheli du Crest's too-early attempt to apply the principles of mea-surability and empirically precise depiction[33] of topographical conditions consistently to Swiss cartography, and his clear renunciation of dragons, fossils, and ball lightning, remained as suspect in the ancien régime as the demographic-statistical works of Jean-Louis Muret or Johann Heinrich Waser. They had either to be forbidden or neutralized in some other way, no matter how.[34] It was probably also for this reason that Micheli du Crest wrote a proposal to begin a trigonometric survey of Switzerland, which he handed over to two members of the federal Diet in 1754.[35] "To have precise, highly detailed, well-drawn maps of the whole Country is a very useful and convenient thing for the government of a State, since one judges infinitely better from the plan of a place than from the place itself, for the plan represents to the eye not just the lands but also everything that surrounds them for farther than the eye can see," Micheli du Crest wrote in his memorandum. The proposal, however, never received careful attention.[36] The transparency of geographical spaces advocated by proponents of the Enlightenment—where "the sun's light streams through fleeing mists"[37]—was threatening to the ancien régime because it implied transparency of political space as well. At stake was the construction of a bourgeois public sphere, which already existed in practice in the scientific, economic, and patriotic societies of the eighteenth century, and which finally began to demand recognition in the political sphere as well.[38]

Political Constellation and Federal Cartography

The example of Micheli du Crest demonstrates not only the political dimension of the cartographic enterprise, it also makes clear that the eighteenth century was not lacking in conceptions for founding a national cartography.[39] Models and instruments for this enterprise were available. Theodolites, surveying poles, barometers, survey tables, thermometers, hygrometers, and telescopes of sufficient perfection were also on hand. Even the problem of projection, or of the curvature of the earth, had been sufficiently solved, thanks to the French survey work in Peru and Lapland.[40] In the ancien régime, at least in Switzerland, the missing element was a national authority interested in creating political instruments to empower the surveyors' tools. A political authority was still lacking that could have identified with a national cartography project strongly enough to provide the financial resources necessary for its realization.

With the downfall of the Helvetic Republic of 1804, that kind of political authority receded far into the distance. The Confederation was nothing more than a loose coalition of states.[41] The resolutions of the Diet had de facto no instrumental force; their realization depended largely upon the cooperation of the individual cantons. Solutions achieved at the federal level came about for the most part under the political and military pressure of Napoleon Bonaparte.[42] With Napoleon's fall, the centrifugal forces already manifest were strengthened even further. The federal treaty of August 7, 1815, was not even an actual constitution, but rather an "agreement that combined the cantons in a league of sovereign states."[43] The reaction led by Karl Ludwig von Haller (the grandson of the Enlightenment figure), wanted to return to the pre-Napoleonic conditions. Everything that could have supported national unity was abruptly abrogated. Each canton had its own currency, its own postal service, and insisted upon the sovereignty of its military and foreign policy affairs. More than 400 customs barriers prevented free commerce; different measures of distance, weight, liquids, and grain applied in almost every canton.[44] Only in the area of military organization were higher, confederal authorities formed.[45]

In these circumstances a unified federal cartography was unrealizable. Until the early 1830s, the budget for confederal surveying remained at an extremely low level. The Diet listened patiently to the reports of Quartermaster Colonel Hans Konrad Finsler (1765–1841), it is true, and repeatedly vouched for petty credits from the confederal military fund, but a substantial contribution to a national project was out of the question.[46] A combination of completely different factors eventually made it possible to secure regular federal outlays for surveying around 1833.[47] First, the Swiss Naturalist Society, an organization dating from the eighteenth century, began to promote a national mapping project. On the occasion of an annual convention at the Great St. Bernhard, the Society commissioned a committee to work out the details of a program and an appeal for subscriptions for a Swiss map.[48] In this "Appel au zèle scientifique," the confederal authorities were above all criticized for having in the past years proven themselves incapable of at least canvassing the entire country with a trigonometric net. The Society resolved to take up the cause energetically and to exert cooperative pressure on the confederal military authorities.

Second, opinion within the army itself changed with regard to the work that had thus far been accomplished.[49] In 1832, Quartermaster Colonel Ludwig Wurstemberger (1783–1862), who had been chosen as Finsler's successor, invited all persons involved in survey work to a conference in Bern, which was to last for several days and to assist in the reorientation of Swiss triangulation.[50] In fact, a fundamental consensus emerged as a result of this conference concerning the current poor state of domestic topography as well as the steps necessary for an improvement of the situation. The protocol ascertained that after 23 years of work there was still no certainty about the longitude and position of the two groundlines near Zurich and Aarburg, and that confederal triangulation did not in any way agree with the triangulation projects of the individual cantons. Finally, one had to admit that it had not yet been possible to complete the triangulation net. What made the conference "one of the most important moments in the history of Swiss surveying"[51] was an agreement on the mapping scale to be used in the future, the selection of a method of projection, and the decision to use the meridian and the parallel latitude through Bern as orientation for the net.[52]

Third, the intensive promotion of national cartography should be seen in the context of a far-reaching transformation of the political system. Encouraged by the Paris July Revolution of 1830, liberal politicians stepped forward in many cantons, demanding recognition of the principle of national sovereignty, elimination of suffrage limitations, separation of powers in the state, as well as creation of a more transparent administration. Eleven cantons, in which altogether more than two-thirds of the entire Swiss population lived, gave themselves liberal constitutions in less than a year and guaranteed the political equality of their citizens, the separation of powers, the right to petition, freedom of the press, and freedom of trade and commerce.[53] The creation of media for expressing popular opinion, led by the *Neue Zürcher Zeitung* and the liberal opposition's *Appenzeller Zeitung*, the revitalization of the Helvetic Society, and the founding of the Confederal Rifle Club (1824), the Swiss Gymnastics Club (1832), and the Officers' Society (1833) created the forums of a bourgeois-liberal public sphere in Switzerland that looked far beyond the boundaries of the individual cantons and were generally receptive to "national" issues. At the same time, liberal politicians energetically set themselves to the task of revising the Federal Treaty of 1815. At the end of 1832, the Diet was presented with a moderate liberal "Federal Charter of the Swiss Confederation," which was to realize at the federal level the liberal constitutional changes achieved in the cantons.[54] One year later, the Diet increased federal outlays for cartography, tripling the sum of the previous annual budget,[55] and continued this policy of increased support for topographic-scientific projects in the following years as well (Fig. 7.1).

The political transformations between 1830 and 1832 resulted in a significant shift in the social composition of the army leadership, which was also beneficial to the national cartography project. The patrician-

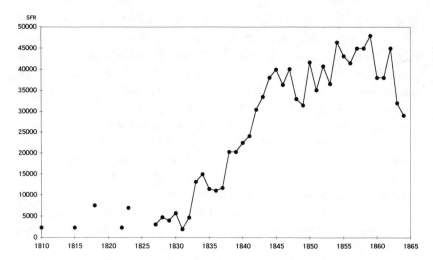

Figure 7.1. Federal expenditures for domestic surveying, 1810–1865. Data from Johann Heinrich Graf, *Die schweizerische Landesvermessung, 1832–1864: Geschichte der Dufourkarte* (Bern: Eidgenössiches typographisches Bureau, 1896), pp. 241–42.

minded Quartermaster Wurstemberger, a member of the Grand Council of the canton of Bern, refused to take an oath to Bern's new liberal constitution and withdrew from all of his offices. That the choice of his successor went to Guillaume-Henri Dufour (1787–1875) was an indication of the military technical reorientation of the 1830s; as "the leadership of the military became more scientific" and "much that earlier had been left up to the heavens was now included in the calculations, the awareness, and the plans of the generals,"[56] the demands on the General Staff, and the Quartermaster Staff officers as well, began to change. Neither Finsler nor Wurstemberger had had the benefit of professional military or technical training.[57] Dufour, on the other hand, was a graduate of the École Polytechnique in Paris[58] and the École du Génie in Metz. From 1810 until 1813 he had fortified Corfu and led the fortification works in Lyons in the service of the French military. After his return to Geneva, he was named canton engineer in 1817[59] and instruction officer at the newly founded confederal military school in Thun in 1819. There he had already been working on problems of the cartography of the region.[60]

For a variety of reasons, the liberal reform movement of the 1830s remained, until the Sonderbund War of 1847–48, ineffectual at the federal level. For my purposes here, a significant development of this period was a clear division of the members of the Diet into a "progressive-liberal" and a "Catholic-conservative" camp. In the common language of the time, national "progressive-liberal" projects could be assured of gaining an extremely narrow majority in the Diet, thanks to the formation of this split.[61] A national mapping project now suddenly gained political opportunity by complementing the ideological goals of this fragile liberal majority. Their goal of establishing a unified Swiss economic space was assisted by a unified cartographical representation; and the borders of a Switzerland conceived as a nation were to be precisely defined not only in military, political, and economic terms, but topographically as well.[62]

Although "bourgeois" projects with an apparent Enlightenment orientation resurfaced during the 1830s—some with significant political effects[63]—this did not mean that the Swiss map now being undertaken had the same political content as Micheli du Crest's project. The political circumstances of 1832, and with them the political function of a map of Switzerland, were not comparable with those of 1754. Dufour's mission was to demonstrate through the transparency of a scientific cartographic representation the political-economic unity of the "Swiss Confederation." Dufour's modern military-technical training and his political stance "entre libéralisme prudent et conservatisme conciliateur"[64] were the best guarantees that this goal could be achieved.

Precision for the Record and Legitimated Surveying Techniques

A discussion between Dufour and his new colleagues in March 1833 made it clear that one practically had to start from scratch: "In order not to leave any legitimate cause for doubt or mistrust hanging over the results as a whole, we found ourselves obliged to start over and conduct the operation as if nothing had been done before our time, from the measurement of a groundline to the final triangle."[65] The old measurements carried out under Finsler were neither verifiable nor internally consistent. Dufour deemed it essential that the new map of Switzerland should be established on precise measurements beyond all scientific doubt. It was therefore decided to take new measurements for the two groundlines near Aarburg and Zurich that had been used for triangulation until then.[66]

The first of the two new surveys at Zurich's Sihlfeld in the summer of

1834 demonstrated that the break with earlier projects was a fundamental one. Dufour not only used them for a systematic training of his colleagues in the use of surveying instruments, he also had the instruments and methods carefully standardized and demanded that the measurements be accurately recorded according to an exacting protocol.

Using three examples, I will demonstrate briefly the meaning precision assumed in the field laboratory at Sihlfeld and later near Aarburg. Of particular interest is how the data received by the engineers was not only recorded but transformed by them in the process, and how both the survey records and the instruments utilized were subsequently subjected to verification in order to insure the foundations for the topographical designation "Switzerland," but also to guarantee that the measurements would be internationally compatible, a key element in legitimating the state.[67]

The laths used for establishing the groundline were eighteen Paris feet or three toise long (5.848 m). On one end was a spherical segment made of steel, at the other a cylinder also of steel, but with a flat transverse section. During the measurements, the rods were situated in a channel and could be moved vertically and horizontally using screws. In order that no collisions could occur when two rods were placed together, rendering the measurements imprecise, a small space was left open between the rounded end of one rod and the flat end of the next, into which a wedge of tempered steel was inserted (Fig. 7.2).[68] The space between the measuring rods then had to be determined by calculation. Because the wedge ABC determines the interval mn instead of ab, the distance ap had to be subtracted in order to determine ab. If r is the radius of the zone of contact, then the formula applies for the correction:

$$\overline{pa} = r(1 - \cos a)$$

or, in Dufour's expression:

$$\overline{pa} = 2r\sin^2 \frac{a}{2}$$

If an extremely tapered wedge was used, as was done during the verification of the measuring rods ($a = 2°30'$), a correction of 0.014277 mm per rod resulted at a radius of 15 mm for the spherical segment. The device was so precise that even small vibrations or slight changes in temperature made it impossible to ascertain the same length twice in a row. A somewhat blunter wedge therefore had to be selected and provided with a vernier, so that in the field measurements could nevertheless be taken to

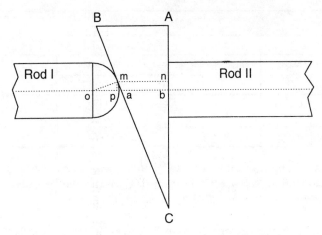

Figure 7.2. Measurement of the space between measuring rods. Based on Graf, *Schweizerische Landesvermessung*, pp. 46–48.

within 1/500 of a rule, or 15/10,000 mm.[69] Measurement results were checked by the engineers several times,[70] then sent immediately to Dufour, who usually checked the correct calculation of the measurements again; he was also the one who instructed the engineers as to the formulas to be used—nothing was to be left to chance.

The unconditional prerequisite for such a multistage verification of the results was the development of a systematic and reliably kept measurement protocol. Figure 7.3 reproduces an excerpt from the protocol of the Aarburg survey at measuring rod no. 571. The sequence of measuring rods, in this case II, III, and I, was given in Roman numerals. The next measurement followed the sequence III, I, and II. The temperature reading was also noted three times, from which the mean was taken each time. In the same way, the distance of the measuring wedge was noted, as determined and corrected by the method described above, and reduced to a theoretical length at a temperature of 10° Réaumur. The fourth column was reserved for measurement with the *T* utilized with differences of elevation. The protocol itself is to be seen as the first step in a routine, complex system of recordkeeping, at the end of which a printed map was produced. In this regard, it is striking that here already transformations were made with regularity (mean and reduction) that obstructed any direct translation back into the original. Survey protocols did not record any landscape, their numbers did not convey any mountains that skilled map readers could then visualize as natural spaces.[71] The deadly boredom

571st Measuring Rod						
II		III		I		
Thermometer	Wedge	Thermometer	Wedge	Thermometer	Wedge	Thermometer
15.5	16.9	37.7	16.1	17.0	41.7	16.1	16.6	

Mean	Reduction to 10° R	Mean	Reduction to 10° R	Mean	
16.2	1.89	16.5	1.97	16.3	

Figure 7.3. Example of a survey protocol on the baseline measurement in Aarburg. From Dufour, "Notice sur la mesure de la base d'Arberg en Suisse," *Bibliothèque universelle des Sciences, belles-lettres et arts* 57 (1834): 383.

from which the survey engineers must have suffered in the face of the profusion of measurements just for the two baseline determinations, points in the same direction. In Zurich and Aarburg, a total of 2,810 survey rods were placed together. Even at this lowest level of the cartographic recording, the reward of delight for the "ever-seeking eye," of which, for example, Haller still dreamed, was hardly conceivable. And even when this "delight" made itself known—as in the sight of a "true gathering of the perpetual snows of Helvetia, who allow nothing foreign in their lap"—an engineer was called back to his charts by his superior: "Immersed in contemplation, I sat in wonder until Mr. Eschmann called for me to write down and reduce his observations."[72]

Transposed data recorded in a protocol had to exhibit regularities validated by other protocols, and not necessarily those of the Swiss engineers. An example is provided by the revision of the distances found at Aarburg and Zurich, the primary reason being a discrepancy discovered in comparison with French measurements of a groundline at Ensisheim. Bringing their measurements into line with French results ultimately required a revision of the temperature tables used for the reduction to 10° R. It is interesting to observe the meticulousness with which the Swiss engineers tried to find an explanation for the difference between their measurements and those of the French; and it is also interesting to see how they revised their protocols in order to mesh with those of the French cartographers. Already during the verification of the rods, it was discovered that their elongation did not change in proportion to the temperature, something that had to do with the sluggishness of the thermometer. The measuring rods expanded more quickly than the thermometer was able to indicate. This problem, which threatened the goal of perfect precision, was tackled with a degree of exactitude that emerges from the report of Johann Eschmann, Dufour's closest colleague:

A trough in the form of a channel closed at both ends was taken and filled with warm water. The measuring rod was placed into it and given time to match the temperature of the water, the thermometers were checked, then the rod was taken out to be placed on the comparator. Here one had to wait for the measuring rod to match the temperature of the surrounding air. Now the comparison was made all over again and the contraction divided by the difference in the temperatures of the warm water and the air. Then the temperature of the water was varied and one proceeded in the same way in order to find out whether high temperatures produced the same elongation as low ones. Because it was impossible to make the comparisons of length at the moment of removal from the water, the first comparison was taken precisely 30 seconds afterwards; then it was checked from 30 seconds to 30 seconds, so that a series was gained that articulated the law of cooling. Then the value at the moment of removal itself was determined by means of interpolation.[73]

The result of these trials was a correction to all measurements of the Aarburg baseline by 0.03647 Paris measures per 1° R difference in temperature. In all, 26.919 rules, or 0.187 Paris foot, had to be added. Afterwards, Eschmann reduced this standard to 13° R, the temperature of the toise of Peru, which meant a total correction of the distance measured at Aarburg (40,189.691 feet) of only 4.484 feet, or 0.01 percent.

Now, it would be completely wrong to assume that Dufour's enterprise would have been satisfied by determining a single groundline as precisely as possible and thus providing the basis for the triangulation net.[74] As the revision of the temperature table suggests, it was equally important to make the Swiss survey internationally compatible. The triangulation work recorded in terms of the groundline measurement had precisely the purpose of guaranteeing connections to other national boundaries. This was also the reason for Eschmann's weeks of work at calculating. Whereas the differences from an older groundline determination at Sihlfeld, taken by Johannes Feer in 1794–97 on behalf of the mathematical-military society in Zurich, seem not to have disturbed him, Eschmann struggled with every imaginable computational art to make the connection with the French network possible.[75] Even given the primacy of measurement, the data remained a construct. In July 1836, Eschmann wrote about this to Dufour:

The doubts concerning the additive correction of the Aarburg baseline have in no way surprised me, because I reviewed the matter several times without being able to explain it to myself. Finally I noticed that the groundline must have experienced an increase at 10° R, because the new elongation coefficient was much larger than the old one, and the average temperature was 12° R, while its reduction at 13° R is even more significant than before; because France's topographical bureau made use of our old warmth table, which is somewhere around 10° R at its minimum, whereas it is now proportional to the number of degrees everywhere and much larger.

Eschmann then made up a table of the expansion coefficients for iron determined by various physicists and used it to calculate the mean (0.00001491 rules per 1° R). His own determination would have given 0.0000141 rules. With the new coefficients, the rods measured up to 10° R (a total of 831) had to be shortened, whereas those with a temperature over 10° R—all 1,400 of them—had to be lengthened: "*There is thus no mistake, either on the part of the French or on mine,* . . . and the results, which I have the honor of sending to you as definitive, I hope will be so forever."[76] The connection with the French network was achieved trigonometrically across two different triangulation nets. Whereas the length measured in Aarburg came to 13,053.740 m, using the French groundline at Ensisheim resulted in an additional 0.035 m. The same method applied to a second triangulation net resulted in a length 0.020 m shorter. With this level of accuracy, the foundation of the Swiss survey project was secured and internationally confirmed.[77]

Transpositions: From Landscape to Map of a Nation

In 1835, the military surveillance authorities reported to the Diet that the protocols, "some from the late Mr. Quartermaster General Finsler, who continued to be entrusted with the verification and calculation of the triangles," were being kept in the best order and contained "154 triangles of the first class, 894 of the second, and 1,446 of the third; altogether thus 2,494 *observed, calculated, verified, and recorded triangles*."[78] The gigantic machinery of a new "national" recording system for the landscape of Switzerland had come to fruition (Fig. 7.4). "Here we are at last, masters of our elements" (*Maîtres de nos éléments*), Dufour commented on December 9, 1836,[79] when the connection with the Lombard triangulation network was reported to him. "Maîtres de nos éléments"—the expression implied mastery of natural spaces through the means of their representa-

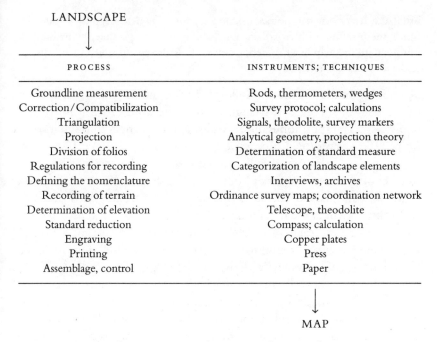

LANDSCAPE

PROCESS	INSTRUMENTS; TECHNIQUES
Groundline measurement	Rods, thermometers, wedges
Correction/Compatibilization	Survey protocol; calculations
Triangulation	Signals, theodolite, survey markers
Projection	Analytical geometry, projection theory
Division of folios	Determination of standard measure
Regulations for recording	Categorization of landscape elements
Defining the nomenclature	Interviews, archives
Recording of terrain	Ordinance survey maps; coordination network
Determination of elevation	Telescope, theodolite
Standard reduction	Compass; calculation
Engraving	Copper plates
Printing	Press
Assemblage, control	Paper

MAP

Figure 7.4. Stages of transposition in the cartographic recording system.

tion in a cartographical recording system, in which the third stage had only just been reached.[80] The extremely complex process of transformation from landscape to the finished "Topographical Map of Switzerland" cannot be described in all of its details here. Now that I have discussed the problems of the instrumental measuring of the groundline and its entry into the protocols, the following pages will examine in closer detail only two of the partial transpositions indicated in Figure 7.4: the regulations for recording and the technical production of the map at the press. In them, the political dimensions of cartography are probably expressed most clearly. They also illustrate the filters translating between "landscape" on the one side and "map" on the other.

Dufour's "Instruction for Recording in 1/25,000" contains the most important rules of transposition for standardized terrain recordings.

In the areas in which assessor surveys exist, the engineer will reduce these plans in order to use the same as a first draft. While he ranges through the villages, he will take note of the changes that have become necessary and newly added objects. For this rectification, he makes use of the survey table or the swivel rule. If no assessor plans exist, the engineer will make a triangulation of the third order by using the trigonometric points assigned to him. Then he records the details, fixing the position with cross-bearings, or if that is not possible, by making an outline with the help of a magnetic compass. It is also permissible to make use of stadia for detail operations.

We might summarize these as follows: Transposition Rule 1: Reduction of the assessor survey or a triangulation of the third order as functional equivalent. Transposition Rule 2: Assessment of changes through inspection of the terrain (field observation). Transposition Rule 3: Use of the survey table or the swivel rule, the compass, or the stadia. The work instructions are reminiscent of Micheli du Crest: reification and retention of the present image of the landscape by means of technical instruments, according to which the remaining landscape elements are to be oriented. This fundamental catalog of rules was followed by a categorization of landscape elements, a definition of what is "of interest." This included "watercourses of the valleys and ravines, the ridges of the mountains, the tops of the hills, as well as the watersheds," transportation routes, lakes, ponds and swamps, peat moors, mines, quarries, rock formations, slopes, moraines, abysses, canyons, scablands, faults, "in a word, all characteristic features of the land." These were, however, "to be expressed according to their real forms, as they would present themselves to an observer situated directly above them, and not with conventional symbols." Pictograms were, however, allowed for castles, factories, chalets or isolated farms, ruins, and "in general all edifices." Observing the interests of the map's military sponsor, everything that created "a serious obstacle for a movement of troops" was to be included. The same went for the exact depiction of roads, which were divided into five categories, in which trade and political interests played a role: Postal roads with four parallel lines, "other good major roads with two lines, one heavy and the other fine, side roads in good condition, where a wagon can pass through easily, with two fine lines, lanes that are only passable for oxen carts with one fine and one dotted line, mule tracks or foot paths with solid or dotted lines." Colors, stroke types, and symbols each received a definite meaning assigned to them: watercourses blue, buildings red, vermilion, or crimson. Stone bridges were drawn with red, wooden bridges with black lines. The anchor points of the map received special attention: "The numbers of the most important [canton] border markers . . . are written in red and in Roman numerals. . . . The trigonometric signal points are indicated with a small triangle, the church towers with a small black circle, the inside of which remains white." Even the form of the captions was specified: "The handwriting of the original recordings is executed with the usual roundness, but carefully executed, and the size [is chosen] in the correct relationship to the importance of the object. It is even better, when the engineer adheres to the pattern supplied to him by the confederal topographical bureau. With the exception of print that refers to rivers, streams, valleys, and mountain chains, the same is always made parallel to the long side of the page, that is, from west to east."[81]

Dufour repeatedly specified which instruments were to be used for transcription of the landscape onto the ordinance survey map. For example, the entry of the "stadia gradations," was defined for the most important points with the "known geodesic methods." "For points of less major importance, as for example those that only serve to determine horizontal contours . . . , differences of elevation can be calculated with an elevation diagram or the logarithmic slide rule." In the mountains, "the terrain is expressed as precisely as possible with horizontal contours that represent the intersecting lines of the ground surface with level planes vertically separated by 30 m. . . . The contours are rendered by unbroken lines in brown (burnt sienna)."[82] In every instance, the engineer was to be wary "of every affectation." The provisions written by Dufour[83] are a comprehensive set of transposition rules for a precise cartographic depiction of natural objects and landscape elements. The need for centralized oversight of the terrain recordings—which recall Latour's "centres of calculation"[84]—was of crucial importance because the recording had for the most part to be delegated to the cantons. Federal projects had to perform a delicate balancing act to survive financially while ensuring the cooperation of the cantons. The active participation of the canton under the more or less gentle central supervision of the federal state was an integral component of the political culture of Switzerland in all other areas of federal policy (with the exception of foreign policy).

Like the specifications for surveying the groundlines at Aarburg and Zurich, the derivation of correctional formulas and projection methods, and the above-mentioned specifications for original recordings, the rules for the reduction to the selected map scale of 1:100,000 also had to be established and monitored, as did the protocols for transferring the map onto copper printing plates. Lacking proper experience, Dufour turned to the Dépôt de la Guerre de France for guidance. It can be assumed that the projects in Dufour's topographical bureau took a form similar to that described by the first draftsman of the Dépôt de la Guerre in his response to Dufour.[85] The instructions provided by the Depôt illustrate the number of complex tasks involved in translating field data to a standard map. According to this, the survey maps of the original recordings, which were colored at the bureau (in the field, a pencil was used exclusively),[86] were transferred onto tracing paper and reduced to the scale map by a draftsman, who worked primarily from the geodesic points provided on the survey maps for orientation. This draft went to the engraver of the "trait," who transferred it onto the copper printing plate.

From the engraving, two proofs are taken on white paper; the text draftsman next enters the inscription in the prescribed manner and afterwards it is engraved by the text engravers. A new proof thus now contains the trait and the inscription. The plate next reaches the terrain engraver, who works from a pattern that has been produced by a draftsman from the second of the two aforesaid proofs by means of quill and who also has construed the contours. The terrain engraver works from these two drafts. The hatching of water is finally entered by an engraver who does nothing other than this.[87]

The production of the engraving is thus a process involving extreme specialization, in which every functional stage follows its own transposition rules. One of these numerous transposition rules concerned the nomenclature of the map—a diplomatic matter of the highest order.[88] The confederal military commission had decided that the titles and commentaries on the folios were to be in German, while the town names would be given in the language spoken by the majority of the local population. This was, especially in the border area between Bern, Fribourg, Neuchâtel, and the Waadt, or in the Bern Jura, politically a very delicate and risky venture. Local opposition to the state's unifying power of definition was practically preordained. The declared policy of the military authorities was to reduce multilingual nomenclature to a single language. "If a mountain or a river has several names, only a single one is to be written, namely the most well-known or the one that is accepted by the local authorities. *Through this the name becomes to a certain degree official.*"[89] Translation protocols were thus required in order to even read the map. Folio V of the "Topographical Map of Switzerland" contains a translation catalog of approximately 450 important or controversial place names in four languages.[90] In it, the map reader was instructed to think of Gebsdorf as Courchapoix, Ueberstein as Surpierre, to translate Marmels as Marmorèra, Tinzen as Tinzogn, and Treiten as Treiteron. Thus, prior to the founding of the federal state in 1848, the multilingualism of Switzerland, according to the slogan "national unity—cultural diversity," had already become an integral component of the future state ideology.[91] In order to ensure that "diversity" did not assume a threatening nature, however, "official" nomenclatures were used wherever possible. The engineers involved in terrain recording asked the local authorities, the village teacher if need be, what a village, hamlet, valley, or hill was "correctly" called. This practice allowed the producers of the map to reduce topographical complexity considerably; and in the process any influences of dialect were purged of local varnish through transcription into written language, while traditional and locally rooted names of town and country were almost entirely ignored.[92]

The state's topographical monopoly on definition was not easy to establish. The publication of the first folios of the map project launched a furious debate, which led Dufour to resign temporarily. The primary complaints made against Dufour in the *Schweizerischer Beobachter*[93] were directed against his system of depiction, against the determination of place names, and against the elevation readings; in short, against the very cornerstones of his cartographic recording system. But most important, the complaints came from the most diverse camps at the same time: a former collaborator of Dufour's claimed that the map had been carelessly and quickly produced: "I find it oddly surprising that the federal authorities have seen fit to allow this map to be published with such serious deficiencies."[94] Colonel Maillardoz, the president of the confederal military council, had already accused Dufour beforehand of insufficient precision in the specification of national boundaries.[95] The article in the *Schweizerischer Beobachter* made it clear that the nomenclature chosen would inevitably meet with the immediate opposition of local residents. Interestingly, in his response Dufour frequently referred to the names ascertained in the original recording maps; he thus used one stage of the recording system in order to deflect criticism from the final map.[96] In other cases he depended upon local informants; but he tended to rely on persons, such as school teachers, who opposed traditional geographical knowledge.[97] Although Dufour dismissed problems of nomenclature with counterattacks, such as on what he termed "cavils" or "chicanery for our bilingual country," in the area of elevation statistics he defended himself exclusively with scientific-methodological arguments. Thus Johann Eschmann wrote as justification to Dufour that "the same instruments and the same methods" had been consistently applied and all accepted elevation readings and calculations had been verified—"those are the fruits of 1,000 combina-

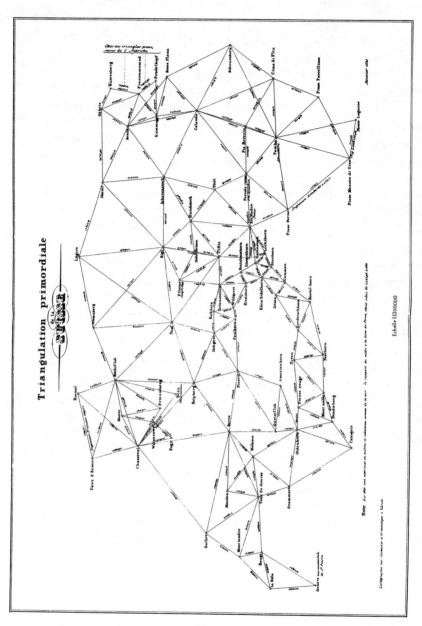

Figure 7.5. *Triangulation primordiale de la Suisse* (Foundational triangulation of Switzerland). Reproduced from Graf, *Schweizerische Landesvermessung*, plate between pp. 96 and 97.

tions." Furthermore, the best instruments had been used in constructing the new map.[98] The recording system itself attained legitimate character.

Assembling Switzerland: Map Popularity and the Loss of Images

What about "nature," the physical spaces captured within the network of national triangulation? At the beginning of 1838, the folio "Triangulation primordiale de la Suisse" appeared as the first publicized result of the topographical survey. The lithograph contained nothing but the bare net, with indications of the names of the intersections, the lengths of the individual segments of the triangulation "expressed in meters and brought down to sea level." Neither rivers nor lakes, valleys nor villages nor canton boundaries were indicated—pure abstraction in trigonometrical terms, with the clear message: The natural alpine expanses and deep valleys were to be transformed into the uniform geographical space of Switzerland (Fig. 7.5).

One then set out to fill in this network, right down to the survey map of the engineer involved in the original recording, to translate it to map scale according to the rules, and in highly specialized and equally regimented sequential processes to produce a total of 25 copper plates that were supposed to represent the country beyond all scientific doubt—the precision of the groundline measurement and its record in the protocol were to vouch for that. There had been a few minor problems during the terrain recording, such as opposition of the population or lack of cooperation from the local authorities. The storm of protest only broke when the concrete results were made public.

But once again political events came to the aid of the project. In 1847, Dufour was elected as commander-in-chief of the army against the Separatist League created by the Catholic-conservative cantons. The clear victory of his troops and the foundation of the federal state with a liberal-republican constitution realized political conditions at the national level, the precursors of which had already made possible the take-off in confederal topography in 1833. After 1848, Switzerland was identified with an institutional political stability that promoted modern economic growth, the creation of a domestic market with a uniform system of measures and currency, and the unhindered boom in the export economy during the second half of the nineteenth century. The Dufour map's bid for identi-

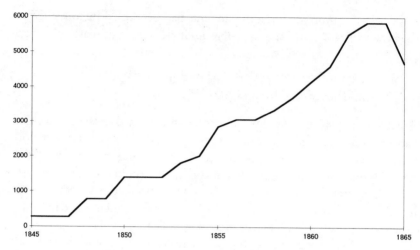

Figure 7.6. Annually printed folios of the Dufour map, 1845–1865. Data from Graf, *Schweizerische Landesvermessung*, p. 237.

fication as a national accomplishment increased under these new political conditions—the map itself became the cornerstone of the federal state's self-representation. It was not without reason that the single folio published in 1848 comprised an "overview" (in French, *assemblage*)—instructions, therefore, on how "Switzerland" was to be pieced together. From then on the printing press ran ever faster until all 25 folios were pieced together in 1864 (Fig. 7.6).

Although the main topographical work on the Dufour map falls within the phase of the structural formation of Swiss liberalism,[99] the actual technical production and diffusion of the map took place in a period of political stability and ideological hegemony of the "liberal" party. I have attempted to explain the effort by Swiss liberals to revitalize the Enlightenment-motivated cartography project of the eighteenth century in terms of the instrumental role played by a "national" cartography in constructing the economically, politically, and ideologically unified space "Switzerland."[100] What became of the map of 1848 appears to confirm this thesis. At first, no opportunity was neglected to present the Dufour map at international exhibitions. The representation of the natural space of Switzerland also contributed to the representation of Switzerland as a politically unified nation. In 1855, the "Topographical Map of Switzerland" won a gold medal of honor on the occasion of the Paris World Exhibition, in 1873 a certificate of honor in Vienna, in 1876 a certificate of award in Philadelphia, in 1878 a certificate of honor once again in Paris, in 1889 a grand prize in Paris. The map was awarded many other distinctions on other occasions as well.[101]

Parallel to this, a cult of personality developed around Dufour as "father of the nation," which was closely related to the cartographical undertaking he had led. In recognition of his leadership of the map project, the state government presented Dufour with a "silver centerpiece, consisting of two fruit bowls, one above the other. . . . Towering above the upper fruit bowl is Helvetia bestowing a wreath, the rim of the fruit bowl being designated for inscriptions. At the base, the corners of the pedestal are occupied by three figures of solid silver, which with their emblems represent topographical surveying, copperplate printing, and technical military work."[102] That a victorious Helvetia and her bountiful fruit is supported by these three figures suggests that together with political and military means Dufour's cartographical work was to be viewed as one of the founding pillars of the federal state.[103]

Nearly three years before the presentation of the centerpiece, the cult of Dufour had been inscribed on the landscape itself: in 1863, the highest Swiss mountain peak in the Monte Rosa group received the name "Dufour Peak" by a resolution of parliament. A popular biography of Dufour—"compiled with special consideration of the service he rendered for the political independence and unity of Switzerland, as well as for scholarship, art, and humanity"—that appeared in 1878 and went through many editions traces a direct line from Dufour Peak across the immense masses of the nation of summits to the precise measurement, in certain respects the graphic representation, of Switzerland.

Dufour Peak that majestic summit of rock is called, which, adorned with everlasting snow, sits upon its throne as lord amid an entire, gigantic army of alpine princes of nearly the same high birth, princes of the Land of the Swiss, who lift up the silver tresses of their heads, in rings closer and wider about him, from out of the immense masses of the lower people of summits and daily receive the first hail and the last kiss of the sun god. It does not carry this name without reason: the representatives of the Swiss nation have named it thus in honor of General Dufour, and indeed with the special purpose of erecting a monument to the priceless accomplishments of this man in the service of the precise measurement and graphic representation, as correct as it is effective and expedient, of the Land of the Swiss.[104]

Senn uses here a metaphor as forced as it is revealing. The senate of these divinely inspired sages, whose primus inter pares and at the same time "lord" is said to be Dufour, is surrounded by the immense masses of the lower people, whose—presumably democratically elected—representatives erect for the precisely measuring cartographer a monument to his "correct," "effective," "expedient" representation of the "Land of the Swiss"; Helvetic triangulation of cartography, politics, and representation—the geographical space of Switzerland has become transparent and uniform, the mists have finally dissipated.

The "correct measurement" was in fact soon corrected,[105] but the map remained "effective" and "expedient" nevertheless—as a planning aid for the construction of railroad lines and streets, for the development of telegraph networks, for the postal administration, for tourism, and for military home defense.[106] The Swiss map was substantially more important, however, as an instrument of national politics. "The popularization of cartography, its application to all branches of public life, is an achievement of the past few decades."[107] It was supplied to cantonal administrations and schools at subsidized prices, and as early as 1853 the state ordered a compact edition on a scale of 1:250,000.[108] Into the middle of the grand century-project of bringing literacy to the entire Swiss population, initiated by enlightened pastors of the eighteenth century and institutionalized by the bourgeois-revolutionary Helvetic, fell a successful project that attempted to teach the "Swiss people" the official, federal manner of reading geographical space. "We can achieve great things when we have learned to use this rich source!"[109] Reified, this claim appeared at the Swiss National Exhibition of 1883 in Zurich. In the exhibit of "Modern Art," one found "in the branch of sculpture the glorification of today's cartography, the best illustration of its contemporary popularity. It is the shepherd lad who studies the map."[110] For Haller, this shepherd lad was still a "student of nature," whose "reason [was bound] to no school law"; in his Alps "rules reason, led by nature."[111] After Dufour and after 1848, one wanted to hear no more of this. The shepherd lad, in 1883 as in the eighteenth century, still stood synechdochically for Switzerland, but this personified Switzerland now grasped the rarified condition of its landscape from a folio of federal cartography. "It is certainly no mere coincidence that led the artist to chose precisely this motif and inspired him to express the far-reaching influence of cartography in such perfect form. It is the ethical influence of cartography that speaks to us out of the marble, that eminent significance of the map as a means always ready to serve the elevation of the spiritual and economic prosperity of solitary individuals as well as the entire society."[112] The transformation of Haller's "He knows his fatherland and with its treasures / Can deify his ever-seeking eye"[113] into the map-reading shepherd lad of marble as he was presented at the National Exhibition of 1883 is itself a breathtaking work in the semiology of mapmaking. "The majority of our readers have probably already seen the Dufour Map pulled out into a single piece," Senn had written in his biography of Dufour only a few years before, pointing on that occasion as well to "our people's developed sense for maps."[114] "It is a mighty work, this map—the fruit of Dufour's 33-year effort and his exceptional field surveyors, geometers, engineers, draftsmen, engravers, copperplate printers, in short, all of those men of science and art whose precise collaboration such a work demanded. . . . *All Switzerland* may thus place its just pride in Dufour's map work." Yet, although Senn here evokes the specialized transposition process of landscape to map once again, he seemed to wish to ascertain the illusory legibility of the map by claiming "that there is no footpath, no bridge, no house, no brook in the entire land from the plains to the highest alps that is not depicted in the right place and in the correct form, not to mention the graphic representation of the lay of the land, which is carried out here in all details so precisely and with such plastic effect that it seems as if one could soar like an eagle high above the land and looking down upon it study with a sharp eye cliff and crevice, hill and slope, ridge and valley floor in reality." Only those who read Senn's hagiographical text closely will discover that the cartographic transposition of the Swiss landscape must have transformed not only the reader, but also the landscape itself according to the principle of measurability. What topographers still found was a "fantastic confusion of ice banks, of precarious glacial towers, of sheer cliffs and yawning crevasses, . . . days away from every human habitation," a world characterized by "snow storms, lightning, and thunder." Not even the most skillful map readers have come to such visions from the "Topographical Map." The cartographic principle of measurability was an "effective" vehicle of the bourgeois mastery over nature in the nineteenth century.[115]

The same principle of measurability, however, also resulted in the subsequent suppression of the world of images. Ulrich Meister, colonel in the General Staff, national president of the left-liberal party, forestry superintendent of the city of Zurich, and author of a work on the national fortification of Switzerland, formulated this loss of images in a short essay on the "current standpoint of Swiss cartography" and the "legibility of our maps": "Under the influence of the state's leadership and guidance of cartography, there has developed a noticeable preponderance of measurable factors or of the mathematical element as opposed to the simply depictive element in the recording as well as in the execution."[116] Little could be done to prevent this. The perspectivist representation, which Meister proposed at least for the smaller cities,[117] so that one could recognize them on sight, belonged to the past. Meister was actually aware of the impossibility of his demand:

We miss the *picturesque element* greatly in the current production of cartography, or better said, the use of *perspective*. This demand must seem somewhat awkward, when it is otherwise demanded of maps that they should only contain material that is *measurable* either directly or by conveyance. Originally, the map was a painting, a pictorial tablet; the cartographer drew or painted the terrain objects on the map in the form in which they appeared to him, and he integrated them into their linear relation as far as this was known to him. In the period of more precise geodesic foundation, mathematical precision was established in the foreground as the most important priority.[118]

The Dufour map had closed off the avenue of perspectivist representation once and for all, reserving it for panoramas and reliefs.[119] "It is not easy, forcing this ponderous, imposing, mountainous land into a suitable likeness and at the same time still preserving the legibility of the map."[120]

Meister was right about the demands on the sense imposed by the requirements of abstraction, precision, and the new technical media. The end point in the development of the cartography of the "long nineteenth century" is formed by a machine that in its functional method resembled the phonograph to an astonishing degree: Von Orel combined the stereocomparator constructed by Pulfrich with a pantographically functioning lever system that was able to transfer all side and depth movements of the comparator onto a firmly attached drafting carriage.[121] "Machines take over the functions of the central nervous system"[122]—before the First World War in the area of cartography as well. Long before this, however, bourgeois-liberal cartography had begun to translate cartographical substitute sensualities into the abstract measurability of the topographical map.

TRANSLATED BY KEVIN REPP

8. GILLIAN BEER

Writing Darwin's Islands: England and the Insular Condition

Long before his name was linked with evolutionary ideas, Darwin was famous for his work as a geologist and theorist of islands. In the 1840s and into the 1850s his reputation rested on his distinction between continental and oceanic islands, on his theory of coral reefs, and on his contribution to the realization that South America had in the early mammalian stage been a vast island continent cut off from North America. All these discussions focused on the means and manner of change: the appearance and disappearance of land forms, the subsidence of ocean floors and the building up of coral, and the extraordinary wealth of extinct mammalian forms then littering the earth in Argentina and Uruguay.

Islands are central to evolutionary theory. Remote oceanic islands produce ecosystems in which the strains and the opportunities of diversification can be seen uninterrupted by the advent of incomers with different strengths. The very large and the very small may survive on islands in a way that astonishes the observing naturalist. Life-forms need not necessarily change if the environment is stable: islands may therefore serve as museums or repositories of early forms that elsewhere have become extinct. The autonomy of island populations provides a measure of distinctive development and a visible graph of change. Neighboring islands set beyond the range of inhabiting species display the burgeoning of counterpossibilities in their endemic species. But swimming and flying inhabitants may create an island network in which the biosystem extends to a group of islands, however isolated each of them may appear to be, and however limited in range may be the other life-forms that inhabit them. Some of these observations had been formulated before Darwin, but his work brought them together into a coherent system whose consequences are still being worked out in later biogeography.

During the late nineteenth and earlier twentieth century, in the wake of Darwin and coworkers such as Alfred Russel Wallace, islands were perceived as providing closed domains in which the processes of variation and of natural selection may be observed working in a high degree of undisturbed historical continuity. They were figured not only as museums and repositories but as gardens, utopias, primal scenes. Now, with airplanes and global markets, to say nothing of plate tectonics, such emphasis on autonomy is out of fashion, at least among biogeographers.[1] In chaos theory the figure of the island has recently provided an instance both of expansion and recursiveness. Mandelbrot argues in *The Fractal Geometry of Nature* for an infinite coastline, crenellation beyond computation: "As increasingly small rock piles become listed as islands, the overall list lengthens, and the total number of islands is practically infinite. Since earth's relief is finely 'corrugated,' there is no doubt that, just like a coastline's length, an island's total area is geographically infinite."[2] Among a wider public, islands have been commodified but the commodity they offer is still nostalgically represented as isolation and insulation (the etymological drift is no accident).

The question of founding populations and how to sustain them, reexamined in MacArthur and Wilson's *Theory of Island Biogeography*, has recently and rapidly been caught up in issues of colonization and postcolonialism being pursued in other fields.[3] In the study from which this essay is drawn I explore these issues extensively and consider the meaning of islands as body, theater, pocket, book, and crucial site of contest in the history of colonization and postcolonialism.[4]

In this essay, however, I concentrate on ways in which island sites formed part of the experiential, political, and natural historical material by whose means Darwin came to frame his new theories. In particular, I explore the changing representations of England in his writing during and immediately after the voyage of the *Beagle*. Moreover, I want to suggest that the categories I have just proposed (experiential, political, and natural historical) were not separated in his thought. The instability of register across which he worked (often the abrasion between different styles of observation) in itself helped to break up the securities from which he set out and brought into question the assumptions and values that would have curtailed his inquiry. This is not to suggest that Darwin extricated himself from his culture's assumptions, but rather that the dubieties he reached provoked new insights—and embarrassments. Uncertainties about boundaries are as much part of his discourse as his theory.

The first version of the work we now ordinarily know as *The Voyage of the Beagle* was written between October 1836 and September 1837 and appeared as the third volume of Robert Fitzroy's *Narrative of the Surveying Voyages of His Majesty's Ships Adventure and Beagle Between the Years 1826 and 1836*. That volume drew extensively on the diary and field notes that Darwin kept during the voyage.[5] A few months later, Colburn reissued Darwin's volume separately as the *Journal of Researches* (without permission or payment). It was a great success. A second, revised edition appeared in 1845, by which time Darwin had already made two attempts at formulating his evolutionary theory and was far into that apparently dissociated long-term enterprise on cirripedes, the barnacle whose capacity for variation and whose transsexuality would become an important control for Darwin in thinking about the processes of evolution in two-sexed systems.[6] Out of the voyage of the *Beagle* also issued the four volumes of geological researches that initially made Darwin's name as a scientist. In particular, his theory of coral reefs (his only theory begun in what he calls "a deductive spirit"—that is, by a chain of reasoning alone before he had ever set eyes on a coral reef) and his distinction between continental and oceanic islands established him as a leading geologist.

Indeed, *The Voyage of the Beagle* can be read as a narrative whose peripeteia is just beyond the scope of the text, has not yet been performed. Yet that same peripeteia is the determining precondition of our reading. To put it another way: Darwin, as he wrote his journal on the voyage and recast and developed it on his return to form the third volume of Fitzroy's history, had not yet quite thought through natural selection; we cannot now think outside it, particularly when we think of Darwin.

Since Frank Sulloway's two important essays, "Darwin's Thinking: The Vicissitudes of a Crucial Idea," and "Darwin's Conversion: The *Beagle* Voyage and Its Aftermath," the tendency among historians of science has been to emphasize the belatedness of Darwin's appreciation of the materials he gathered on the *Beagle* voyage and of his theoretical formulations.[7] Instead of a *coup de foudre* insight while on the Galapagos, we are offered the emergence of his views from sober conversation with his professional coworkers such as Gould, Lyell, Owen, and Henslow back in England during 1837, the year after the voyage was completed. This has been a useful corrective to the eureka view. But it harbors a eurekism of its own. Concentrating only on the "natural historical" or "evolutionary" topics of the voyage contracts the process by which ideas emerge and hold together. An implicit privileging of observation and immediate record is part of a romantic legacy for the discipline of the history of science even while it proposes itself as a severely empirical methodology. That is to say, a reading that concentrates solely on "scientific" material does not take account of the shape-shifting of ideas, their way of emerging in contexts apparently remote from the main drive of professional inquiry. Recognizing key ideas is often a matter of having the confidence to place at the center of attention what has already been present on the periphery or as part of another inquiry.[8]

A different picture emerges if we take into account, for example, Darwin's descriptions of his encounters with other ethnic groups, his method of comparing what he sees with England and the English, and the anthropomorphism of his meetings with animals. After the actual moment of encounter all the experiences of the voyage were revivified as writing, reencountered as reading. If we examine Darwin's writing and rewriting, his copying-out of his writing, and his musing on scenes and observations present to his eyes now only *as* writing and rereading, we shall uncover material constitutive of theory though not directed toward abstraction.

Such acts of writing, whether in letter, diary, field notes, or later redaction, are themselves prototheory. These acts are intensified in the process of reading again (self-reading), particularly for the purposes of revising or recasting. And because language is never owned by the single writer, it will be traceried with the contradictions and the history active in the society from which the writer emerges.[9] Those contradictory histories we know to have beset Darwin. Theory formation and creativity imply that the writer no longer lies quite level with the assumptions embedded in the initiating communal discourses, but pulls across them as she or he writes. One trope deeply embedded in English self-identity is that of the island. This essay will argue that Darwin's coming to awareness of the English idea of "the island" gave him insights to work with and the liberty to move askance, to institute change.

To many people the Galapagos Islands still remain an emblem of Darwin's initiating insight. The islands seem to represent a primal moment of observation that led to new construals of the world. And though we may agree that in fact that observation was constructed retrospectively by means of writing and displacement as much as presence and engagement, Darwin's initial and later writing of the encounter yields a medley of material that bears on related issues of closed domain, empire, intruder, and autochthonous inhabitant. The young Darwin describing himself encountering for the first time the gigantic indigenous tortoises of the Galapagos provides a vignette at once comic and revealing. It turns out that the famous version of that passage occurs in the 1845, not the 1839, account. In much of my argument I shall be pointing to different theoretical, ideological, and experience-filled meanings of islands in Darwin's thought, but let us start with this famous encounter.

Darwin recognized the absurdist comedy of human beings entering ecological systems in which they had no previous place: take, for example, his first landing on one of the Galapagos, Chatham Island. The British name imports a supposed naval good order to the scene (Chatham is the site of a naval dockyard in England), but what Darwin finds there is disquieting both in its likeness to and difference from his home territory. His description is a surreal mingling of the scenes of industrial modernity, classical literature, the domestic, and the primordial:

The entire surface of this part of the island seems to have been permeated, like a sieve, by the subterranean vapors: here and there the lava, whilst soft, has been blown into great bubbles; and in other parts, the tops of caverns similarly formed have fallen in, leaving circular pits with steep sides. From the regular form of the many craters, they gave to the country an artificial appearance, which vividly reminded me of those parts of Staffordshire, where the great iron foundries are most numerous. The day was glowing hot, and the scrambling over the rough surface and through the intricate thickets was very fatiguing; but I was well repaid by the strange Cyclopean scene. As I was walking along I met two large tortoises, each of which must have weighed at least two hundred pounds: one was eating a piece of cactus and as I approached, it stared at me and slowly stalked away; the other gave a deep hiss, and drew in its head. These huge reptiles, surrounded by the black lava, the leafless shrubs, and the large cacti, seemed to my fancy like some antediluvian animals. The few dull-coloured birds cared no more for me, than they did for the great tortoises.[10]

Darwin is a small Gulliver in the world of black lava, vehement heat, and Brobdingnagian creatures assured of their own centrality ("As I was walking along I met two large tortoises": the casualness and continuousness of the imperfect tense, "was walking along," persuades the reader to accompany his motion and to accept the tortoises as on equal social terms with humans: "I met"). Darwin is a time traveler, moving from the iron foundries of Staffordshire (in which county his kin and future in-laws the Wedgwoods had their potteries) back in instantaneity to the "antediluvian," the world before the flood now still present. In his *Diary* entry at the time, he goes further in the oscillation between the domestic and the exotic in his language of encounter: "Surrounded by the black Lava, the leafless shrubs & large Cacti, they appeared most old-fashioned antediluvian animals or rather inhabitants of some other planet."[11] The turtles are at once domesticated ("most old-fashioned") and exotic ("inhabitants of another planet").

He works at this scene several times: in his diary at the time, in the 1839 account, and again for the 1845 edition. Intriguingly, there is no description of the encounter with the tortoises in his letters home. There, the Galapagos are "that land of craters."[12] The 1845 version, with its narrative imperfect tenses and its intensitive "strange Cyclopean scene," enhances the dreamlike immediacy of the experience and may indeed have been provoked by a rereading of his original diary entry, to which it is closer in tone than the intermediate published version. The 1839 form is less dramatic: "In my walk I met" instead of "As I was walking along"; "when I approached" instead of "as I approached"; the tortoise "looked," not "stared," at me and the reptiles "appeared," not "seemed," to my fancy.[13]

Subjectivity is enhanced in the revised 1845 description as though the scene has become more meaningful, not less, in the longer retrospect. This might be that Darwin now realizes (but still secretly) the implications of those creatures for the theory he is formulating. The intensity of revisioning may conceal a new message that he does not wish yet to communicate. Certainly, the linguistic register has been elevated. In 1839 he writes of the craters in a separate paragraph and remarks with humorous emphasis: "From their regular form, they gave the country a *workshop* appearance, which strongly reminded me of those parts of Staffordshire where the great iron-foundries are most numerous" (Fitzroy, p. 455). The "workshop" has become "artificial"—retaining the old sense of "artifice" alongside the sense of "produced by man," "against nature."

In other important ways, however, the emphasis of the passage remains unchanged. In the moment of encounter Darwin was more impressed by the size and tameness of the birds and animals than by the differences of the inhabitants from island to island, so much so that he even mingled the materials collected from each island and had to rely later, to a large extent, on other people's collections. But what he did observe and describe proved crucial to his later work: "The birds are Strangers to Man & think him as innocent as their countrymen the huge Tortoises. Little birds, within 3 or four feet, quietly hopped about the Bushes & were not frightened by stones being thrown at them. Mr. King killed one with his hat & I pushed off a branch with the end of my gun a large Hawk" (*Diary*, p. 334). Tameness, he discerned for the first time, is not a concept confined to culture; rather than being an outcome of domestication, it may imply the absence of the human. Wildness may be produced by the intrusion of new forms, he realizes as he observes Indian tribes being forced from a settled to a nomadic life by the predations of the Spaniards. If nineteenth-century ethnography posits the nomadic as the more primitive organization, Darwin nevertheless observes that these categories are not fixed and "natural" but adaptive. These realizations of the mid 1830s control the crucial opening paragraphs of chapter four, "Natural Selection," in *The Origin of Species* (1859). Culturally established polarities of wild and tame, inhabitant and intruder, past and present, near and far, are the first opposites to lose their stability in his thinking. Re-thinking them is fundamental to Darwinian evolution.

Darwin came to and passed the brink of his key formulations by means of the processes of writing and rewriting. Crucial also were the shifts and juxtapositions of place in which the work was written: first, diaries, journals, letters, and field notes day by day or in bunches around the world, producing a series of cross-hatched images as he wrote of one site while at another; then back home at a desk in a forever altered England. He reached his new insights, too, by observing other human societies and by destabilizing England from its normative position in the language with which he had set out.

For Darwin the continued imaginative presence of England through-out his travels was not only culturally inevitable but methodologically essential, a means of tabulating new experience, giving form to the intel-lectual project in which he was engaged, and providing evidence from an island-base for comparison with the many islands that he visited on his journey round the world.[14] But as he went on, that base changed from being always foundational to being an instance among many. One of the most important discoveries that Darwin made on the voyage of the *Beagle* was that islands are *systems of difference*: they are extraordinarily different from each other and they harbor difference within their economy. In-deed, extinction is the inevitable consequence where insufficient differ-ence is maintained among the population.

It was necessary for him to learn this because England to the English tended then (and at a less acknowledged level still tends now) to be seen as both an ideal and a normative island. Of course, the identification of England with an island is already, and from the start, a fiction. The island is equaled with England in the national discourse of assertion though England by no means occupies the whole extent of the geographical island: Scotland and Wales are suppressed in this description, and only part of Ireland can now be corralled even within that very different group, "the British Isles."

Some of Darwin's favorite reading as a schoolboy was Shakespeare's history plays. He pictures himself in memory perched in a window seat immersed in the honors and triumphs of Shakespeare's national apologia. Many English people have imbibed a sense of collective self from the language of these works that often survives their own later skepticism about patriotism. Shakespeare's play *Richard II* provided the initiating communal self-description, alluringly emblematic and topographical at once. Gaunt calls England:

> This royall throne of Kings, this sceptred Ile.
> This earth of maiestie, this seate of Mars,
> This other Eden, demy Paradice,
> This fortresse built by Nature for her selfe,
> Against infection and the hand of warre,
> This happy breede of men, this little world,
> This precious stone set in the siluer sea,
> Which serues it in the office of a wall,
> Or as a moate defensiue to a house,
> Against the enuie of lesse happier lands.
> This blessed plot, this earth, this realme, this England.[15]

England is seen as supremely and reflexively natural; built by Nature for herself. It is a "little world" containing everything sorted and selected, at once a miniature cultivated place, "this blessed plot" and an extensive realm. In this ideal description it is body, fortress, and moated country house. It is also, as the cycle of history plays makes clear, a site of struggle, of factions at war, of food chains, and of disputed hierarchies. Within dominant ideological constructions countermeanings are always lying in wait.

When Darwin set out on the voyage of the *Beagle* in 1831, Linnaeus had only 45 years before been translated into English as saying that "If we therefore enquire into the original appearance of the earth, we shall find reason to conclude, that instead of the present wide extended regions, one small island only was in the beginning raised above the surface of the waters."[16] The place of islands in the natural history of the earth was a site of controversy, as Janet Browne demonstrates in *The Secular Ark*.[17] Darwin, writing in *Voyage*, is skeptical about the paradisiacal imagery of the primal island: visiting St. Paul's Island, he observes that "not a single plant, not even a lichen, grows on this islet." The "terrestrial fauna" consists of a "fly living on the booby, and a tick which must have come here as a parasite on the birds; a small brown moth . . . a beetle . . . and a wood louse . . . and lastly numerous spiders. . . . The often-repeated description of the stately palm and other noble tropical plants, then birds, and lastly man, taking possession of the coral islets as soon as formed, in the Pacific, is probably not quite correct; I fear it destroys the poetry of this story, that feather and dirt-feeding and parasitio insects and spiders should be the first inhabitants of newly formed oceanic land."[18]

The issues concerning indigenous inhabitants and the myths of the noble savage and the enlightened explorer alike haunt Darwin's language. In the *Diary* entries he made while on the Galapagos, Darwin writes with disgust at the clumsy lizards: "They are as black as the porous rocks over which they crawl and seek their prey from the Sea." Somebody calls them "imps of darkness" (p. 334). The reference is to Caliban in Shakespeare's play most concerned with colonization, *The Tempest*.[19] A week later, describing the conditions of the British settlement on Charles Island, Darwin comments on the lack of water and continues: "Generally the islands in the Pacifick are subject to years of drought and subsequent scarcity; I should be afraid this group will not afford an exception. The inhabitants here lead a sort of Robinson Crusoe life" (*Diary*, p. 337). We do not have reading lists for Darwin's time on the *Beagle* (though we can piece together a good deal of his more studious library) but among his reading in the latter part of 1840 (June 10–Nov. 14) Darwin noted together on one line "Some Arabian Nights Gulliver's Travels Robinson Crusoe."[20] The group is recorded among Horne Tooke's work on language theory, Humboldt's *Personal Narrative* of his scientific explorations, Harriet Martineau's novel on the uprising of the slaves under the leadership of Toussaint L'Ouverture, *The Hour and the Man*, Carlyle's *Sartor Resartus*, Hume's *History of England*, and a variety of geological, biogeographical, and travel literature.

Darwin was meticulous about recording his reading even of books he had read before, so we have no need to suppose that this was his first acquaintance with the *Arabian Nights* or *Gulliver's Travels*, and we know it was not for *Robinson Crusoe*. Indeed, that would be most unlikely. Rather, these literary texts suggest very diverse interpretative models for understanding the business of sailing round the world and visiting strange populations, which was what Darwin had spent the years between 1831 and 1836 doing and writing about. It is suggestive to find him reading them in the midst of his work for the geological volumes issuing from the voyage: *The Structure and Distribution of Coral Reefs* (1842), *Geological Observations on the Volcanic Islands Visited During the Voyage of the HMS Beagle* (1844), and *Geological Observations on South America* (1846). Edward Said has alerted us in *Orientalism* to the permeating effects in Western culture and writing of the self-projection figured onto the East. Sinbad the sailor can become a repository for guilt and self-exculpation. But the two eighteenth-century works, *Robinson Crusoe* and *Gulliver's Travels*, present salutorily conflicting views of the Englishman abroad and of the relationships between self-possession and the possession of territory. Crusoe, the castaway, becomes at last king of his island, gratified to be able to substitute imported human subjects for the domestic animals from whom he had earlier formed his court and family. Gulliver takes leave of his readers by defending himself against the charge that he has shown a lack of zeal for enlarging his Majesty's dominions by his discoveries:

To say the truth, I had conceived a few scruples with relation to the distributive justice of princes upon these occasions. For instance, a crew of pirates are driven by a storm they know not whither; at length a boy discovers land from the topmast; they go on shore to rob and plunder; they see a harmless people, are entertained with kindness, they give the country a new name, they take formal possession of it for the king, they set up a rotten plank or a stone for a memorial, they murder two or three dozen of the natives, bring away a couple more by force for a sample, return home, and get their pardon. Here commences a new dominion acquired with a title *by divine right*.[21]

Of course, the *Beagle* was on no such unruly piratical voyage.[22] It was one of a great number of admiralty-encouraged and part-financed surveying enterprises: not so much exploring as mapping, part of a system of commercial knowledge-seeking that would in time serve the turn of colonizers and empire makers as well as the safety of seamen. The purpose of the enterprise was to annex information, not land. The journals of writers such as Jukes and Webster are typical of the genre of report that these journeys brought forth. Taking part in such a journey seems also to have been something of a rite of passage for Victorian natural scientists (Huxley on the *Rattlesnake*, Hooker on the *Chanticleer*, for example).

By 1880, when he published *Island Life*, Alfred Russel Wallace (the codiscoverer of natural selection) could see very clearly another kind of ransacking and predation for which these numerous voyages had become responsible:

I cannot avoid here referring to the enormous waste of labour and money with comparatively scanty and unimportant results to natural history of most of the great scientific voyages of the various civilized governments during the present century. They have brought home fragmentary collections, made in widely scattered localities, and these have been usually described in huge folios, whose value is often in inverse proportion to their bulk and cost. . . . The result of this wretched system is that the productions of some of the most frequently visited and most interesting islands on the globe are still very imperfectly known, while their native plants and animals are being yearly exterminated.[23]

But when Darwin set out in the early 1830s those conservationist fears were not yet active. Nevertheless, the voyage of the *Beagle* was to some degree a voyage of reparation and return. Darwin was not implicated in that earlier voyage on which Fitzroy took away a group of young Fuegian people from their homes, kidnapping them or taking them hostage, arguing always that he intended to return them as educators and missionaries to their people. Fitzroy did return them, after a sojourn in London and elsewhere during which they were subjected to display, education, sartorial transformations, and Christian doctrine. One of the purposes of the *Beagle* expedition was to set them down near the place they had left (the English misjudged by about twelve miles and so landed the Fuegians among people who spoke a different language). So, significantly, among the 70-odd people on board ship for half the journey was a group of Fuegians, dressed in western style and apparently to some degree assimilated to British culture.[24] Though it would be overstating things to suggest that they were Darwin's constant companions (he and Fitzroy shared the captain's cabin, along with twenty chronometers), Darwin did talk with them and learned from them, particularly from Jemmy Button, whose information he appeals to as authoritative. Questions of migration, rafting, cultural adaptation, and ecological imperialism were thus daily enacted within the ship's company and trouble the language of Darwin's records.

In the introduction to their useful abbreviated edition of *The Voyage of the Beagle*, Michael Neve and Janet Browne go some way to counter the solely scientistic reading of the *Beagle* voyage.[25] Explaining their principles of selection, they remark that they have retained Darwin's descriptions of political and military campaigns he encountered and his descriptions of "the social mores of the groups of people that he met." The reason they give for doing so is that "natural history and coastal surveys were . . . heavily politicized nationalistic pursuits." They continue: "Darwin and Fitzroy were never neutral observers, and this text shows just how far Darwin, at least, saw foreign society through English eyes and endorsed the social order and parliamentary structure of his native country" (p. 10). But that formulation threatens to divide the political from the scientific even while it acknowledges the material base of the scientific expedition. Darwin's encounters with political and military campaigns and his observation of the peoples that they met in fact became important constituting elements in the idea of natural selection. In particular, they bore on the problems of extinction and diversification. Although, as Browne and Neve observe, Darwin certainly never was a neutral observer, it is not the case that he "saw through English eyes" only. Instead, he more and more looked back with exotic and awakened eyes on the taken-for-granted England from which he had set out. The process begins on the first page of the 1839 version. Arriving at St. Jago, Darwin remarks:

The scene, as beheld through the hazy atmosphere of this climate, is one of great interest; if, indeed, a person, fresh from the sea, and who has just walked, for the first time, in a grove of cocoa-nut trees, can be a judge of any thing but his own happiness. The island would generally be considered as very uninteresting; but to any one accustomed only to an English landscape, the novel prospect of an utterly sterile land possesses a grandeur which more vegetation might spoil. (Fitzroy, pp. 1–2)

That capacity imaginatively to *enjoy* difference proves to be Darwin's central intellectual gift.

Darwin muses on his native space and society satirically, fondly sometimes. By this means the tumult of impressions is held and sorted, and some of the pleasure in his thought is in comparisons and gradings, such as his observation that the Chilean horses "appeared to recover from any injury much sooner than those of our English breed" (Fitzroy, p. 25) or that the South American partridges do not "conceal themselves like the English kind. It appears a very silly bird" (p. 51). Sometimes such comparisons point to underpinning and reassuring likenesses: the brine-lakes throughout Patagonia, in Northern Chile, and at the Galapagos Islands, as well as the brine-pans at Lymington at home in England, all harbor life: "Thus we have a little living world within itself, adapted to these little inland seas of brine. . . . Well may we affirm, that every part of the world is habitable!"[26] The younger Darwin tends to the sanguine.

Other passages, however, show him working at the questions of difference with real conflict and skepticism. The effects are visible in the often discordant assessments he will rally within a sentence. Had he really looked only with English eyes and "endorsed the social order" of his country he could never, in my view, have reached the radical insights of his evolutionary theories.

Traveling, recording, rereading those records, and writing afresh all combined to make him aware of social construction and of difference. They made him aware, too, of the instability of observation and the value judgments always implicit in description: "a party of soldiers being sent" to recapture a group of runaway slaves, "the whole were seized with the exception of one old woman, who sooner than again be led into slavery, dashed herself to pieces from the summit of the mountain. In a Roman matron this would have been called the noble love of freedom: in a poor negress it is mere brutal obstinacy."[27] That insight came readily enough, it could be argued, to one raised in a family traditionally opposed to slavery and active in the Wedgwood "Am I Not a Man and a Brother" campaign. More striking is Darwin's analysis of the degree to which St. Helena has become, ecologically, a parody of England and its social injustices:

St. Helena, situated so remote from any continent, in the midst of a great ocean, and possessing a unique Flora, this little world within itself—excites our curiosity. Birds and insects, as might have been expected, are very few in number; indeed I believe that all the birds have been introduced within late years. Partridges and pheasants are tolerably abundant: the island is much too English, not to be subject to strict game-laws. I was told of a more unjust sacrifice to such ordinances, than I ever heard of even in England. The poor people formerly used to burn a plant, which grows on the coast rocks, and export soda; but a peremptory order came out prohibiting this practice, and giving as a reason, that the partridges would have nowhere to build! (Fitzroy, p. 584)

Perhaps most striking of all are passages where the writing registers a deep unease, a disequilibrated reach for a vocabulary that in the event cannot steady itself along one axis of judgment. That disturbance is mani-

fest in Darwin's description of his visit to the penal settlement Van Diemen's Land (now Tasmania). His judgment of colonial proceedings vacil-
lates between an attempt to justify the treatment of the native inhabitants
and a disturbed recognition of its cruelty:

> All the aborigines have been removed to an island in Bass's Straits, so that Van
> Diemen's Land enjoys the great advantage of being free from a native population.
> This most cruel step seems to have been quite unavoidable, as the only means of
> stopping a fearful succession of robberies, burnings, and murders, committed by
> the blacks; but which sooner or later would have ended in their utter destruction.
> I fear there is no doubt that this train of evil and its consequences originated in
> the infamous conduct of some of our countrymen. Thirty years is a short period,
> in which to have banished the last aboriginal from his native island,—and that
> island nearly as large as Ireland. (Fitzroy, p. 533)

Almost unavoidably, the name of Ireland enters the colonial discourse,
overstepping its function here of providing a land measurement. The
young Englishman shifts from seeing the absence of a native population as
"a great advantage," to a bleak recognition of the monstrous speed of
extirpation, the guilt of the colonists, and the natural claims of the aboriginal inhabitants to their native island. The uncertainties in the passage are
epitomized in the single change Darwin makes for the 1845 edition—a
change that is effectively no change but registers his sense of the awkward
moral fit between the two halves of the sentence. Instead of "*but* which
sooner or later would have ended in their utter destruction," he substituted "*and* which sooner or later would have ended in their utter destruction" (emphasis added).

The equivocation in the word "native" works painfully here. The
phrase "native intruders" is frequent in nineteenth-century writing,
without any sense of its scandalous paradox. As he writes *Voyage* Darwin
has not yet thought his way through this ideological dilemma, but he is
amassing material concerning indigeneity and autochthony that is fundamental to his particular formulation of evolutionary theory. His increasingly disenchanted observation of the processes of colonization and
colonialism among human populations contributes quite as much to that
argumentative process as does his visit to the a-human territory of the
Galapagos islands. "Wherever the European has trod, death seems to
pursue the aboriginal. It was melancholy at New Zealand to hear the fine
energetic natives saying, they know the land was doomed to pass from
their children."[28]

His tone toward the Falkland Islands in his letters in April 1834 is positively Swiftian:

Here we, dog-in-the-manger fashion seize an island, leave to protect it a Union Jack; the possessor has been of course murdered; we now send a Lieutenant, with four sailors, without authority or instructions. . . . This island must some day become a very important halting place in the most turbulent sea in the world. . . . In other respects it is a wretched place.[29]

The struggle between inhabitants and incomers continues to haunt Darwin and becomes part of the argument about adaptation and sudden environmental change in *The Origin of Species* twenty years later. There, he opens the chapter on natural selection by imagining an ideal intellectual and ecological space, "an island" that is secure from the predations of "improvement" from outside, from those he describes as "intruders": "had the area been open to immigration, these same places would have been seized on by intruders."[30] Instead the aboriginal inhabitants or lifeforms discover more and more ecological niches within the island economy solely by the slow means of diversification: "In such case, every slight modification which in the course of ages chanced to arise, would tend to be preserved: and natural selection would thus have free scope for the work of improvement." As I have argued elsewhere, this description does not—as many commentators have asserted—provide a language for colonialism, but it does, incipiently, provide a language for the exclusion of immigrants—an equally damaging extension.[31] In their major biography, Adrian Desmond and James Moore hold to the view of Darwin as an unperturbed English patriot and point to Darwin's support of missionary activity in New Zealand.[32] "Fitzroy and Darwin looked through Christian spectacles at the heathen hordes. Both used an inflexible 'scale' of civilization, with progress measured against the European ideal" (p. 175). They note Darwin's pleasure, for example, at seeing "an English farm house & its well dressed fields, placed there as if by an enchanter's wand" (Desmond and Moore, p. 175). One might set against that Darwin's diary entry of August 9, 1835, where he contrasts the current state of misrule in Christian Lima, with its "extraordinary number of churches," with seeing

one of the very numerous old Indian villages, with its hill-like mound in the centre. The ruins in this plain of houses, enclosures, irrigating streams & burial mounds, give one a high idea of the ancient population. When the Earthen ware is considered; the woollen clothes, the utensils of elegant forms cut out of the

hardest rocks, the tools of Copper & ornaments of precious metals, it is clear they
were considerably advanced in civilization. (*Diary*, pp. 332–33)

In any case, my argument does not deny the supremacist pressure in
Darwin's responses. Here, for example, he wincingly concedes more to
developmental views of race in the 1845 revision: "It is impossible not to
respect the considerable advance made by them in the arts of civilization"
(*Journal*, p. 368). Instead, I want to emphasize that Darwin's language
manifests the difficulties he experienced in sustaining the orthodox acceptable point of view. That struggle between normative values and in-
surgent insights makes for difficulties—and those difficulties are productive of fresh awareness. There is no unidirectional movement toward or
away from orthodoxy between 1831 and 1845 in his views on English
colonization or the English self-image. The disturbance never settles.

Any long sea voyage makes the traveler aware of islands: most of the
places visited in such conditions are certain to be islands. The ship itself,
moreover, becomes an intensified form of national floating island, preserving rigorously the forms of the home community, exaggerating them
in the tight hierarchies and lack of privacy of shipboard life. In one sense,
every time Darwin touched shore on that voyage he was leaving England
anew. Telling comparisons sprang up from this tight conjunction of distant homeland and present place. What had been exotic was now the
everyday. The alternation between ship and shore dramatized the process
of contact, observation, record, and discovery.

Darwin started out from and returned to an island, in small loops
repeatedly to the *Beagle*, but also encircling the globe to perform the
narrative of awakened recognition. Not only the distant places visited but
the homeland is seen as if for the first time. That narrative of travel and
return is familiar from much story; it was particularly important in the
romantic period through the form of Bildungsroman, typified in Goethe's *Wilhelm Meisters Wanderjahre*, and through the parabolic, such as
Novalis's *Die Lehrlinge von Sais*, where the familiar becomes erotically
valued and known only when the furthest antipodes of experience are
reached. So Darwin's initiating and homeward island is England (or, more
correctly, Scotland, England, and Wales, in all three of which he spent
prolonged formative periods of his early life). But Darwin's is not a story
of complete reassimilation or maturing into accord with the home culture. It is rather one of silent resistance pursued over years to come and
issuing in fundamental changes to that culture.

A trope of nineteenth-century historiography was to emphasize the essentially law-abiding and secure nature of island kingdoms, especially if they happen to be England. Macaulay opens his *History of England* with, first, a dark picture of an early England as barbaric as the Easter Islands. But after the Norman settlement of the land, argues Macaulay, the thirteenth century established the character of England: "Then it was that the great English people was formed, that the national character began to exhibit those peculiarities which it has ever since retained, and that our fathers became emphatically islanders, islanders not merely in geographical position, but in their politics, their feelings, and their manners. Then first appeared with distinctness that constitution which has ever since, through all changes, preserved its identity . . . [and of which] all the other free constitutions in the world are copies."[33]

From time to time in *Voyage* Darwin allows himself a cheerful and pacific patriotism, calling on the French writer Monsieur Lesson to provide the necessary cover to his boast: Of Ascension Island he remarks, "The whole island may be compared to a huge ship kept in first-rate order. . . . M. Lesson has remarked with justice, that the English nation alone would have thought of making the island of Ascension a productive spot; any other people would have held it without any further views, as a mere fortress in the ocean" (Fitzroy, p. 586). The ship here stands in for a tightly bounded, intricately taxonomized—and fertile—concept of society.

But a newly learned sense of the frailty of English society and custom is figured in his reaction, at another point in the voyage, to the great earthquake at Concepción. Darwin was lying down in a wood on the island of Quiriquina when the quake occurred. To his horror, seasick as he always was, the earth began to behave like the sea: "It was something like the movement of a vessel in a little cross-ripple" (Fitzroy, p. 369). "The island itself as plainly showed the overwhelming power of the earthquake, as the beach did that of the consequent great wave" (p. 370). "A bad earthquake at once destroys our oldest associations; the world, the very emblem of all that is solid, has moved beneath our feet like a crust over a fluid" (p. 369). (The 1845 version has "earth" for "world" and other changes.)

When he reached the town Darwin was confronted with the human enormity of the event. His observations of the appalling distress of the inhabitants ends with a paragraph in which he broods upon the seeming stability of England, in social as well as geological terms:

If, for instance, beneath England the now inert subterranean forces should exert those powers which most assuredly in former geological ages they have exerted, how completely would the entire condition of the country be changed! What would become of the lofty houses, thickly packed cities, great manufacturies, the beautiful public and private edifices? If the new period of disturbance were first to commence by some great earthquake in the dead of the night, how terrific would be the carnage! England would at once be bankrupt; all papers, records, and accounts would from that moment be lost. Government being unable to collect the taxes, and failing to maintain its authority, the hand of violence and rapine would go uncontrolled. In every large town famine would go forth, pestilence and death following in its train. (Fitzroy, p. 373)

The frailty of social order, the inevitability of breakdown, the loss of the processes of inscription and communal memory that sustain government, come home to him as he recreates first in his diary and then in his published writing those experiences of which he was part observer, part participant, while on his bildungsroman journey. His writing is imbued also with the fears of social upheaval that were already in the 1830s gripping the English middle classes, as Carlyle's essay *Chartism* demonstrates. In the *Diary* entry for March 5, 1835 (p. 282), an extra sentence appears at the end of his earthquake fantasy of England undermined: "Who can say, how soon such will happen?" That sentence is dropped in the published version. Perhaps in England in 1837 the question pressed too close.

Darwin had seen several revolutions by the time he returned home and knew that England would not rely on immunity either from earthquake or violent social change. Indeed, his training as a geologist habituated his thinking of England as "troubled," an unstable island.[34] Connections between geology and revolution were never far to seek. Charles Lyell's theory of uniformitarianism affirmed stability, but stability of explanation, not of event—and as Lyell remarked in his opening sentences, geology is a theory of change.

Darwin grew up an inland islander more often in country houses than by the sea, and in that inland dwelling he shared the particular condition of many inhabitants of Britain. English people tend to be unaware of how unusual the population distribution of the island is, and how fortunate they are to be able to use almost its whole surface. By way of contrast, consider islands of equivalent size such as Japan, where extensive tracts of mountainous country force the population to cluster in pockets, or an island continent of enormous extent such as Australia, where most of the population live on the littoral strip, fronting the sea, with vast unoccupied land spaces at their back. Darwin grew up—as do many of us—taking for granted the densely netted sociology and ecology of the British Isles. Notable among its characteristics is the possibility of being an inland dweller, a midlander far from the shore, though claiming that shore ideally. This is a privilege much taken for granted in nineteenth-century fiction, for example, in George Eliot and even Dickens.

While on the voyage of the *Beagle* Darwin was much of an age with a present-day graduate student; when he set out he was almost 23; when he returned he was 28. During these years he undertook grueling land journeys, often accompanied by only one local guide. He spent quite long periods away from the *Beagle*, trekking across unfamiliar and lightly charted or unmapped terrain. He was a figure utterly different from the myth of the sedentary Darwin, based on a considerably later period of his life. One thing that such a journey teaches you, he wrote in the conclusion to *The Journal of Researches*, is that "the map of the world ceases to be a blank; it becomes a picture full of the most varied and animated figures. Each part assumes its proper dimensions: continents are not looked at in the light of islands, or islands considered as mere specks, which are, in truth, larger than many kingdoms of Europe" (p. 505). He learned to rescale England and also the class community out of which he had emerged to take account of the scope and difference of population across the world.

The other thing such extensive travel teaches is the overwhelming amount of water in the world: crossing the Pacific, "moving quickly onwards for weeks together we meet with nothing but the same blue, profoundly deep, ocean. Even within the Archipelagoes, the islands are mere specks, and far distant, one from the other. Accustomed to look at maps drawn on a small scale where dots, shading and names are crowded together, we do not judge rightly how infinitely small the proportion of dry land is to the water of this vast expanse" (Fitzroy, p. 446).

The study of geology begins to seem a counterenterprise, seizing land back into analytical memory from the infinite forgetfulness of ocean. His visit to the Falkland Islands and—particularly through the doubled presence of Jemmy Button as savage and as European companion—to Tierra del Fuego were quite as important imaginatively as his visit to the Galapagos. They dramatically strengthened Darwin's awareness of how fundamental are the processes of adaptation and descent for the generations of life-forms dwelling in an enclosed environmental system. Islands are dramatic evidence of how slight environmental differences privilege and require different characteristics.

Island sites also tax the powers of observers by the hope they seem to hold out of providing a complete record. Darwin writes disconsolately in 1845 to J. D. Hooker that Hooker and Macrae's collections of plants do not correspond: "Thus though you both collected on one small Isld. your collections are very different, 11 out of his 15 you did not collect & he only 5 out of the 60 that you got, this is terribly unsatisfactory."[35] Darwin concludes that the boundaries of islands are in ecological terms never as determined as the land form may suggest: "There is still an enormous deal to be done with the materials—a comparison of the Islets with all the extra Galapageian species eliminated . . . the proportion of driftable & portable xtra Galap. plants in each:—those that fly or *are flown* by birds, those that salt water does & does not kill: that birds do & do not digest &c&c&c" (p. 221).

How to distinguish indigenous from interchanged species becomes a taxonomic issue that goes to the heart of his inquiry—and to the heart of much of the subsequent nineteenth-century social theory concerning immigration and colonialism that emerged from Darwin and Hooker's apparently autonomous and technical study of "Teucrium inflatum, Salvia tiliaefolia, Scoparia dulcis, Lantana canescens, Verbena littoralis, Bousingaulita baselloides, Brandesia echinocephala" and other plants on Charles Island.

The *reading out* of Darwinism into other domains that he sought to debar by his exclusion of humankind from the argument of *Origin* had already been written into *Voyage* more than twenty years earlier. The writing and revising, the miniaturization of the world back into book, obliged him to redraft the mental maps he and his compatriots shared. So in 1837 the writing self is back home, at a desk in England, an England mercifully steady underfoot, but now also an England forever destabilized by the observation and experiences of that earlier traveling self. Writing *brings home* the exotic. It cannot be left behind. Inscription forges new connections between earlier observations; writing renews experience and changes it. The "other" provides allegories for the familiar and may allow alarming perceptions to flourish without their needing to bare themselves to scrutiny.

Darwin had seen the processes of manmade selection at work between peoples: in Van Diemen's Land, in Tierra del Fuego, between the Spaniards and the Indians, between the British and the Aborigines. In his formulation of the terms of "natural selection" he sought to describe a very different process than these ethnic clashes: unconscious, working for the good of each species, extended gradually over tracts of time. But he was driven to use also the perceptions he had gained, and the language of recourse, from his political and social experiences during his five years circumnavigating the globe, as often on land as sea.

Howard Gruber and, more recently, James Moore and Adrian Desmond have argued that Darwin profoundly feared the religious implications of the evolutionary ideas he had already reached before the end of the 1830s and that dread, as much as the desire for further evidence, delayed the publication of his findings by twenty years. That may well be so. But further, an awakened cultural relativism, natural historical comparisons, and a sense simply of the *variety* of possible life-forms, possible social groupings, meant that Darwin henceforth could never be sanguine about the centrality or the permanence of his own society, nor the efficacy of its hierarchies or laws.

Perhaps, in conclusion and in a different guise, the concept of the island can help us better to understand the processes of Darwin's creativity in those later years. The "island state" of Down House, inland, paternalistic, in a village community where the unknown had little obvious place, grounded Darwin in his imaginative journeys into complexity, profusion, and evanescence. His role as what Jim Moore calls squarson—a condensation of squire and parson—gave him a prominent role in the ordering of the small community. He pacified the unruly forces of his imagination. At the same time, he relished the rich monotony of family life, with its endless proliferation of difference within the biota of the household, its shift of power as the generations move on. He knew, and vehemently, that it is possible for "one second of time [to create] in the mind a strange idea of insecurity, which hours of reflection would never have created" (Fitzroy, p. 369). Change and variation are not to be avoided: that he also knew. The peaceable circumstances of his later ordered life—a contracted island—were the necessary fiction for an imaginative and argumentative revolution.

Illustration as Strategy in Charles Darwin's 'The Expression of the Emotions in Man and Animals'

On November 21, 1826, the night he petitioned to join the Plinian Society at the University of Edinburgh, Charles Darwin attended a lecture refuting professor Charles Bell's treatise, *The Anatomy and Philosophy of Expression as Connected with the Fine Arts* (1806). Bell proposed that a rigorous understanding of the anatomy of expression could be useful to visual artists. Advocating that scientific knowledge should inform artistic practice, he reasoned that

anatomy [is] the examination of the structure by which the mind expresses emotion, and through which the emotions are controlled and modified; it introduces us to the knowledge of the relations and mutual influences which exist between the mind and the body. To the painter, therefore, the study is necessarily one of great importance; it does not teach him to use his pencil, but rather it teaches him to observe nature, to see forms in their minute varieties [and] to catch expressions so evanescent that they must escape him, did he not know their sources.[1]

Bell believed that emotional expressions reflect an internal formula that can be captured and communicated by artists. As a result, he argued, a comprehensive understanding of the physiology of expression could help artists to effectively render the emotional states of their subjects and to extrapolate details of movements too fleeting to study in situ.

By the time Charles Darwin published *The Expression of the Emotions in Man and Animals* in 1872, the logic of Bell's argument had been reversed. The invention of photography allowed scientists to examine rapid movements without relying on the expertise of visual artists. For the first time, a technology existed that could depict physiological movements without the direct intervention of a painter or draftsman, and the camera had emerged as an authoritative source of visual information. Whereas Bell had been forced to theorize about transitory expressions without precise visual data, Darwin was among the first scientists to attempt to use photography to record moments of activity previously invisible to the human eye. Darwin believed that the objectivity of photographic evidence could be used to challenge Bell's analysis of expression. With this in mind, he commissioned and collected hundreds of photographs depicting various aspects of human expression. Although technical limitations compromised the utility of these photographs, Darwin used many of them to illustrate *The Expression of the Emotions*. *The Expression* thus became one of the first photographically illustrated scientific treatises, and the only work in which Darwin reproduced actual photographs. Far from scientifically factual, these photographs formed part of a narrative strategy designed to advance his theoretical concerns.

Darwin's return from the voyage of the *Beagle* in 1836 coincided with William Henry Fox Talbot's introduction of the first photographic process in Britain.[2] Talbot's method entailed the production of a paper negative, which was printed in contact with a piece of chemically sensitized photographic paper. Other early photographic techniques, such as the popular daguerreotype process from France, resulted in the production of a single photographic image that could not be reproduced. Talbot's method allowed the photographer to produce an unlimited number of relatively inexpensive prints from a single negative. This meant that even in cases where specimens were rare or unique, representations of those specimens could be distributed easily for critical analysis. As a result, by the 1840s a number of British photographers began to use Talbot's process for the documentation of scientific samples.

Talbot's academic associations were crucial to the development of his photographic process. Both Darwin and Talbot had studied at Cambridge University, where they had met the eminent professor John Herschel. Herschel became a mentor to Talbot, and was instrumental in working out the chemical procedures necessary for producing photographic prints. He is credited with the invention of sodium hyposulfate as a photographic fixative, and he devised much of the basic vocabulary of the medium, such as the terms *photography*, *snapshot*, and *fix*. Darwin admired Herschel, and listed his *Introduction to the Study of Natural Philosophy* as one of the most compelling books he had ever read.[3] Darwin maintained his acquaintance with Herschel until Herschel's death in 1871, and kept in touch with him through their mutual friend, the photographer Julia Margaret Cameron.

Darwin had no formal training in the visual arts, although he did maintain a lively interest in painting, printmaking, and photography. While a student at Cambridge University in 1831, Darwin frequently visited the Fitzwilliam Galleries, where he met with curatorial staff to discuss the pictures on display.[4] He read Sir Joshua Reynolds's *Discourses at the Royal Academy* (1790), and began a small collection of engravings under the tutelage of Charles Thomas Whitley, an associate at Cambridge. He also made occasional trips to the newly opened National Gallery of Art in London, where he admired the works of Renaissance masters.[5] Nevertheless, Darwin considered himself unable to prepare even rudimentary sketches.

Although Darwin did not produce his own photographs, he would have been acutely aware of developments in photographic technology. The cataloguing of botanical speciments was one of the first scientific applications of the medium. As early as 1835, Talbot had discovered that by simply placing a botanical specimen on top of a piece of photographic paper, then exposing it to light, one could create a perfect life-sized silhouette of that specimen. Such prints were aesthetically pleasing and rich in morphological detail. Talbot maintained that photography was uniquely capable of producing representations of flowers and leaves, for example, "with the utmost truth and fidelity, exhibiting even the venation of the leaves, the minute hairs that clothe the plant, etc."[6] He is known to have exhibited such illustrations before his colleagues at the Linnean Society, where Darwin was frequently in attendance.

Photography rapidly became a dominant means of visual communication in nineteenth-century Britain. The purported objectivity and relative ease of production afforded by the medium proved irresistible to Victorian consumers. As early as 1843, *The Edinburgh Review* reported that

the art of photography [allows us] to obtain perfect representations of all objects, whether animate or inanimate, through the agency of light which they emit or reflect. . . . It has called to its aid the highest resources of chemistry and physics;

and while it cannot fail to give a vigorous impulse to the fine arts, it has already become a powerful auxiliary in the prosecution of physical science; and holds out no slight hope of extending our knowledge of the philosophy of the senses.[7]

By the 1860s, commercial mass production of photographic materials insured that even individuals of modest income could afford to own and commission photographic prints. The effect of the new medium was profound and extensive. The quantity and quality of imagery that could be produced using photography was unparalleled in the history of representation. Soon after its invention, photographic imagery was understood to be a standard of truth in a wide variety of applications, from the popular to the scientific and documentary.

The growing reputation of photography for producing engaging evidence must have excited Darwin. To be effective, all scientists must convince an audience of the validity of their work, and must both attract and persuade their readership. These are substantial obstacles to the acceptance of any research development, but were particularly onerous to Darwin. The broad paradigmatic shift inherent in Darwinian evolutionary theory required significant cultural changes in Victorian Britain. A substantial departure from the ideas that preceded it, Darwin's theory of evolution did not feature the cognizant activity of a divine being. Darwinian theory was deeply troubling to natural theologists, who sought to describe biological phenomena in religious terms. Evolution by natural selection challenged the canon of natural theology, which maintained that there could be nothing imperfect or unnecessarily complex in the works of God. Darwin's success in gaining acceptance for his theories is remarkable given the sensitivity of his research and its religious implications.

Contemporary theorists have suggested that the rapid assimilation of Darwinian theory into Victorian culture is in part due to Darwin's effective literary style. Gillian Beer characterizes Darwin's approach as metaphorical:

It is the element of obscurity, of metaphors whose peripheries remain undescribed, which made *The Origin of Species* so incendiary—and which allowed it to be appropriated by thinkers of so many diverse political persuasions. It encouraged onward thought: it offered itself for metaphorical application and its multiple discourses encouraged further acts of interpretation. The presence of latent meaning made *The Origin* suggestive, even unstoppable in its action upon minds.[8]

Edward Manier has pointed out that this quality of Darwin's work was persuasive as well as polemic.[9] By adopting nomenclature that suggested but at the same time replaced the theories he sought to supplant, Manier argues, Darwin was able to create a new scientific vocabulary, and in turn a new way of describing and perceiving nature. This may have facilitated the acceptance of Darwinian theory by establishing a neutral framework in which his observations could be considered.

Historians now acknowledge the importance of Darwin's narrative techniques to the acceptance of Darwinian science. Darwin is understood to have been more than a clever theorist; he was also a skilled writer with a persuasive literary style. Darwin produced works that engaged his audience. At the same time, his narrative methods helped to popularize his work, for his writing was highly accessible, imaginative, and persuasive. Darwin used an innovative lexicon to describe his evolutionary worldview. The effectiveness of crucial terminology such as *natural selection, struggle*, and *chance* in *The Origin of Species* may in part account for the widespread acceptance of Darwinian science. In addition to their metaphorical resonance, the terms Darwin used to describe his theory of evolution were simple and distinct. These terms served to make Darwin's arguments more palatable to his Victorian audience, and helped his evolutionary theory gain popular acceptance.

From this perspective, the photographic illustrations Darwin used in *The Expression of the Emotions in Man and Animals* invite closer scrutiny. At the time Darwin published *The Expression*, photographic representation had become accepted currency in Victorian popular culture. Nevertheless, photographic illustration had seldom been used in scientific discourse. Darwin suggested that the photographic illustrations used in *The Expression* were more objective than the drawn and painted illustrations that had been used in other works. In fact, his photographic illustrations were carefully contrived to present evidence Darwin considered important to his work. Whether or not Darwin consciously recognized the suspect nature of his photographic evidence, he applied it with sophistication in *The Expressions*. He knew that photography could assume more than a straightforward documentary role in scientific writing, and recognized its powerfully persuasive potential.

The Expression of the Emotions contains 29 photographic illustrations and a number of engraved plates made from photographs. A representative plate from *The Expression* is reproduced in Figure 9.1. The success of these images derived from their ability to persuade the reader to accept Darwin's theoretical arguments. Given his knowledge of the work of Talbot, Herschel, and others, Darwin was uniquely situated to take ad-

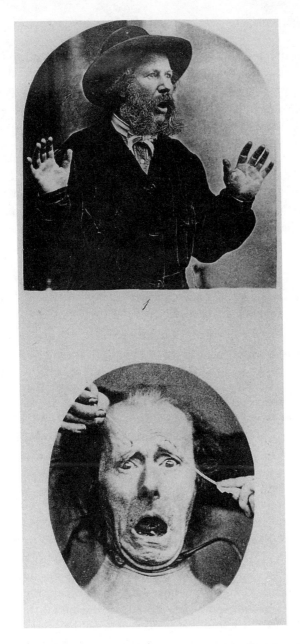

Figure 9.1. Plate 7 from Darwin's *The Expression of the Emotions in Man and Animals* (Chicago: University of Chicago Press, 1965 [1872]). Like many of the illustrations in the second half of *Expression of the Emotions*, this plate combines a posed photograph by Rejlander (above) with a photograph of a Duchenne de Boulogne experiment (below).

vantage of the most up-to-date developments in photographic technology. If, as some theorists have suggested, Darwin used verbal constructs that facilitated the acceptance of his theories, it stands to reason that he may also have used photographic illustration to his advantage.

Although contemporary theorists have attempted to delineate the compatibility of Darwinian narrative with Victorian popular culture, they stop short of attributing his success to the intentional manipulation of his published material. Analysis of the illustrations used in *The Expression of the Emotions* suggests a more intriguing vision of Darwin. In composing that work, Darwin strategically applied illustrations to augment the effectiveness of his textual arguments. The photographs he selected for inclusion in *The Expression* were designed to interest and engage his readers, even at the expense of scientific objectivity. Consideration of the photographic illustrations in *The Expression* demonstrates that Darwin had the capacity to act as a shrewd strategist, who carefully structured his work to have maximum effect on his readers.

The Evolutionary Significance of the Expression of the Emotions

Upon his return from the voyage of the *Beagle* in 1836, Darwin professed to have been overwhelmed by the expressions of his family and friends as they greeted him after a five-year absence. With the birth of his son William in 1839, Darwin began to take notes on the first expressions of his own infant children. He was convinced that even in infancy children would instinctively exhibit the complex and subtle shades of expression visible in adults.[10] Although it was not until some 30 years later that Darwin would publish these observations, it is clear that the subject of the expression of the emotions had long fascinated him personally.

Originally, Darwin had intended to devote only a chapter of *The Descent of Man, and Selection in Relation to Sex* to the question of the expression of emotions.[11] Later he decided that he had gathered sufficient material to merit the creation of a separate treatise on the subject. Darwin composed *The Expression of the Emotions* after the completion of *The Descent of Man, and Selection in Relation to Sex*. It was finished one year later and published in 1872.

The connections between *The Descent of Man* and *The Expression of the Emotions* are complex. In *The Descent of Man*, Darwin argued that the close parallel between the characteristics of humans and lower animals suggests that humans have evolved from animal ancestors. Those characteristics, Darwin argued, have been subject to evolution by natural selection as described in *The Origin of Species*. However, Darwin believed that a "simple" model of natural selection may be insufficient to account for all evolutionary developments. Darwin argued that more complex selective forces may account for some animal characteristics, including social behaviors. He proposed that evolution is driven in part by sexual selection, which exerts unique pressures on individual animals and affects the dynamics of animal populations.

The Descent of Man contained numerous references to the relationship between human and animal behavior, but avoided a detailed discussion of expression. In the sixth chapter of volume one, Darwin referred to the similarity of human emotional expression to that of various primates. Darwin characterized these as "various small points of resemblance between man and the quadrumana." Among these, Darwin described how the various emotions "are displayed [in primates] by nearly similar movements of the muscles and skin, chiefly above the eyebrows and below the mouth."[12] To Darwin, these behavioral similarities constituted evidence of evolution.

In fact, the question of behavioral adaptation was crucial to Darwin's understanding of evolution. In *The Origin of Species*, Darwin noted that the social behaviors of honeybees may have adapted collectively to benefit the hive:

The most wonderful of all known instincts, that of the hive-bee, can be explained by natural selection having taken advantage of numerous, successive, slight modifications of simpler instincts. . . . The motive power of the process of natural selection having been the economy of wax; that individual swarm which wasted least honey in the secretion of wax, having succeeded best, and having transmitted by inheritance its newly acquired economical instinct to new swarms, which in their turn will have had the best chance of succeeding in the struggle for existence.[13]

As Darwin explained in *The Origin of Species*, such behavioral adaptations may be extremely complex and difficult to decipher. Nevertheless, he remained convinced that adaptive behaviors were intrinsic to evolution by natural selection. Like Jean-Baptiste Lamarck before him, Darwin associated morphological differences between organisms with habitual behavior. Darwin believed that anatomical adaptations may facilitate certain types of behavior, which may in turn result in the selection of organisms with certain beneficial characteristics. In addition, he argued that behaviors themselves may be inherited, and are therefore responsive to evolutionary pressures.

In *The Expression of the Emotions*, Darwin argued that facial and other expressions originated in precisely this type of habitual behavior, although they may or may not serve a discernible purpose for the animal that displays them. He theorized that some expressions are the visible result of movements of muscle tissues in preparation for certain activities. The baring of teeth in anger, for example, may have developed as a biological preparation to bite or attack with the mouth. Conversely, Darwin believed that the derivation of some expressions may not be obvious to the analyst, as many behaviors have evolved strictly in opposition to others. To express satisfaction, for example, it may be necessary to do the opposite of express anger, that is, *not* to bare one's teeth. Thus the expression of pleasure may have developed in contradistinction to anger, and not from any obvious biological need. Finally, Darwin argued that the elaborate structure of the nerves themselves may be responsible for some expressions. The nervous system is required to perform divers functions that overlap and compete in human and animal bodies. As a result, Darwin argued, the complexity of some expressions may have developed as a mixture of voluntary and involuntary nerve action. He suggested that the movement of some tissues, which may have evolved in response to overt biological pressures, inadvertently stimulates other tissues to act in reflex, thus producing an overall expressive posture.

These ideas build on the principles set out both in *The Origin of Species* and *The Descent of Man*. While maintaining that behavior is responsive to natural and sexual selection, in *The Expression of the Emotions* Darwin suggested that some behaviors may be extraordinarily difficult to understand. He maintained that residual and associated habits may account for some forms of expression, and described some of the psychological principles that may precipitate other types of expressive behavior. Darwin proposed that nervous action, and in particular the action of the mind, may be so complex as to obfuscate completely the forces that produce any given behavior.

For this reason Darwin maintained that lower animals are ideal subjects for the observation of "emotional" expression. Lower animals, according to Darwin, are less susceptible to the highly complex associative psychological behaviors that humans are prone to. Moreover, they are excellent controls. Animals are unaware of the scientist's purpose, and are unlikely to modify their behavior in reaction to experimental observation.

This focus on lower animals was consistent with the approach Darwin had taken in *The Origin of Species*. In *The Expression of the Emotions* Darwin extended his principle of evolution by natural selection in arguing that humans and their behaviors respond to the same evolutionary pressures as all other animals. By identifying human expression, perhaps the most personal and emotive of human behaviors, with the biological instincts of lower animals, Darwin argued for a continuum of animal life: "We have seen that the study of the theory of expression confirms to a certain limited extent the conclusion that man is derived from some lower animal form, and supports the belief of the specific or subspecific unity of several races; but as far as my judgement serves, such confirmation was hardly needed."[14] At the foundation of Darwin's argument was the supposition that humans themselves, including their psychological makeup and their aesthetic sensibilities, have evolved from lower life forms.

In *The Expression of the Emotions* Darwin attempted to clarify and extend the workings of his principle of natural selection. In so doing, he found himself at odds with Christian theological doctrine, which held that humans have not evolved at all, but are the creation of an omnipotent God. Darwin's argument that human and animal life share the same origins contradicted the Church's literal interpretation of biblical events. That argument was also in sharp contrast to previous treatises on expression, which had posited since classical antiquity that expressions were the manifestation of divine energies, and the result of spiritual forces arising from the soul.

In the preface to *The Expression of the Emotions* Darwin noted his indifference to the disciplines of physiognomy and phrenology. He referred to physiognomic theory in the first line of the work, stating that he was not concerned "with Physiognomy,—that is, on the recognition of character through the permanent form of the features."[15] Nevertheless, it appears that Darwin may have conceived of his work as a corrective to dubious works on expression and character, such as Lavater's extremely popular turn-of-the-century *Essays on Physiognomy* (*Physiognomische Fragmente*, 1772). The principles of adaptation by natural and sexual selection precluded the physiognomic argument that character is expressed in the permanent fixation of the features. Instead, Darwin argued, one must look at the transitory effects of emotional expression to observe the workings of evolution.

Antecedents

The Expression of the Emotions derived from a tradition of investigation into the nature of human expression that had its inception in classical discourse, but was repeatedly explored in a number of works dating from the early Renaissance. The question of physiognomy, the relationship between character and appearance, dates from antiquity. To an author in the circle of Aristotle, one of the earliest known writers on physiognomy, it seemed "that the soul and body react on each other; when the character of the soul changes, it changes also the form of the body, and conversely, when the form of the body changes, it changes the character of the soul."[16] Subsequent writers perceived in this observation the foundations of a science of physiognomy. By the second century, the Greek physician Galen had proposed that such a science, predicated as it is on observations of links between psychology and behavior, requires medical training. Galen observed that the expression of the emotions was a physiological phenomenon, and should be considered using biological research methods. This argument was precocious, for the biological character of expression would not become a theoretical focus again until well into the seventeenth century.

Sixteenth-century artists devoted considerable effort to the study of both physiognomy (the external appearance of inner character) and pathognomy (the external appearance of the passions). As early as 1501, Jacopo de' Barbari had complained that contemporary paintings were "improbable and false" because painters lacked the knowledge of astronomy necessary to understand facial expression: "Physiognomy and chiromancy have their being by the force of the stars, which exert their greatest influence over man's outward appearance in the face and hands. And it is necessary that painters should understand this, in order to record the celestial influences appropriately in their painted histories."[17] Books were produced that depicted a variety of idealized pathognomic and physiognomic types. These were intended as teaching aids for students to follow in learning to communicate properly the precise character of people in their compositions. This effort was seen to be closely allied to the sciences, for it was only through accurate understanding of the mechanism of human expression that an artist could hope to paint a true likeness. Albrecht Dürer, Leonardo da Vinci, and G. B. della Porta are a few of the artists who devoted themselves to the documentation of ideal expressive types.[18]

Table 9.1 lists several of the most influential of these treatises on human expression. Descartes's *Les Passions de l'âme* (The Passions of the Soul), published only a few weeks before his death in 1649, was one of the first comprehensive scientific attempts to understand the subject. Descartes subordinated the arts to the dominion of scientific reason, and was one of the first to elaborate an analytical interest in human psychological theory.[19] Descartes theorized that the soul is united with the whole body, and that it exercises its functions by the mediation of a "little gland" in the middle of the brain.[20] He argued that this little gland orients the animal spirits, which in turn affect the nerves and muscles. The Cartesian conception of the passions was foremost a physiological one, arguing that emotional expressions could be rationally explained in terms of chemical and physical interactions.

Descartes's *Les Passions de l'âme* was illustrated with representative schematic drawings. A more elaborate illustrative presentation was devised by the seventeenth-century artist Charles Le Brun, whose *Method to Learn to Design the Passions* was published in 1667. Le Brun accepted the Cartesian view that expression should be understood in terms of the relationship between mind and body. By contrast, Le Brun considered artists as foremost in the physiognomic debate, lamenting that despite the ongoing efforts of philosophers and physicians, "no one has ever thought of making it his particular study with an eye to painting."[21] Le Brun's *Method to Learn to Design the Passions* was illustrated extensively with engravings from his own hand.

Arguably the most popular treatise on expression was published by Johann Caspar Lavater in 1772. His *Essays on Physiognomy* were elaborately illustrated by a number of influential artists, including the American Benjamin West and the Britons Henry Fuseli and William Blake. Lavater was principally a phrenologist, who based his study on the correlation of behavior with the external resemblance of humans to various animal types. Followers of Lavater maintained that one could deduce an individual's personality from his or her appearance. Ironically, the influence of Lavater's essays on physiognomy nearly prevented Darwin from joining the voyage of the *Beagle*: "Afterwards, on becoming very intimate with [Robert] Fitz-Roy [*sic*], I heard that I had run a very narrow risk of being rejected on account of the shape of my nose! He was an ardent disciple of Lavater, and was convinced that he could judge of a man's character by the outline of his features, and he doubted whether one with my nose could possess sufficient energy and determination for the voy-

TABLE 9.1

Selected Treatises on Expression, and Techniques Used to Illustrate Them

Author	Publication	Illustration techniques	Contributing artists	Illustration layout
Charles Darwin	The Expression of the Emotions in Man and Animals (1872)	Wood engravings, engravings from photographs, and heliotype photographs	Oscar Rejlander, G. B. Duchenne de Boulogne, T. W. Wood, Joseph Wolf, Briton Riviere, et al.	Full-page images grouped to follow text
Guillaume-Benjamin Duchenne de Boulogne	Mécanisme de la Physionomie Humaine (1862)	Photographs	G. B. Duchenne de Boulonge with Adrien Tournachon	Quarter- to full-page images interspersed with text
Charles Bell	Anatomy and Philosophy of Expression (1806)	Soft-ground etchings	Charles Bell, Giulio Romano, et al.	One-eighth to full-page images interspersed with text; on the top or bottom of each page.
Johann Caspar Lavater	Physiognomy (1772)	Etchings. Divers techniques	Benjamin West Henry Fuseli William Blake (Hunter ed.)	Mostly full-page images, printed opposite a page of text, with filigree decorations in text margin. The last eight illustrations follow text as appendix
Charles Le Brun (John Williams, ed.)	A Method to Learn to Design the Passions (1734)	Line engravings	Charles Le Brun	Five full-page illustrations interspersed with text
René Descartes	Les Passions de l'âme (1649)	Engravings. Outline drawings rendered from three separate perspectives, arranged to grid		

Figure 9.2. "Wonder and Astonishment," from Charles Bell's *Anatomy and Philosophy of Expression* (London: G. Bell, 1912). Note the similarity to Rejlander's illustration of "Surprise" in Fig. 9.1.

age."[22] Fortunately, Darwin's scientific talents prevailed over his unsatisfactory appearance.

The immediate antecedent to *The Expression of the Emotions in Man and Animals* was the University of Edinburgh professor Charles Bell's *Anatomy and Philosophy of Expression as Connected with the Fine Arts*, first published in 1806. As the title of the work suggests, Bell's argument focused on the application of human expression in drawing and painting. In the introduction to the third edition of the text, Bell explained the purpose of his undertaking:

I am not without hope that a new impulse may be given to the cultivation of the fine arts, by explaining their relation to the natural history of man and animals. . . . Till he has acquired a poet's eye for nature, and can seize with intuitive quickness the appearances of passion, and all the effects produced upon the body by the operations of the mind, he has not raised himself above the mechanism of his art, nor does he rank with the poet or historian.[23]

The Anatomy and Philosophy of Expression was illustrated with fanciful soft-grounded engraved plates depicting artistic interpretations of human emotion, as in Figure 9.2.

Darwin recorded his unfavorable reaction to Bell's work in his "C" notebook of 1838. He noted sarcastically Bell's observation that the practice of grinning was designed to expose the canine teeth, "no doubt a habit gained by formerly being a baboon with great canine teeth. . . . Laughing modified barking, smiling modified laughing. Barking to tell other animals in associated kinds of good news [sic].—discovery of prey.—arising no doubt from want of assistance.—crying is a puzzler. Under this point of view, expression *of all animals* becomes very curious."[24]

Though these notes do express some antipathy for *The Anatomy and Philosophy of Expression*, they also reveal a crucial difference between the approaches of Darwin and Bell. Whereas Bell had tried to interpret all expressions for their utility, Darwin maintained that the reasons for certain types of expression may not be immediately obvious to the contemporary observer. Bell believed the expressions to be gifts from God, designed to equip the recipient with tools appropriate to the machinations of the earthly world. He argued that

as the Creator has established this connexion between the mind and external nature, so has He implanted, or caused to be generated, in us, various higher intellectual faculties. In every intelligent being He has laid the foundations of emotions that point to Him, affections by which we are drawn to Him, and which rest in Him as their object. In the mind of the rudest slave, left to the education of the mere elements around him, sentiments arise which lead him to a Parent and a Creator.[25]

This argument paralleled William Paley's assertion that the rationality of God is evident in the structure of organisms.[26] Indeed, Bell had collaborated with Paley in the production of his classic *Natural Theology; or, Evidences of the Existence and Attributes of the Deity*. Although the prevailing atmosphere at Edinburgh was radical and Lamarckian, Bell was a natural theologist. He saw in human expression the designs of a divine being.

Darwin, by contrast, believed chance variation to be the engine of adaptive change. But he also understood behavior in terms of instinct, which he considered to be inherited individual habits. Thus Darwin was prepared to admit the possibility of vestigial and residual habits, a possibility not accounted for in Bell. Moreover, Darwin was able to explain the full range of human and animal characteristics in empirical terms, without having to appeal to theological arguments. Darwin's goal in *The Expression of the Emotions* was to explain some of the social behaviors that appeared problematic in a straightforward reading of *The Origin of Species*.

The works of Descartes, Le Brun, Lavater, and others form part of an ongoing historical endeavor to analyze the nature of human expression. Published considerations of human expression involved a mixture of both artistic and scientific theorists. These theorists considered both the empirical study of drawing and painting, and the accumulated knowledge of physiologists and physicians. Some, like Le Brun, argued that the artist's eye should provide guidelines for the perfect expression of the passions. Others, like Descartes, argued that expression was the product of physiological interactions, and was not governed by any abstract aesthetic ideal. That Darwin cited much of this material in *The Expression of the Emotions* indicates that he had studied works that contained extensive references to pictorial considerations of expression. His understanding of this work informed his effort to illustrate *The Expression*.

These works had grown in visual complexity over time. Table 9.1 describes the various illustration methods used in Descartes, Le Brun, Lavater, Bell, and Darwin. Each of these authors included illustrations to support his arguments. The increasing quantity and sophistication of these illustrations may have been a result of technological advances in printing methods. As printed books became more common, illustrations became more economical. But these illustrations also proved popular with the consuming public. Viewing the illustrations to Lavater's *Essays*, for example, was considered an entertaining parlor distraction in eighteenth-century Britain. By 1810, twenty different English-language translations of Lavater's *Essays* had been prepared, many of which were widely available through newly constituted book clubs.[27] To compete with work such as Lavater's for the attention of the consuming public, it was necessary for Darwin to produce a book that was more appealing to his potential audience. Darwin's challenge was to produce illustrations that clarified and advanced his theories on expression, but also persuaded and engaged his readers. To accomplish this, Darwin chose to use photographs to illustrate *The Expressions*.

Illustration Strategies

Early in the nineteenth century the British photographer Roger Fenton described the challenge of capturing the fleeting features of human expression on paper: "Those natural attitudes of the human form which come unbidden—is there any pencil so rapid that it can depict them before meaning has departed from the pose?"[28] By mid-century, photography could provide such a pencil, but only barely. Advancements in photochemical technology had reduced exposure times, and made it possible for the camera to capture relatively small moments in time. By contemporary standards exposures were still long: in the order of tens of seconds. Nevertheless, this was considerably shorter than the exposure times of several minutes that the earliest materials required. Technology had progressed to the point that photographers felt equipped to produce images of singular events in time.

Illustrating a book with such pictures was expensive, however. Darwin had been advised that the photographic illustrations for *The Expression of the Emotions* could make the book a money-losing proposition, and cost an additional £75 per 1,000 sets.[29] Despite this financial obstacle, he insisted on proceeding with seven photographic plates to illustrate the volume. That he would insist on these illustrations despite the risk to the economic viability of the publication indicates the extent of Darwin's commitment to photographic imagery. Darwin was impressed with the unique ability of photographs to effectively communicate facial expressions. He had witnessed the persuasive impact of photographs he had shared with colleagues in researching the work: "That the expression is true, may be inferred from the fact that out of fifteen persons, to whom the original photograph was shown, without any clue as to what was intended being given them, fourteen immediately answered, 'despairing sorrow,' 'suffering endurance,' 'melancholy,' and so forth."[30]

The first edition of *The Expression of the Emotions in Man and Animals* was illustrated with heliotype plates. Heliotyping was a photomechanical printing process in which images were transferred onto printing plates, then inked and printed onto paper. Previously, books printed with photographic illustrations contained actual photographic prints, which were attached to leaves of paper and inserted into the book. Books produced this way were tedious and expensive to produce. Moreover, they were often relatively impermanent, as the photographic materials used had to be meticulously prepared in order to remain chemically stable.

It seems likely that Darwin chose the heliotype method through his association with the London photographer Ernest Edwards (1837–1903). Darwin had modeled for Edwards in 1867 for Edward Walford's *Representative Men in Literature, Science, and Art* (1868).[31] Edwards was a pioneer in the production of photographically illustrated books, who specialized in portraiture and topographical landscapes. He was also an inventor, responsible for the creation of a new kind of pocket camera, and in 1869, the invention of the heliotype photomechanical reproduction technique. *The Expression of the Emotions in Man and Animals* was the first book ever to use heliotype illustrations. Heliotypes were slightly more detailed and cheaper to produce than previous photographic reproduction methods, and the reddish-purple tone achieved using this process had a distinctive, modern look. Although the economic advantages of the process must have appealed to Darwin, it may also be that Darwin was attracted to the technique for aesthetic reasons. The heliotype illustrations in *Expression of the Emotions* gave the book an appearance unlike any book that preceded it. The use of heliotypes assured that *The Expression* would be perceived as a unique work, distinguished from contemporary books by its novel illustrations.

Darwin organized the illustrations for *The Expression of the Emotions* to complement the logic of his written arguments. The first half of the book, through page 145, is illustrated with wood-engraved vignettes, according to the convention of the day. This technique had been used in Charles Lyell's *Principles of Geology* (1830), one of the few references Darwin brought with him on the voyage of the *Beagle* from 1831 to 1836.[32] Darwin considered Lyell's *Principles* extremely persuasive: "The science of Geology is enormously indebted to Lyell—more so, I believe, than to any other man who ever lived. . . . I am proud to remember that the first place, namely St. Jago, in the Cape Verde Archipelago, which I geologized, convinced me of the infinite superiority of Lyell's views over those advocated in any other work known to me."[33] Whether Darwin consciously emulated Lyell's illustration methods is not known. The engraved illustrations in each work cover roughly one quarter of the pages they are printed on, and are interspersed with the text to which they refer. This illustration method was advantageous in that it allowed the author to insert illustrations directly into relevant sections of the text. As a result, Darwin was able to structure the engraved illustrations in *The Expression* to coincide with developments in his written arguments and visually reinforce information presented in the text.

The second half of *The Expression of the Emotions* contains a remarkably different set of illustrations. These are the photographic plates, as depicted in Figure 9.1. Each of these plates contains reproductions of between two and seven separate photographs, which were assembled for presentation on a single page. One photographic plate appears in nearly every chapter. Unlike the engraved illustrations, which Darwin inserted to coincide with the introduction of written information, the photographic plates are referred to at various points in Darwin's narrative. They are of a variety of shapes and sizes, and depict different aspects of human expression.

In broad strokes, Darwin's visual narrative followed the hierarchical progression of evolution itself. The use of photographic illustrations in the second half of the book coincides with a transition in Darwin's argument from a discussion of animal to human behavior. The engraved illustrations Darwin used in the first half of the text began with anatomical drawings of facial musculature, then proceeded to depict smaller animals such as dogs, cats, and birds. By the end of the section, however, Darwin included illustrations of the facial expressions of quadrumana, as in Figure 9.3.

The progression of photographic plates in *The Expression* reflects the chronology of the human life cycle. In these, Darwin began by illustrating the expressions of infants, then proceeded to the expressions of children, and ended with the expressions of adults. At the end of the book, Darwin returned to drawn illustrations, and included three engraved images interspersed with the photographic plates. Thus the book began with lower order animals, which are depicted in engravings, and progressed from simple to more complex animals. With the introduction of human expression, the book switched from engravings to photographs, which were maintained until the final sections of the text. At the end of the book Darwin reintroduced the engraved illustrations, as if to unify the early illustrations with the late ones.

This use of engravings at the beginning and end of the text may also have been designed to make the book seem less radical than it would have if photographs alone had been used. By introducing photographs midway through the text, then returning to engravings at the end, Darwin managed to integrate the photographs in the work without appearing to have departed substantially from long-standing traditions of scientific illustration. Although photography had gained acceptance in the objective representation of events, Darwin may have felt that the singular presence of photographs would have been overwhelming to his readers. The placement of engravings in *The Expression* may have helped legitimize the work, inasmuch as it helped the book seem more conservative. Although photography was understood to be an objective implement of tremendous documentary usefulness, it had only just begun to emerge in scientific treatises. By combining old and new techniques of scientific illustra-

136   SPECIAL EXPRESSIONS :   CH AP. V.

Fig. 16. *Cynopithecus niger*, in a placid condition. Drawn from life by Mr. Wolf.

Fig. 17. The same, when pleased by being caressed.

Figure 9.3. "Cynopithecus niger, in a Placid Condition" and "The Same, when Pleased by Being Caressed." Figs. 16 and 17 from Darwin's *Expression of the Emotions*. Engravings are used to illustrate animal expression through the first half of the book. Humans are the only subjects depicted photographically.

tion, Darwin gave his work a cutting-edge appearance without relying too heavily on untested forms of illustration.

Darwin seems to have been determined to separate the photographic from the engraved illustrations in *The Expression*. Darwin could have used photography for all of the illustrations in the work. He seems to have consciously avoided this, even though some of the engraved illustrations were actually based on original photographs. The engraved image of a small dog in the first chapter of the text, for example, is annotated "from a photograph by Oscar Rejlander." Other images are described as drawings "from life." The documentary "truth" of all the illustrations in *The Expression* was important to Darwin. Nevertheless, he avoided the potentially objective capacity of the camera and lens to delineate the expressions of animals. The difficulties inherent in managing animals may have contributed to Darwin's decision not to attempt to use photographs to illustrate their behaviors. However, by reserving photographs for the depiction of human subjects, the work achieves a visual climax at precisely the point where human expressions are introduced.

The disjunction between the illustrations of humans and animals in *The Expression of the Emotions* has the effect of distinguishing human from animal expression. This distinction has a dramatic effect, as the striking appearance of the photographic illustrations corresponds to the move from consideration of animal to human subjects, and punctuates an increase in the complexity of Darwin's subject matter. The reintroduction of the engraved illustrations at the end of the work invites the reader to associate the arguments of the first section with the arguments of the second. At the same time that Darwin may have been compelled to use photography to add excitement to his narrative progression, he also felt the need to return to engraved illustrations to unify the work. Darwin could not make the break between human and animal subjects too severe, or he would have caused readers to doubt his evolutionary argument that humans and animals share the same origins.

Darwin positioned the illustrations in *The Expression of the Emotions* to augment the effectiveness of his textual arguments. The carefully considered placement of images in the text reinforces the logic of Darwin's narrative progression. But the impact of the illustrations in *The Expression* is not limited to their layout. The images themselves were chosen individually for their capacity to successfully communicate aspects of Darwin's thesis.

The engravings used in *The Expression* were carefully selected both for their artistic pedigree and for their visual appeal. The engravings were produced by James Davis Cooper (1823–1904) after drawings by Thomas Wood (fl. 1855–72), Arthur Dampier May (fl. 1872–1900), Joseph Wolf (1820–99), and Briton Riviere (1840–1920). These artists were renowned for their illustrations of animals. Wood was a painter and watercolorist, who exhibited at the Royal Academy and who had achieved notoriety for illustrating the popular *Bible Animals* of 1869. Wolf worked as staff artist at the British Museum, and was one of the most prolific animal illustrators of the nineteenth century. Philip Henry Gosse's *Romance of Natural History* (1862) was heavily populated with Wolf's images depicting various species of animals in spectacular action scenes. Darwin was keen to emphasize the contribution of Briton Riviere to two of the engraved plates, although original drawings in the Darwin Collection at Cambridge University Library demonstrate that these were almost completely reworked by Wolf prior to publication.[34] Riviere was perhaps the most celebrated of the draftsmen Darwin enlisted to illustrate *The Expression*. Despite the fact that his drawings were not quite suitable to the text, Darwin made it a point to list Riviere as one of the artistic contributors.

Darwin's choice of well-known and respected artists to illustrate *The Expression* helped him to market the book to his Victorian audience. That these artists had proven themselves successful in other projects indicated to Darwin that they could prove valuable assets to the success of *The Expression*. Each of the engraved artists had shown himself capable of capturing the Victorian imagination. In addition, the fact that all of them contributed to *The Expression* allowed Darwin to link their reputations to the authenticity of the engravings, affording the latter a legitimacy that would have been absent with lesser draftsmen.

The photographic illustrations also were selected for their ability to contribute to the success of *The Expression*. Nearly all of these images were produced by Guillaume-Benjamin Duchenne and Oscar Rejlander,[35] two distinctive photographers now recognized as giants in the history of photography.

Collaboration with Duchenne

Guillaume-Benjamin Duchenne (1806–75), called Duchenne de Boulogne, was one of the pioneers of French neurobiology. Mentor to Jean-Martin Charcot, and he in turn to Sigmund Freud, Duchenne's contributions to European neurological science cannot be overestimated.

Duchenne had been a resident clinician at La Salpêtrière, a public hospital for the poor in Paris. It was an unusually rich source of clinical information. A seventeenth-century account described the hospital as "a city within a city, consisting of about forty-five buildings with streets, squares, gardens, and a beautiful old church."[36] The hospital was noted for its extraordinary number of patients with "incurable" neurological disorders, including the mad, epileptic, spastic, and paralytic. In the early 1840s Duchenne began to catalog these patients, and is credited with a variety of pathological discoveries, including previously unknown forms of motor neuron disease, muscular dystrophy, palsy, and paralysis. In early 1862 he published what is considered to be the first photographically illustrated medical book, a treatise cataloguing some of the more acute cases he observed in La Salpêtrière titled the *Album des Photographies Pathologiques*. It comprised seventeen plates of clinical photographs, together with twenty pages of explanatory text.

The problems of the mentally ill were familiar to Darwin from his experience as a student in Scotland. While in Edinburgh, Darwin had been nominated for membership in the Plinian Society by the radical William Browne, who had a specialist interest in madness. Browne graduated from Edinburgh in 1826, having studied the inmates of Montrose Lunatic Asylum to prove that religious fanaticism was a kind of insanity.[37] According to Browne, many of the charismatic religious thinkers who had been canonized by the church in prior centuries would have been diagnosed as insane in nineteenth-century Britain. Browne believed that this demonstrated the fallacies of church doctrine, and used this as an argument for anticlerical politics. In fact, it was Browne who gave the lecture refuting Charles Bell's pious *Anatomy and Philosophy of Expression* on the night of Darwin's petition for admission into the Plinian Society.

Some 45 years later, Darwin turned to Browne's son, Dr. James Crichton Browne (1840–1938) for photographic evidence of distinct expressions among the insane. Dr. Browne served as Director of the progressive West Riding Lunatic Asylum in Wakefield from 1866–76. He was also an amateur photographer, who had begun to document the physiognomic character of patients in his care. Although Browne sent more than 40 photographs to Darwin for his consideration in writing *The Expression*, Darwin chose to reproduce only one, a photograph of an insane woman with bristly hair, reproduced by engraving as Figure 19 in the text. Nevertheless, Darwin cited Browne's research heavily in *The Expression*, and even suggested that Browne should be credited as coauthor of the book.[38]

In 1853, the English photographer Dr. Hugh Welch Diamond (1809–86) had initiated photography of the mentally ill by documenting female patients in the Surrey County Asylum. Diamond photographed patients in the asylum to catalog symptoms of various forms of insanity. In a paper delivered to the Photographic Society of London in 1856, Diamond reported: "The photographer catches in a moment the permanent cloud, or the passing storm or sunshine of the soul, and thus enables the metaphysician to witness and trace out the connexion between the visible and the invisible in one important branch of his researches into the philosophy of the human mind."[39] Diamond reported therapeutic success using photographs of patients he examined. He claimed that he had cured several of his patients of delusional tendencies by showing them their photographic likenesses.

In the 1840s, Duchenne de Boulogne began to experiment with neurological treatments using electrical stimulation. In 1847 he presented a paper describing a method whereby electricity could be used to stimulate individual muscles to contract without having to cut or pierce the skin.[40] He applied this principle to artificial resuscitation techniques, and is known to have developed a successful procedure for reviving patients suffering from chloroform intoxication and asphyxiation.

These endeavors culminated in the publication of his classic work, *Mécanisme de la Physionomie Humaine ou Analyse Électro-physiologique de l'Expression des Passions Applicable à la Pratique des Arts Plastiques* (The Mechanism of the Human Face, or an Electrophysiological Analysis of the Expression of the Passions Applicable to the Practice of the Fine Arts), in 1862. In that work Duchenne documented a series of experiments he performed to electrically induce facial expressions in human subjects. To do this, Duchenne attached electrodes to facial muscle groups, then stimulated them with electrical charges to contract in various postures. As Duchenne explained,

Using electrical currents, I have made the facial muscles contract to speak the language of the emotions and the sentiments. This careful study of isolated muscle action showed me the reason behind the lines, wrinkles, and folds of the moving face. These lines and folds are precise signs, which in their various combinations result in facial expression. Thus by proceeding from the expressive muscle to the spirit that set it in action, I have been able to study and discover the mechanism and laws of human facial expression.[41]

Mécanisme was illustrated with 84 individual photographic plates of Duchenne's experiments, produced in collaboration with Adrien Tournachon (1825–1903), nephew of the pioneering French photographer Félix Nadar (1820–1910). It also contained nine synoptic plates providing a comparison of the plates in series, and approximately 150 pages of textual analysis.

Although Duchenne's previous work had been purely scientific in content, *Mécanisme* included an analysis of facial expression for the use of artists, and Duchenne's observations on anatomical aesthetics. In this Duchenne placed himself firmly in the tradition of the physiognomic debate, and may have helped sensitize Darwin to artistic concerns about photography. In his introduction Duchenne praised the work of Le Brun, Lavater, and Bell, but criticized them for having made numerous factual errors. Duchenne reasoned that his study was a necessary antidote to these early works, for "without sacrificing all past knowledge for the greater glory of modern progress, we may affirm today that before my electrophysiological research, we had very incomplete ideas of the actions of the muscles."[42]

Through the use of photography, Duchenne was able to offer detailed analyses of facial movements despite their momentary nature. Duchenne claimed that the contractions of the facial muscles were so transient that previously it had not been possible for "even the greatest masters to grasp the sum total of all their distinctive features."[43] Photography, combined with his electrical stimulation techniques for producing expressive postures, afforded Duchenne unprecedented control in the study of expression. In effect, he had devised a means to "activate" expression with electricity, then "fix" it with photography.[44]

Mécanisme was structured in three parts. The first part was introductory, providing a discussion of the background of Duchenne's experiments, an overview of the principles of facial musculature, and a statement of purpose. The second part Duchenne termed the "scientific" section. In this, Duchenne described the specific physiological processes involved in such expressions as joy, benevolence, attention, sadness, and surprise. This section concluded with a criticism of anatomical errors made by artists depicting expression in several classical sculptures. The illustrations for this section depict only the face and torso of subjects, excluding hand and other body movements. The third section Duchenne called "aesthetic." In this part Duchenne discussed a series of photographs he produced of a young blind woman dressed in fanciful costume and acting out theatrical poses. The illustrations for this section were three-quarter length illustrations, and included depictions of the movements of the hands, arms, and torso.

Duchenne used six models to produce the illustrations for *Mécanisme*. The majority of the illustrations are of an old man whom Duchenne describes as having "common, ugly features."[45] This subject, Duchenne explained, was chosen because he had a relatively insensitive face that did not react to the mere presence of the electrodes. Duchenne was concerned that the expressions he documented should reflect the results of electrical stimulation alone, and not any of the unsolicited expressions of the sitter. He reasoned that any discomfort experienced in the placement of the electrodes or in the laboratory environment might produce unwanted spontaneous expressions. As a result, he selected "inexpressive" and "insensitive" patients who, he believed, would react mechanically to his experiments. These patients included an elderly man and woman, an opium addict, and a blind girl. He considered only one of his subjects to be "normal," a sculptor he had met at the Académie.

Duchenne's choice of subjects was unusual, and his portrayals are at times brutally ironic. The expression of ecstasy, for example, on the face of an old simpleton, or the look of joy on the face of a drug addict, are highly evocative and at times disturbing. For this reason Duchenne was widely criticized at the time of the publication of *Mécanisme*. Perhaps not surprisingly, it is for this same reason that contemporary artists have lauded Duchenne's illustrations as enigmatic and insightful commentary on the state of nineteenth-century society.

The matter-of-fact portrayal of the subjects who posed for *Mécanisme* belies the personal histories of the sitters themselves. Duchenne performed his experiments on a series of social outcasts who appear vulnerable in his photographs. Perhaps inadvertently, Duchenne captured some of the pathos of the characters he selected. In most of these images the subjects were photographed facing straight into the camera against a neutral background. The subjects appear to stare out of the page, and at times seem to confront the viewer. The lack of environmental information in these images leaves the subjects floating in a vacuous negative space. As a result, the artificially induced expressions recorded in *Mécanisme* seem strangely out of place and have a chilling universal quality.

Duchenne claimed that he did not intend the illustrations in *Mécanisme* to evoke emotional reactions from his audience. To the contrary, Duchenne sought to demonstrate that expression was a purely physiological phenomenon that might be observed independently of beliefs about the spiritual condition of the individual. Darwin, however, would not have been bound to interpret them so clinically.

While paying tribute to Duchenne for his contribution to *The Expres-*

sion of the Emotions, Darwin explained the mixed reaction he received when he showed Duchenne's photographs to friends and colleagues:

> Dr. Duchenne galvanized, as we have already seen, certain muscles of the face of an old man, whose skin was little sensitive, and thus produced various expressions which were photographed on a large scale. It fortunately occurred to me to show several of the best plates, without a word of explanation, to above twenty educated persons of various ages and sexes, asking them, in each case, by what emotion or feeling the old man was supposed to be agitated; and I recorded their answers in the words which they used. . . . The most widely different judgements were pronounced in regard to some of them. This exhibition was of use in another way, by convincing me how easily we may be misguided by our imagination; for when I first looked through Dr. Duchenne's photographs, reading at the same time the text, and thus learning what was intended, I was struck with admiration at the truthfulness of all, with only a few exceptions.[46]

Two elements of this commentary are crucial to understanding the inclusion of the Duchenne illustrations in *The Expression of the Emotions*. The first is the element of "truth" Darwin perceived in these photographs. Certainly Darwin was attracted to the objective power of the photographic lens to record factual information normally invisible to the human eye. But more provocative is the enigmatic quality of the *Mécanisme* imagery. Darwin recognized that Duchenne's photographs demanded interpretation and explanation. Taken out of context, it is unclear what these photographs were meant to signify.

In this sense, the Duchenne illustrations exhibit the same metaphorical quality that Edward Manier posited in his analysis of Darwin's narrative strategies. They are eye-catching and even shocking, but they are also pliable. They allow the viewer to project onto them what he or she wants to see, and thus engage the viewer on his or her own terms.

The photographic illustrations Duchenne published in *Mécanisme* were substantially edited versions of his actual experimental results. Figure 9.4 provides an example of the extensive cropping of the images Duchenne used in *Mécanisme*. In Darwin's copy of the image of "Discontent, Bad Humor," for example, a substantial portion of the original image has been masked from view. Duchenne chose to eliminate information contained in the areas peripheral to the central figure, and blacked-out all but a small oval portion containing the subject's face. The full negative would have been the same size as the extensive border that surrounds the remaining image, and would have contained information such as the dress of the

Figure 9.4. "Discontent, Bad Humor." Plate 51 from Darwin's copy of *Mécanisme de la Physionomie Humaine*. The original image was the size of the extensive black border surrounding the subject, but the negative has been substantially masked. Courtesy of the Syndics of Cambridge University Library.

subject, the position of the experimenter and his equipment, and the appearance of the laboratory environment. Instead, Duchenne edited the photograph to focus attention solely on the face of the sitter and the application of an electrical probe to the bridge of his nose.

Daston and Galison have argued that nineteenth-century scientific imagery was characterized by an increasing preoccupation with objectivity.[47] They argue that the capacity of photography to objectively record experimental events motivated scientists to include photography among their documentary tools. The editing of photographs in *Mécanisme* demonstrates that scientists may have adopted a more interventionist role in cultivating the appearance of objectivity in their illustrations. Although it may be argued that the unedited versions of the photographs in *Mécanisme*

Figure 9.5. "Horror and Agony." Plate 6 from Darwin's copy of Duchenne's *Mécanisme*. Note Darwin's cropping notation horizontally across the lower center of this image. Courtesy of the Syndics of Cambridge University Library.

contain more factual information about Duchenne's experimental methods, Duchenne preferred his readers to consider an abstracted version of his experimental results. Apparently preferring an idealized representation of the experiments performed to the reality of the scientific laboratory, Duchenne eliminated much of the evidence of human involvement in these images.

By limiting evidence of human intervention in the *Mécanisme* photographs, Duchenne created the illusion that his images had been produced with mechanical objectivity. Duchenne edited out evidence of human activity in his illustrations that might have caused readers to question the clinical control of his electrophysiological experiments. Despite the fact that the experimenter's probe is still visible in the edited image, it is depersonalized and its size substantially reduced compared to the full

Figure 9.6. Darwin's interpretation of Fig. 9.5. Fig. 21 from Darwin's *Expression of the Emotions*. By transcribing Duchenne's photograph to an engraved plate, Darwin was able to edit out the intrusion of the scientists' hands and probes.

printing. Duchenne's edits cause the viewer to focus attention on the expression of the sitter himself, without questioning the experimental environment. Although Duchenne ostensibly had used photography for its objective capacity to record experimental information, ultimately he felt compelled to remove disconcerting information recorded by the camera.

Ironically, Charles Darwin may have been concerned that Duchenne's photographs were still not objective enough for reproduction in *The Expression*. Most of the photographs Darwin appropriated from Duchenne for inclusion in *The Expression* were directly copied from images published in *Mécanisme*. Two of these images were not reproduced photographically, but transcribed onto engraved plates, as shown in Figures 9.5 and 9.6. A comparison of Duchenne's original image of "Horror and Agony" (Fig. 9.5) and Darwin's engraved interpretation of that image (Fig. 9.6) reveals that Darwin further edited Duchenne's photographs for inclusion in *The Expression*. In Darwin's engraved version of "Horror and Agony," evidence of human intervention in the image was removed entirely. Darwin instructed the engraver to remove the hands of the experimenters and the electrical probes visible in Duchenne's photograph from his engraved copy. Areas of the subject's forehead and neck that had been obstructed in the Duchenne original were extrapolated by the Darwin engraver. Darwin may have felt that the experimental apparatus captured in Duchenne's photographs was distracting to the viewer. Duchenne's photographs were too honest, in that they recorded the actual situation of the sitter in his laboratory environment. To engage his readers, Darwin cultivated an appearance of objectivity that actually misrepresented experimental events.

Collaboration with Rejlander

With few exceptions, the rest of the illustrations for *The Expression of the Emotions* were commissioned by Darwin from Oscar Gustave Rejlander (1814–75), an expatriate Swedish photographer living in London. Rejlander was highly controversial in photographic circles, and had alternately enjoyed both extreme critical acclaim and ruthless disapproval. He is most often identified with the composite printing process, in which two or more negatives are combined to create a photographic print with elements of several pictures. As a result Rejlander was able to manipulate his images, and produce convincing photorealistic images that were actually artificially assembled in the darkroom.

Little is known of Rejlander's early life. He is believed to have been born in Sweden, although his birthplace has never been traced. When he died in 1875, his death certificate gave his age as 61, yet his self-portraits in *The Expression of the Emotions* suggest he may have been older. He was trained as a painter, and claimed to have been influenced by the old masters. As a youth he had studied in Rome, where he supported himself by painting portraits, copying the work of old masters, and making lithographs. He claimed to have been particularly attracted to the allegorical nature of the work of the painters Raphael and Jan Steen.[48]

Rejlander maintained that the use of photography could make painters better artists and more precise draftsmen. Of Titian's *Venus and Adonis* he remarked that "Venus had her head turned in a manner that no female could turn it and at the same time show so much of her back. Her right leg is also too long. I have proved the correctness of this opinion by photography with various shaped models. In *Peace and War* by Rubens, the back of the female with the basket is painted from a male, as proved by the same test."[49] Rejlander recognized the capacity of photographs to provide objective information to critique the work of visual artists.

Rejlander produced both straightforward and manipulated photographs, although his reputation was built on the creation of his composite-printed works. In 1856 he submitted his masterpiece, an elaborate composite print titled "The Two Ways of Life" (Fig. 9.7), to the Manchester Art Treasures Exhibition. It was a large, frescolike work made from 32 separate photographic negatives depicting sixteen separate groupings of models. The composition was organized in two halves. On one side, allegorical figures of scantily clad and naked women enacted scenes of moral depravity and despair. On the other side of the image, righteous women portrayed the virtuous way of life, and engaged in acts of piety and kindness. Poised between the two was a likeness of Rejlander himself, listening and considering the merits of each way of life. Rejlander explained that he had made the photograph "to show the plasticity of photography. I sought to bring in figures draped and nude, some clear and some rounded with light, others transparent in the shade; and to prove that you are not, by my way of proceeding, confined to one plane, but may place figures and objects at any distance, as clear and distinct as they relatively ought to be."[50] "The Two Ways of Life" was the most ambitious photographic collage ever undertaken, and its effect was completely convincing. Reviews were mixed however, and the piece became highly controversial. Rejlander had designed the piece to demonstrate the capabilities of photography beyond straight documentation. As a result, he brought into question the reliability of photographs as objective representations of nature.

Darwin commissioned the photographs for *Expression of the Emotions* in 1871, asking Rejlander to produce images to illustrate fear, anger, pleasure, surprise, indignation, and so on. Rejlander himself posed for five of these images, which appear to have been taken at various sittings. His interpretations of "disgust" and "helplessness" appear to have been taken at the end of his commission, as Rejlander appears stooped, as if weakened by the diabetes that ultimately killed him in 1875. In some of the images Rejlander's fingertips are stained, probably with the silver nitrate he used to sensitize the emulsions of his glass plate negatives.

The technical difficulties involved in making Rejlander's "spontaneous" images were formidable. Exposure times were long, and artificial light was not generally available. Depending on the weather, a single image might have required an exposure of ten to twelve seconds.[51]

Figure 9.7. Oscar Rejlander's "Two Ways of Life," 1856. Courtesy of the Royal Photographic Society.

Although *The Expression of the Emotions* was only moderately successful for Darwin, it was extremely successful for Rejlander. The first run of
The Expression sold completely, but the reprint of 1873 remained largely
unsold throughout Darwin's lifetime. One illustration, titled "Mental
Distress" in the text, depicted a male infant (in Darwin's words) "howling
his head off." (Fig. 9.8) An enlargement of that image was displayed in the
popular press, under the title "Ginx's Baby" after the name of a popular
novel. Rejlander objected to this title, but the name was popular, and he
was soon overwhelmed with requests for reproductions. In total Rejlander sold approximately 60,000 24 × 30 cm. prints and a quarter of a
million cartes de visite prints (approx. 5 × 8 cm.) of the image to the
general public.[52]

"Ginx's Baby" was hailed as one of the first "momentary" images
widely distributed in Victorian England. Rejlander appeared to have
succeeded in capturing the child's appearance while his body and facial
muscles were in vigorous movement. This was unusual in photography of
the time, causing one critic to suggest that prior artistic portrayals of the
emotions had "finally been scattered to the winds by the wonderful photographs of the late O. G. Rejlander, especially noting the illustrations in
Dr. Darwin's work on expression, and 'Ginx's Baby,' as examples."[53]

Those who admired the photographic quality of "Ginx's Baby" may
have been surprised to discover that it was not, strictly speaking, really a
photograph. Rather, it was a photographic copy of a drawing *after* an
original photograph. The prototype (Fig. 9.9) on which the drawing was
based is a photograph remarkably similar to the drawn version. Rejlander
carefully copied the original photograph onto paper, then photographed
the resulting drawing to produce the illustration used in *The Expression*.
One can see Rejlander's talent as a draftsman in his extreme sensitivity to
the original, including the minutiae of the child's clothing and the intricate curls of his hair. By transcribing the photograph into a drawing,
Rejlander would have been free to highlight elements of the image Darwin sought to express. Although the child's hair, cheeks, and brow do
seem slightly more lively and energetic in the drawn version, Rejlander
does appear to have remained largely faithful to the original photograph.
It is probable that the image was redrawn in an effort to enlarge the image
to assist the printer in preserving some of the subtlety of the original
photograph which would have been lost in reproduction. The heliotype
process produced images high in contrast, which would have resulted in
the loss of some of the detail in the original photograph of the child's face.

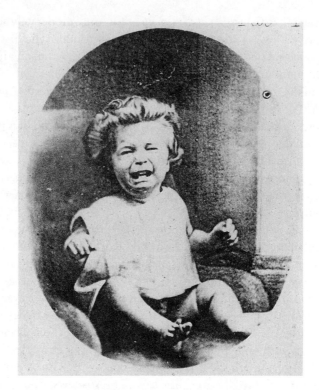

Figure 9.8. "Mental Distress." Plate 1, fig. 1 from Darwin's *Expression of the Emotions*.

Redrawing the image allowed Rejlander to augment those aspects of the child's expression that would have been lost in reproduction.

Nevertheless, it is interesting to note that in the drawn copy Rejlander rendered the child perched in a conspicuous padded chair that is completely missing from the original photograph. The chair itself seems unnaturally small for the child, giving him a larger-than-life appearance. It also serves to ground the child in a domestic setting that would have been instantly recognizable to Darwin's audience. It helps to make the image seem plausible, and provides it with a personal intimacy lacking in the photographic original. By inserting a chair into the drawn rendition of the image, Rejlander created an illustration that would have seemed persuasive to Darwin's readers. It appeared to capture a perceptible moment from everyday life that readers could sympathize with and mentally verify.

Darwin refers to "Ginx's Baby" as a photograph in the text, but fails to explain that it is actually a drawn and altered copy of a photograph. He would certainly have known of Rejlander's alterations of this image, as he had both versions in his possession prior to publication of *The Expression*, one of which Rejlander annotated as an "enlarged chalk" and "my enlarged drawing."[54] As Darwin was satisfied that Rejlander's image was factually correct, he was not inclined to explain the machinations behind the production of "Ginx's Baby." Yet the composition of the image had changed substantially from the photographic original.

The techniques Rejlander used to produce the images for *The Expression* were far from scientific. He selected his sitters from among his friends and neighbors, but would occasionally take a stranger from off the street.

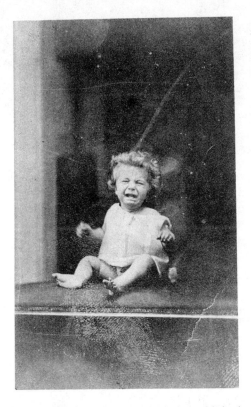

Figure 9.9. The original photograph from which the published version (Fig. 9.8) was drawn. Collection of the author. Although Rejlander's transcription was generally faithful to the original, he decided to render the child sitting in a small, padded armchair.

Figure 9.10. "Sneering." Plate 4, fig. 1 from Darwin's *Expression of the Emotions.* Darwin described the subject as a woman "who sometimes displays the canine on one side, and who can do so voluntarily with unusual distinctness." The sitter was actually Rejlander's wife, Mary.

Rejlander's wife, Mary, modeled for the illustration of "Sneering" (Fig. 9.10), which Darwin described as a depiction of a "lady who sometimes unintentionally displays the canine on one side, and who sometimes can do so voluntarily."[55] Rejlander's personal involvement with his sitters afforded him considerable control when producing the photographs Darwin had commissioned. Many of the photographs were enacted by Rejlander himself, who posed in melodramatic interpretations of the expressions Darwin sought to illustrate. In each of the emotions acted out by Rejlander, the photographer is dressed in a plush velvet jacket, and in some he poses with a fashionable top hat. Such attire would have been appropriate for a theatrical performance of the day.

The question of photographic responsibility was a topic of lively discussion in the Victorian era. Darwin's close associate Thomas Henry Huxley had been one of the main critics of photographic manipulations such as those practiced by Rejlander. In the 1860s he had devised a system for documenting individuals with the intention of "the formation of a systematic series of photographs of the various races of men comprehended within the British Empire."[56] Huxley argued that "[although] great numbers of ethnological photographs already exist . . . they lose much of their value from not being taken upon a uniform and well-considered plan. The result is that they are rarely either measurable or comparable with one another and that they fail to give that precise information respecting the proportions and the conformation of the body, which [is of paramount] worth to the ethnologist."[57] In an effort to produce photographic documents that would permit the subsequent recovery of reliable and comparative morphological data, Huxley recommended that all subjects be photographed naked, according to uniform poses, as in Figure 9.11. Additionally, he suggested that each perspective should be accompanied by a plainly marked measuring instrument placed in the same plane as the subject.[58]

As a scientist, Darwin was concerned about sources of error in the observation of facial expression. He considered the fleeting nature of expression to be the principal difficulty in observing facial movements. Darwin was concerned that an observer might taint a subject's actions through his or her own reaction to the emotion displayed, and that an observer might read into his or her observations because of the circumstances of the occasion.[59] He was acutely aware of the fragility of the expressive moment, and hoped that through the use of photography he would be able to effectively communicate the true character of expression.

Darwin's patronage of Rejlander may seem curious, given Huxley's formula for the production of scientifically useful documentary photographs. Rejlander's illustrations are simulations of expression, not actual documents of spontaneous events. But the limitations of photographic technology precluded objective recording of facial expressions. With exposure times in the tens of seconds, a "true" rendering of emotional expression would have required the subject to achieve naturally a representative emotional expression, then hold it unadulterated for the photographer to record on film. This could not be done reliably or with precision. Given the technical limitations of the photographic medium, Rejlander may have been an ideal choice to illustrate the expression of the emotions. He was a master of the construction of realistic artificial imagery (Fig. 9.12). Though the photographs he submitted for *The Expression of the Emotions* were contrived, they were also persuasive.

Rejlander's ability to convincingly simulate natural events was precisely what Darwin needed for *The Expression of the Emotions*. Rejlander was known for the theatrical nature of his photography, and had built his reputation on the seamless manipulation of special effects as in the composite printing process. His working methods would have been well known among photography aficionados, but would not have seemed incongruous to Darwin's general audience. Despite Darwin's ostensible desire for error-free photographic documents of actual human behavior, he had to settle for realistic simulations of expressive postures. Darwin

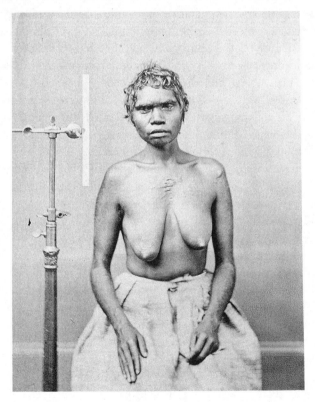

Figure 9.11. Anonymous, "South Australian Aboriginal Female," ca. 1870. Courtesy of the Royal Anthropological Institute, RAI 1747.

Figure 9.12. "Disgust." Plate 5, fig. 3 from Darwin's *Expression of the Emotions*. Darwin eschewed Hùxley's photometric method in favor of Rejlander's looser, more personalized approach.

included Rejlander's photographs not for their ability to document per se, but to provide points of discussion by creating the illusion of documenting spontaneous human expression. The Rejlander illustrations contributed little of scientific value to *The Expression of the Emotions*. Yet they were an integral part of the work, for they contributed to Darwin's narrative effect.

Conclusion

Recent theorists have attempted to decode Darwin's narrative strategies in an attempt to understand the massive public appeal of his work in Victorian Britain. The sophistication of the illustrations in *The Expression of the Emotions in Man and Animals* suggests that Darwin's narrative expertise was not limited to the written word. His deep awareness of developments in the photographic community and his recurring involvement in questions of aesthetics suggest that Darwin was both master and product of a wide range of knowledge.

Darwin's profound sensitivity to design questions, along with his personal involvement with a variety of important figures in the arts, sheds light on the breadth of his influences. His work was tempered with concerns both scientific and artistic, as well as personal and cultural. As we attempt to describe Darwin's position in the Victorian social milieu, it is important to remember that his personality derived from a rich and deeply complex interaction between his individual character and the tapestry of influences that surrounded him.

Darwin was one of the first theorists to use photography to freeze transitional movements for scientific evaluation. Darwin's work in *The Expression of the Emotions* paved the way for subsequent experimenters to study motion more thoroughly. The proto-cinematographic motion studies of Eadweard Muybridge in the United States[60] and Etienne-Jules Marey in France are indebted to Darwin's effort to document discrete moments in time. As Marey observed in 1894: "It is instantaneous photography in particular which has exercised a noticeable influence upon the arts, because it allows one to fix in one authentic image phenomena of short duration, like the movements of waves or of the attitudes of men and animals in their most rapid motions."[61] Like many of the scientists who have used photography since Darwin, Marey expressed confidence in the unrivaled ability of photography to objectively record scientific events. Analysis of the photographs contained in *The Expression of the Emotions* casts serious doubt that Darwin considered his photographic illustrations truly to have been rigorous, objective documents. Although Darwin lauded the "fidelity" and "accuracy" of the photographic process, the materials and methods available to him in 1872 were insufficient to capture facial expressions precisely by mechanical means alone. As a result, Darwin was compelled to use photographs that were both enigmatic and contrived.

Although photography was perceived as a cogent evidentiary tool when *The Expression of the Emotions* was published in 1872, it is clear that the ability of photography to capture rapid movements was severely limited at the time. As data, the illustrations in *The Expression* are deeply flawed. Darwin knew this, but instead of avoiding the use of photographs in his writing, he devised creative ways to use them convincingly. The inclusion of photographic illustrations in *The Expression* helped to authenticate Darwin's arguments, even though the methods used to produce them would now be considered dubious. The use of photography in *The Expression* constitutes a narrative strategy insofar as Darwin thoughtfully commissioned and applied photographs to legitimize his verbal arguments. Darwin used photographs to communicate meaning that was not actually contained in the photographs he presented. In so doing, he took advantage of the capacity of photographs to signify meaning without explaining the procedures he and his colleagues had used to create them.

Notwithstanding the tremendous flexibility photographers now have in altering their images electronically, photography is still used authoritatively in contemporary theoretical discourse. The capacity of photographs to objectively communicate information is unrivaled in the history of visual representation. However, as the illustrations in *The Expression* attest, the mere fact of photographic documentation is not sufficient to insure true scientific impartiality. Darwin used creative methods to reinforce the appearance of objectivity in *The Expression*. The photographic illustrations Darwin used in *The Expression* served to persuade his readers of the truth of his arguments. That a scientist of Darwin's stature chose to use photography early in the history of the medium contributed to the legitimization of the medium itself. As a result of his efforts in *The Expression of the Emotions*, Darwin promoted the institutional use of photography in scientific discourse. Darwin helped to define the language of scientific representation in *The Expression*, even as he sought to harness it.

10. SIMON SCHAFFER

The Leviathan of Parsonstown: Literary Technology and Scientific Representation

❖

Social studies of scientific knowledge portray scientists' writing in rather technical, material terms. The term "literary technology" indicates how scientists use texts as knowledge-producing tools and connects literary with other scientific work. The design and use of instrumentation, the social organization of the scientific community and the assignment of credit and status all materially interact with textual production in the formation of natural knowledge.[1] These technologies help fashion the bearer of knowledge as authoritative and competent, and the item of knowledge as independent from the contingencies of human judgment. They make authors and they make facts. In this sense, the array of technologies helps order nature and society.[2] How general is this account of the literary technology of the sciences? Here are some ways in which the extension of literary technologies might be questioned:

Literary technologies are sociohistorically local. Because the craft of verbal performance is so closely linked with appropriate material and social formations, conventions of literary technique are highly localized. Robert Boyle's literary technology of circumstantiated, first-person, past tense, indicative accounts was designed to match codes of gentlemanly honor and the machine philosophy in mid-seventeenth century Britain. Historians have documented the varying literary conventions available to Boyle's contemporaries, such as the essays of Montaigne, the texts on mixed mathematics developed by the Jesuit professoriate, Newton's projects in demonstrative certitude in optics and rational mechanics, or the chemical memoirs of naturalist Academicians in Paris.[3] The proliferation of literary genres in early modern natural philosophy accompanied the crisis of authority in the Baroque republic of letters. So natural philosophers, stationers, editors, and their colleagues strove to develop and then propagate acceptable conventions of polite and learned discourse to which many milieux could subscribe. The process through which a relatively well defined literary technology emerged amidst the academicians and naturalists of the Enlightenment awaits its historian. Resources are available in the compelling studies of print culture and of the learned academies already produced by historians of the book.[4] Social historians of the sciences attend to the enterprises that multiplied the contexts in which the technologies of circumstantiation, of the passive voice, of assignment of authorship, and of the representation of distant or unattainable phenomena were put into action. Part of this attention may center on the textual effects of the interaction between the material technologies of the sciences and those of literary production. A history of the press, of conventions of note-taking and reportage, of the relative powers of editors, booksellers, printers, and savants, will help clarify the emergence of new textual forms and the maintenance of their authority.

Many different representatives compete for the right to produce scientific representations. Literary technologies are crafted to make representations of the natural and social worlds stand independently of their authors. Sociologists of scientific knowledge have pointed to the significance of this representational work. Working scientists make, handle, and translate representations through networks of machines and humans. Because of this process, such representations can take many, varying forms. Just as literary technologies are sociohistorically local, they are also specific to particular representational practices. Manuscript notes, copper engravings, daguerreotypes, chart recorder graphs, and computer printouts all take part in different regimes of representation and different labor processes. These differences have political consequences. Scientists strive to make themselves reliable representatives of nature by qualifying their representations and disqualifying others. Disqualification is often achieved by pointing to the labor involved in making a representation. Hobbes drew attention to the labor process of Boyle's experiments in order to make them part of Boyle's world, not part of nature. Similarly, the invisibility of technicians and support staff is often a prerequisite of making an artifact count as a representation.[5] So representational regimes provide the grounds on which authority is contested. Representatives seek not merely to disqualify rivals but to guarantee the future interpretative stability of the lessons their representations are designed to teach. By propagating conventions of literary, material, and social order, these workers strive to diffuse the cultural formations that help make meanings stable. Literary technologies are limited by the boundaries of these formations. Membership of specific communities thus helps qualify those who are competent, and define those who are incompetent, within a given representational regime. These regimes are not limited to textual practices, but to careful manipulations of instruments and their graphic outputs.[6]

Visual displays are not substitutes for textual description. Literary technologies are bounded by their embodiment in material and social complexes, and their deployment in communities of representatives. These boundaries are especially apparent in the management and production of visual materials. Talk of "representation" is necessarily ambiguous, for the term refers both to visual depiction and political delegation. Contemporary philosophies of science trade on this nuance, in order to challenge the simplistic model of scientific vision as picturing, and to draw attention to the interventionist work of patterning and organization that any representation demands.[7] A picture may be worth a thousand words but it cannot be substituted by them. The work involved in making pictures is a fundamental aspect of the labor process of the sciences.[8] It is necessary to combine an account of such technologies as engraving, photography, printing, and painting with an analysis of the rival means through which such images display the world. The notion of "virtual witnessing" helps capture an aspect of this combination. The authority of the printed textual report aided the multiplication of those who could be counted as present at an experimental performance. But we must not be misled: this authority was always fragile and never unquestioned. The "author" of an image was a socially constituted and conventional matter. Limner, scribe, engraver, mapmaker, and stationer combined in differing ways to make and propagate pictures of the world. Despite recent insistence on the immutability of mobile inscriptions, pictures were always embedded in rather complex technologies that were not easy to translate, and their evident meaning relied on interpretative conventions that were by no means robust. The extension of the account of literary to visual technologies is troubled, and demands new analytical tools drawn from iconography and cultural history.

These factors suggest the problems of an identification between scientist and writer, and the prejudicial consequences of assuming a stable role for the bearer of knowledge and the author of texts. To what extent is it possible to evoke the historical interaction between these unstable and conventional categories—the techniques through which pictures and texts are produced, the roles assigned to authors and representatives, and the communities that credit rival representations? In what follows, I tell a story designed to illuminate the conflictual, local character of this interaction. The period is that of the United Kingdom of the 1840s, when new print and picturing technologies were rapidly developed, authorial categories were unstable, and the definition of the proper bearer of knowledge was not well established. This was when our favored term "technology" was specialized to mean the practical arts and the word "technologist" appeared. The science in question is astronomy, whose status depended on a special, allegedly close relation between observer and nature, unmediated by the vagaries of laboratory or workshop labor. Astronomy of the period is a good case for our purposes not least because despite this positivist reputation it then involved some spectacular technological and experimental enterprises.[9]

A prized emblem of this technological astronomy was the great reflecting telescope constructed under the orders of the Earl of Rosse in western Ireland during the 1840s. The story of this telescope helps show how specific literary-technological processes work. The "Leviathan of Parsonstown," as the instrument was known, was widely seen as a triumph of engineering. In the later 1840s, for example, the economic journalist Harriet Martineau included the Leviathan in the final, celebratory chapter of her account of the three decades of peace following the Napoleonic Wars. Significantly, she juxtaposed the huge reflector with other material means of communication developed since 1815: the electric telegraph, photography (which she called "sun painting"), steam railways, and Brunel's amazing tunnel under the Thames. So the Leviathan took its place with other "processes of virtual education," as she labeled the ways knowledge had been diffused through the populace. Rosse's heroic if fragile telescope, she suggested, "is not yet so manageable as it will be, and errors derived from its use are as enormous as its powers. But it is a vast new opening into science, through which wise men are learning to look and which may hereafter stand wide to the peasant and the child."[10]

During Martineau's "thirty years' peace" the public status and use of such novel technologies by sages and plebs were in question. In traditional accounts of skill, workshop artisans claimed the right to their labor's whole product. Such discourses were challenged by utilitarian reform and factory management. Workshops were surrounded with rites of secrecy to exclude surveillance by managers and thus protect the property and culture of artisans against fashionable doctrines of political economy. British astronomers such as those who founded the Astronomical Society of London in 1820 were intimates of the aggressive commercial culture of the capital. Artisan skill and proper management were indispensable in the making of their new large-scale mirrors and telescopes, inaugurated in a series of very big reflectors built for the naturalist-astronomer William Herschel from the later eighteenth century on. In 1820 his son, John, arranged for an image of Herschel's largest instrument to appear on the Astronomical Society's seal.[11] Such emblematic machines posed major challenges to labor, engineering, and cosmology alike.

In western Ireland during the 1830s, Rosse assembled a team of workmen to build a reflector of unprecedented complexity and scale. A later Victorian historian of astronomy, Agnes Clerke, observed that initially the Earl "had no skilled workmen to assist him. His implements, both animate and inanimate, had to be formed by himself. Peasants taken from the plough were educated by him into efficient mechanics and engineers." Rosse's property rights were therefore crucial aspects of his telescope's meaning and status. All his human and nonhuman resources "issued from his own workshops."[12] There were some important contemporary resources for explaining how an inventor could make such "implements" and thus become an expert and a discoverer. Technologies such as the telescopes had simultaneously to be the unambiguous property of their masters, yet their lessons must be accounts of Nature's properties. Patent laws and rights of intellectual property helped define owners and the machines and machinists they ran. Attempts to repeal these laws in the mid-nineteenth century were resisted by those who celebrated the heroic power of the inventive genius and thus rewrote histories of technical invention to legitimate the property rights of such masters. Tales of invention and discovery, such as those told about the big reflecting telescopes between 1800 and 1850, provide good examples of literary technologies designed to make material technologies into items of individual property and authority.[13]

Such tales counted in the public commercial sphere, where the features of Nature and of makers were simultaneously established. As Iwan Morus has clearly demonstrated, shows of machines and scientific artifacts at such new places as the National Gallery of Practical Science (opened in London in 1832) theatrically connected owners, makers, and technologies in the capitalist market.[14] Telescopic astronomy was a well-established part of this British technological culture, and the Leviathan of Parsonstown was also put on show in print and through innumerable images. Its size and significance precluded easy transport to the metropolitan showrooms, but encouraged display throughout the public world of Victorian engineering and commerce next to other technical marvels. Unprecedentedly dramatic and detailed pictures of the nebular heavens were made by workers at the Earl's observatory at Parsonstown. These representations concerned many different constituencies. The telescope and its output were soon the topic of an avalanche of print. Parsonstown was not a "centre of calculation" through which other interests had to pass and where the labels of "truth" and "conjecture" could be assigned, but rather a stage on which various mid-nineteenth century groups played out claims to power. Literary and visual technologies were used to make authoritative representatives of Rosse's telescope. But power and authority were not easily distributed from the great instrument toward its commentators, from the esoteric scientific device to more popular realms. Journalists, critics, astronomers, priests, engineers, and wits also gave authority to the Leviathan. The history of this Leviathan is a story of how some nineteenth-century material communication systems made sense out of technologies of metal and paper.

"Seeing Is an Art Which Must Be Learnt"

New instruments can embody new disciplines. The emergence of stellar astronomy in the early nineteenth century is an example. The huge reflectors built in Britain from the 1780s on became the emblems of the unprecedented enterprise of investigation into the construction of the stellar heavens. The Leviathan of Parsonstown was a reflector 70 feet in length, carrying a six-foot diameter mirror (see Fig. 10.1). It cost £12,000 to build and was developed between spring 1842 and spring 1845. It was in many ways the last of its generation, never quite being matched for the rest of the century. The series of which the Leviathan was a culmination began with the work of William Herschel. Following a ten-year experimental program inaugurated in Bath in the mid-1770s, he set up a 20-foot reflector at Datchet in summer 1782 and developed an extraordinary 40-foot instrument with a 48-inch mirror at Slough between 1786 and sum-

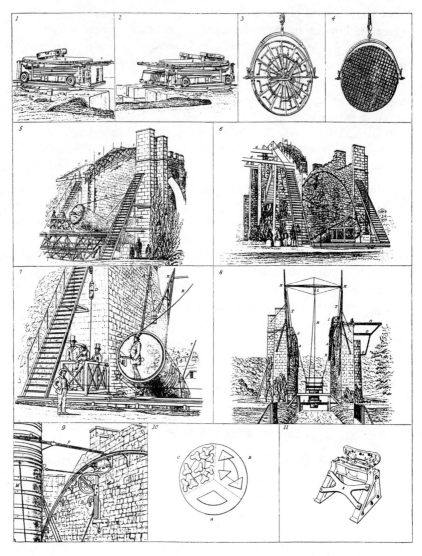

Figure 10.1. The Leviathan of Parsonstown, completed in 1845, described in 1861. Reproduced from Charles Parsons, *Scientific Papers of the Earl of Rosse* (London: Lund, Humphries, 1926), p. 150.

mer 1789. This latter was widely judged a failure, but Herschel developed a relatively lucrative trade in telescope building. In 1796–97 he fitted up a 25-foot reflector for the Spanish monarch, and in the following decade kept up his pursuit of better mirrors and light-grasp. John Herschel inherited the family monopoly over giant reflectors. From 1825 to 1833 he used the improved twenty-foot telescope to revise his father's great nebular catalogues of 2,000 such objects. In a famous extension of this program he reerected the twenty-foot instrument in South Africa in early 1834 to survey the southern skies. Some London makers tried to match the scale of these achievements, but the technology transferred very badly. During the 1820s, a Scottish amateur, John Ramage, developed a series of unsuccessful 25-foot instruments under the patronage of his colleague, the natural philosopher David Brewster, while the wealthy and eccentric Earl of Stanhope projected a utopian telescope of 384 feet with a six-foot mirror. Mirrors greater than one foot remained Herschel's prerogative. This was the situation when the Earl of Rosse commissioned his first great instrument, carrying a three-foot mirror, erected at Parsonstown in 1839, followed by the stunning 72-inch mirror cast in 1842 and first used in 1845.[15]

Building big reflectors was big technology. A great reflector was not simply an isolated and homogeneous emblem of new astronomy. It was an ensemble of interacting technical systems—metal-casting and polishing, construction of the telescope carriage, development of new conventions of recording and transcription, cartography, publication, and dissemination. Each component was both fragile and potentially controversial. To make representations of the heavens was just to secure the working and credit of each system and its interactions. Thus William Herschel's innovation was simultaneously technical, social, and disciplinary. Herschel had been a successful professional musician in fashionable Bath, where he cultivated literary and philosophical contacts in the spa town's polite society. These contacts prompted him to pursue astronomy and natural history, a project that included the discovery of a new planet in spring 1781. He used the status he gained and the patronage he won to propose a vast telescopic enterprise. Herschel's "sweeps" of the stars were prompted by a novel program in the "natural history of the heavens," the picturing of the stellar and nebular populations as natural historical series arranged in temporal species according to their outward form. Key to this stellar botanizing was the distinction between nebulae that could be resolved into stars and those that exhibited what he called "true" nebulosity. During the 1780s, he gradually constructed a temporal and spatial series in which the naturalist of the heavens could see a development from uncondensed nebular fluid to condensed masses of stars.[16]

Herschel's natural history of the heavens was idiosyncratic by the standards of Georgian astronomy. It was not obvious that large reflectors held the key to astronomical advance. Better lenses, such as the fine achromatics made by Josef Fraunhofer in Bavaria from 1812 on, pointed more securely to better precision measurement. Surely no true astronomer would concentrate on nebular shapes at the expense of celestial mechanics. The Astronomer Royal, Nevil Maskelyne, was astonished that "Mr. Herschel looks at the appearances of bodies more than the times."[17] Herschel seemed to his contemporaries a possibly insane and certainly unique investigator. But however isolated on the public stage of Georgian astronomy, his works at the eyepiece and in the casting-shop were collective efforts. In his accumulation of nebular catalogues, Herschel stood in the gallery of a twenty-foot reflector while workmen moved the tube and his sister Caroline took careful notes and observations. In 1785, Herschel began to plan a much larger telescope, "grasping together a greater quantity of light, and thereby enabling us to see further into space." He used friends at court, such as the president of the Royal Society, Joseph Banks, to lobby the monarch for finance. He won £4,000 and an annual stipend of £200 for his workmen. Between summer 1785 and spring 1787 Herschel's team of 40 at Slough built an unprecedented instrument, a telescope of 40 feet with a series of four-foot mirrors (see Fig. 10.2). Caroline Herschel recalled that "the garden and workroom were swarming with labourers and workmen, smiths and carpenters going to and fro between the forge and the forty-foot machinery, and I ought not to forget that there is not one screw-bolt about the whole apparatus but what was fixed under the immediate eye of my brother."[18] The relation between collective labor and personal control was crucial for the development of this important new technology.

Herschel's project was his own property. He recognized this fact and its consequences for his remarkably individual artisan skill. In response to skepticism of his unlikely observational claims to remarkable resolution and magnification after the discovery of Uranus, the former musician and current naturalist of the heavens wrote that "I do not suppose there are many persons who could even find a star with my power of 6450, much less keep it, if they had found it. Seeing is in some respect an art which must be learnt. To make a person see with such a power is nearly the same

Figure 10.2. William Herschel's 40-foot reflecting telescope at Slough. Reproduced from Henry King, *History of the Telescope* (High Wycombe: Griffin, 1955), p. 130.

as if I were asked to make him play one of Handel's fugues upon the organ. Many a night have I been practising to see, and it would be strange if one did not acquire a certain dexterity by such constant practice."[19] Just the same strictures applied to mirror-making, telescope mounting, and the interpretation of images. The artisanal quality of Herschel's complex of theoretical and practical work was in part a consequence of his role as telescope-maker—he could not let his commercial skills become public. His advertisements stressed the advantages of the big instruments but did not specify how they were made: "Here an observer may sit for many hours, with constant entertainment, continually expecting new objects to present themselves, which he never could have perceived in common telescopes."[20] But it was also a fundamental aspect of his account of the natural history of the heavens. He needed great light-grasp, so he needed parabolic mirrors of great size and a vocabulary of his own devising for accounting his results.

Herschel's observatories defined a private space for his new science.

Marvels of polishing, design, observation, and resolution were claimed there. How could such local artifacts be made compelling elsewhere? Mirror polishing, for example, was a private art. Herschel challenged the conventional method—a series of circular, eccentric strokes that aimed for a direct paraboloid—and tried combinations of eccentrics and "glory" strokes, linearly across the center, so as to achieve a spherical surface and then move carefully toward "the so much wished for parabola."[21] There was much debate on these matters with his colleagues. The expert mirror-makers John Michell and Jonathan Edwards both challenged Herschel's recipes. Edwards had his own recommendations when Herschel reported a failure in casting a 6.5-inch mirror: "Here I am certain is the cause of your Failure, I will therefore be particular. Before I melt the Metal in my Air Furnace, I make up a very large Fire in my Kitchen with common small coal." The recipe also called for the right support staff: Edwards's brother-in-law "was a most ingenious young Man, & the best practical mechanic I ever met with; but he is lately dead; so that at present I have no one to assist me."[22] Herschel tended to pursue his own technical repertoire. He favored an unusual recipe for speculum metal, a proportion of copper of seven to three, but he still had problems with brittle metal, with cracking, and with getting parabolas. By 1789 he had designed a powered machine to polish four-foot discs at a rate of eight strokes per minute, the system he used for the 40-foot giant. Some other makers avoided mechanization. In 1833 Ramage explained that his mirrors had been "ground and polished simply with his hand, without the aid of any machinery or mechanical power—a circumstance which, he said, astonished the opticians of London, when it was stated, and which they considered as almost incredible."[23] So any mirror was the personal property of its polisher. Mechanization did not remove it from the immediate locale of the maker. The puzzles of artisan workshop privacy were widely recognized. In 1830 the future Earl of Rosse publicly pointed out that "the practical optician will rarely give you the slightest intimation of the process of working specula which he finds the most successful, nor is it perhaps to be expected." The young Irish nobleman announced that his goal was to ensure that "the art might no longer be a mystery known to but few individuals." But to this extent he failed, for that was what it remained.[24]

Such local features mattered because of the importance of speculum performance in any assessment of a telescope's power and thus the credit of the results reported by its user. Material technology was linked with disciplinary politics and the distribution of social authority. A fundamental trouble characterized the emergence of this technical and political complex in the new science of stellar astronomy. William Herschel's great claim, insisted upon from at least 1789 until his death, was the existence of true nebulosity in space. His cosmology proposed a developmental scheme for stars: stellar systems were condensed out of a widely diffused, luminous nebulous fluid. Herschel's cosmology carried great prestige—the prestige of the discoverer of a new planet and the maker of unprecedentedly large instruments with hitherto unknown light-grasp. However, the status of true nebulosity depended on the astronomer's ability to discriminate clouds of light that were really irresolvable from those that were generated by very distant congeries of stars. So Herschel had to bolster his system's credibility by pointing to his instruments' unrivaled power; then he had to stipulate that some luminous objects would always resist this power. He had to make true nebulosity a property of the heavens, not of weak machines or his own predilection. The trouble was exacerbated by the impossibility of replicating Herschel's machines. Publicly available calibrations were therefore indispensable. He needed means through which true nebulosity and potentially resolvable stars could be distinguished, a substitute found for Herschel's private confident judgments at the eyepiece, and these distinctions made manifest to his extramural community.[25]

Literary, material, and social technologies were developed to make these distinctions robust and public. Herschel's strategy integrated these three repertoires. He announced his care with neologisms. His early work on stellar parallax was marked by this concern. Herschel reckoned that a good catalog of double stars would show which were optical and which physical doubles. Only the former could be used for parallax measures. So the term "double star" embodied the aims of his program. He preferred the term "to any other, such as Comes, Companion or Satellite; because in my opinion, it is much too soon to form any theories of small stars revolving around large ones, and therefore I thought it adviseable carefully to avoid any expression that might convey that idea."[26] Just the same strictures applied when mapping the solar surface. Herschel's view was that sunspots were holes in bright clouds through which the cool, dark, inhabitable solar surface could be seen. So, again, "I have found it convenient to lay aside the old names of spots, nuclei, penumbrae, faculae and luculi, which can only be looked upon as figurative expressions that may lead to error. . . . The expressions I have used are openings, shallows, ridges, nodules, corrugations, indentations and pores."[27] In both cases, new theory-laden claims were bolstered by making language realistic, scotching metaphor, and insinuating much new stellar and solar knowledge through the literary technology itself. Herschel often found himself in this situation. Uranus had first been recognized as a comet. The term "new planet" was oxymoronic. It took two years for the label "planet" and the name "Georgian star" to appear. These terms had disciplinary and political uses. When he detected two small bodies between Mars and Jupiter he remarked that "it appears to me much more poor in language to call them planets than if we were to call a rasor a knife, a cleaver a hatchet &c. . . . Now as we already have Planets, Comets, Satellites, pray help me to another dignified name as soon as possible." Calling these bodies "asteroids" fitted them into a large-scale investigative scheme centered in Germany in which Herschel then took part, and fitted them, too, into the pattern of the solar system.[28] Herschel's work demonstrated that naming novel objects required appeal to networks of familiar concepts, and extended that network in one direction rather than another. Double stars were fitted into the program of parallax measures; asteroids into the regime of celestial mechanics.

Parallax measurement and celestial mechanics were recognized disciplinary programs. So terms like "new planet," "double star," and "asteroid" were extensions of networks of practice and meaning that already commanded allegiance. Trouble occurred in Herschel's radical new program of the natural history of the heavens. Here big new instruments, unprecedented classifications, and unrepeatable observations were all common. Herschel insisted on the link between his new technology and new criteria of competence. A new art of seeing would need a new way of making credit and a new vocabulary for recording the contents of the heavens. The art of seeing would become *experimental* and no longer self-evident. In his notorious paper on parallax delivered to the Royal Society in December 1781, he announced that "too much has hitherto been taken for granted in optics: every natural philosopher is ready enough to allow the necessity of making experiments, and tracing the steps of nature; why this method should not be more pursued in the art of seeing does not appear."[29]

By making seeing an art, and thus both susceptible to and in need of disciplinary revision, Herschel was in danger of severely limiting his possible audience. Astronomy posed as the perfect science because of its empiricism. Herschel proposed making seeing subject to deliberate experiment. He had to develop ways of communicating the techniques of vision to others. His doctrine of scientific language played a key role. He told the Royal Society that his taxonomy of double stars was designed to relate the reality of doubles to the capacities of different observers. First class doubles were just those "which indeed require a very superior telescope, the utmost clearness of air, and every other favourable circumstance to be seen at all. . . . They seemed to me on that account to deserve a separate place, that an observer might not condemn his instrument or his eye if he should not be successful in distinguishing them." Herschel insinuated that he had the tribunal before which other instruments and other observers must be judged, and that his language and his taxonomy could adequately embody the precepts of that tribunal.[30] This was not innocent. Herschel's models of stars and space were conveyed, often tacitly, through these terms and judgments. A good example was his work on star color. He reckoned that the presence of an interstellar ether, crucial for his nebular model, was evidenced by the fact that smaller, dimmer stars were red, suggesting that other colors were absorbed by this medium. He had to give his audience instructions for judging star color. But "different eyes may perhaps differ a little in their estimations." One star "by some to whom I have shewn it has been called green, and by others blue." So "it is difficult to find a criterion of the colours of stars." Herschel resorted to calibrating colors by reference to well-known standard stars, thus begging the question of any way of calibrating for color, other than his own telescope's and eyes' deliverances.[31]

The point of Herschel's literary technology was to stabilize these claims and make them acceptable to others. Good language and calibration would make his reports representatives of the heavens, not of Slough observatory. New instruments helped the task, provided they, too, were accompanied by careful terminology and made acceptable and credible. The aftermath of the Uranus triumph was a debate between Herschel and his London colleagues about the powers of his telescopes. He tried to explain to Maskelyne the means of assessing telescope quality. His comments reveal the difficulty of making this calibration: "I believe that a telescope which does not shew the apparent body of a star very round and free from rays will hardly shew the small double star of 3 Cancri." When Herschel visited Maskelyne's observatory at Greenwich in spring 1781 he found that the Astronomer Royal's best instrument failed this test, "but a good deal must be allowed for my not being used to that telescope; nor did I think the evening a very good one." A year later Herschel had fewer doubts. He told his sister that Maskelyne had conceded Herschel's instruments' superiority: "Double stars they could not see with their Instruments I had the pleasure to shew them very plainly. Among opticians & Astronomers nothing is now talked of but *what they call* my great discoveries. Alas! this shews how far they are behind, when such trifles as I have seen and done are called *great*."[32] Direct comparisons with mobile instruments remained possible. But Herschel designed telescopes and accompanying support techniques that were by no means so easy to translate. His descriptive language began to play its role. In his double star catalog of 1782 he introduced a new precision micrometer and tried to show how its measures of star position should be represented. His instructions simultaneously revealed the personal basis of precision judgment and the need for a means of calibrating novel phenomena: "when I have added *inaccurate*, we may suspect an error of 3 or 4 seconds. *Exactly estimated* may be taken to be true to about one eighth part of the whole distance; but only *estimated*, or *about*, &c. is in some respect quite undetermined, for it is hardly to be conceived how little we are able to judge of distances when, by constantly changing the powers of the instrument, we are as it were left without any guide at all."[33]

Lack of guidance was a trouble for the natural history of the heavens. Between the 1780s and the 1810s Herschel compiled huge nebular catalogs in an effort to make stable categories recognizable by others, then to suggest a temporal and spatial order that divided these types between earlier nebulous matter and later stellar clusters. Most of this project needed a scale against which distance, and thus time and aggregation, could be judged by Herschel, and by others who lacked his instruments. His valedictory discussion of this problem was a paper of June 1818 on the question of "how far the power of our telescopes may be expected to reach into space when directed to ambiguous celestial objects." Here Herschel provided some recipes for discriminating between clouds of light that were aggregations of stars and those that were truly nebular. First, he asked his audience to accept that "we have hitherto only been acquainted with two different principles, the nebulous and the sidereal." Then, an astronomer must find a star cluster resolvable with an instrument of given power. Call this the "gauging power" of that object. Were the power to be progressively diminished, the star cluster would gradually be reduced to "an ill defined star surrounded by nebulosity." This way, the novice would learn the subtle differences between irresolvable star clusters and true nebulae. But these differences were hard to spot. Herschel confessed that he had often used "the expressions *resolvable*, or *easily resolvable*, when, from their appearance, I could not decide whether they belonged to the class of nebulae, properly so called, or whether they might not consist of an aggregation of stars at too great a distance from us to be distantly perceived." Hence his coinage of the term "ambiguous": "When the nature or construction of a celestial object is called *ambiguous* this expression may be looked upon as referring either to the eye of the observer, or to the telescope by which it has been examined."[34]

So Herschel recognized the instability of his scheme. Solutions lay in material or theoretical technique. Materially, if a celestial object could not be resolved into stars, he could call it ambiguous, then find a gauging power with a greater telescope. He envisaged instruments greater than his forty-foot, instruments to which that would act as a mere "finder." This was the invitation accepted by Rosse. Alternatively, he could try to reinforce his theoretical model. The Milky Way, for example, was not ambiguous. Every cloud of light there was in principle resolvable, and "when our gages will no longer resolve the Milky Way into stars, it is not because its nature is ambiguous, but because it is fathomless."[35] Herschel left a story about the origins of stars and planets that demanded the existence of true nebulosity, a criterion for true nebulae that depended solely on the resolving limits of large telescopes, and a tacit understanding that some objects, such as clouds of light in the Milky Way, were stellar, not nebular. These were resources for the new discipline of stellar astronomy, developed in Britain and Germany from the 1830s, but they were scarcely the foundations of a stable and credible world-picture to which all his audience could subscribe.

In 1803 the canny Edinburgh lawyer Henry Brougham, future apostle of useful knowledge, reacted with fury to Herschel's neologistic defence of his cosmology. Terms such as "telescopic sweeps," "natural history of the heavens," and "space penetrating power" were objectionable because they did not convey new knowledge. Brougham, apostle of Edinburgh philosophy of language, reckoned that any who accepted them would be seduced into seeing Herschel's conjectural visions as robust discoveries. "Knowing, as we do, the great power of words in misleading and perplexing our ideas, we cannot allow the unnecessary introduction of a new term to escape unnoticed. Where a new object has been discovered, we cheerfully admit the right of the discoverer to give it a new name; but we will not allow a needless multiplication of terms, or an unnecessary alteration in the old classification of things to be either justifiable or harmless, a substitute for real discovery or a means of facilitating the progress of invention."[36] Brougham's critique represented a Regency commonplace. Language's stability represented a commonsense, ordered world. Terms should refer to unproblematic objects. So Herschel's neologisms and ambiguities, his desperate coinage of a literary technology designed to stabilize irreproducible and unreachable cosmological classes, scarcely carried conviction among the polite society of the literati.[37] In order for nebular astronomy to become public property William Herschel's work had to be developed and recast.

"The Most Wonderful Object in the Heavens"

William Herschel never quite managed to develop his cosmology as a program for a new astronomical discipline. Critiques such as Brougham's told against the speculative character of a general developmental picture of nebulae, stars, comets, and planets. Herschel's image remained that of the great solitary, just as his telescopes proved irreproducible and his 40-foot instrument almost useless. This image was reinforced by the publicity Herschel cultivated. Tours through the telescope tube itself were common among polite visitors. George III led the Archbishop of Canterbury through the tube: "Come, my Lord Bishop, I will show you the way to Heaven!" The instrument was open to elegant callers. The theories it helped develop were closed. In December 1787, for example, while the 40-foot instrument was in the process of being set up, the young Fanny Burney, court lady and celebrated diarist, paid a visit to Slough. "Already with that he has now in use, he has discovered fifteen hundred *universes*! How many more he may find, who can conjecture? . . . By the invitation of Mr Herschel I now took a walk which will sound to you rather strange—it was through his telescope!"[38]

Such episodes set the tone for telescopic astronomy's place in culture. It attracted public attention, became a tourist attraction, and was subject to popular and literary commentary. At the same time, the technologies of esoteric stellar astronomy remained the private property of a few. The space opened between the private observatory and the public culture of astronomy provided resources for a variety of genres. The openness of Slough to visitors reinforced a stereotypical account of the astronomical watchtower, the privileged site of isolated heavenly contemplation. What happened there, visitors implied, was distant from earthly concern, too elevated for public consumption, and the sole property of a solitary astronomer.

So during the Regency the stellar astronomers, especially John Herschel, doyen of the new Astronomical Society, tried to use these resources to give a better account of nebular cosmology. This required a clarification of astronomical language and imagery. Crucial in this project was the definition of the most important objects of William Herschel's natural history of the heavens. Like his colleagues among the Bath Linnaeans, the senior Herschel had fixed a series of type objects with which to model species. Most important amongst these was the Orion nebula (Messier 42), "altogether the most wonderful object in the heavens," as he described it in October 1784. This nebula was the first he recorded in his observing book in March 1774, and he studied it throughout his career.[39] It became one end of a scale of nebulosity, stretching from items which seemed truly nebular to those which could certainly be resolved into stars. The truly nebular character of Orion depended on its resistance to high telescopic power. It also depended on marked changes in printed and manuscript drawings of the object. In his first observation, Herschel noted that he "saw the Lucid Spot in Orions Sword, thro' a 5½ foot reflector; its Shape was not as Dr Smith has delineated it in his Optics; tho' something resembling it. . . . From this we may infer that there are undoubtedly changes among the fixt stars, and perhaps from a careful observation of this Spot something might be concluded concerning the Nature of it."[40] Robert Smith's *Opticks* (1738) was a standard text in Newtonian light theory, and the picture Herschel found there was a drawing made by Huygens in 1656. In December 1778, October 1779, January 1783, and September 1783 Herschel noted major changes in its shape. These changes suggested that the Orion spot was nearby and truly nebular. Thus, before 1784 he had good evidence, so he reckoned, that there was true nebulosity from which stars might develop by condensation.[41]

Herschel's comparisons between his own perceptions of the heavens and those he and others had previously recorded made the Orion nebula the most important object in the skies. His son shared this view, and interest in the means through which pictures of nebulae might be more reliably recorded. The status of Orion as the exemplary nebular object was institutionalized in the speech the new Astronomer Royal, George Airy, gave to the Royal Astronomical Society in 1836, on the occasion of the award of the Society's medal to John Herschel for his work in extending and revising his father's great nebular catalogs. Airy emphasized that "the subjects of nebulae and double stars appear to have been considered" by John Herschel as "a hereditary possession, and the familiarity with the instruments necessary for the observations was also hereditary." He pointed out that William Herschel's attitude to true nebulosity had wavered, but that he had finally fixed on Orion as the evidence of true nebulosity and the condensation of stars from nebular fluid. He told his audience that anyone who had seen Orion "in a telescope of great light" must be convinced that it was "like a pile of cumuli-clouds tossed together in the same capricious manner in which we see them in our summer skies," and could not doubt that "these can be anything but masses of nebulous matter." A print published in the Society's *Memoirs* confirmed this interpretation.[42]

Airy's speech summarized but limited Herschel's nebular cosmology. Airy himself was the principal apostle of precision meridian astronomy. His headquarters at Greenwich avoided nebular observation and the imperial network that Airy forged in the subsequent decades did not see its own function as that of discovery and resolution. John Herschel was, if anything, even more publicly careful with this work. In the papers that won him the medal he avoided complete endorsement of the implications of Orion's true nebulosity. In his observing books, however, he frequently noted evidence that its shape had changed and that it continued to resist resolution. He could see that "stars of the 7th or 8th magnitude and irresolvable nebula may coexist within limits of distance not differing in proportion more than as 9 to 10." This would imply that true nebulosity did exist but that his father's model of condensation of nebular clouds might be wrong. He could also see that Orion's shape continued to change. Yet in 1833 he publicly insisted that "we have every reason to believe, at least in the generality of cases . . . a nebula [is] nothing more than a cluster of discrete stars." The inheritor of nebular astronomy, therefore, provided no very clear message to his audience on the key issue of the nebulae.[43]

This uneasy compromise between the lauding of William Herschel and skepticism about his cosmology changed during the 1830s. In early 1834 John Herschel arrived at the Cape of Good Hope, and in his four years' observation there he revised many of his father's premises. His earliest observations were of the Orion nebula and, on the same night, of the Magellanic Clouds, the first with a big reflector. Herschel found stars and nebulae in the same object, revising the notion that true nebulosity must be separated from the stars it later produced. The drama of his South African tour made him an imperial hero, the symbol of early Victorian science.[44] In the same years the highly popular works of the Cambridge don William Whewell and the Glasgow astronomy professor John Pringle

Nichol changed the meaning of nebular cosmology. The two authors invented a "nebular hypothesis" out of materials they found in William Herschel's papers and in comments by the master physicist Pierre-Simon Laplace. This hypothesis was given many moral, local messages. True nebulosity showed that this solar system had evolved from clouds of luminous fluid. By bringing the nebulae into the center of political and moral debate, Nichol and his contemporaries changed the conditions that had hitherto governed work on stars and nebulae.[45]

Nichol's "nebular hypothesis" first appeared in October 1833, six months after that of Whewell and at the same time as Herschel's masterly textbook on astronomy. Here, in an unprecedented manner, Nichol deployed the fiery metaphors of Scottish pulpit oratory in the service of nebular astronomy. He reckoned that true nebulosity taught the lesson of progress. In the radical *Westminster Review* of July 1836, Nichol cited John Herschel's recent news that the Orion nebula changed its shape and was truly nebular, then immediately criticized him for daring to suggest that much other apparently condensing material might be the result of an optical illusion: "in a case of *demonstration*," Nichol advised the statesman of British astronomy and natural philosophy, "everything must be clear; but when we treat of conjectural subjects, it is not philosophical to refuse to rest on the superior amount of *probability*."[46] Nichol led a challenge to the rights of the clerisy over the judgment of "philosophy." In his paper for the *Westminster* and his subsequent popular triumph, *The Architecture of the Heavens* (1837), Nichol claimed the right to define the bounds of astronomical truth. The "nebular hypothesis" was philosophically sure because in such matters as development and progress the mere grounds of probability were sufficient. Herschel had told the Astronomical Society that since no visible evidence of condensation was available, the nebular hypothesis in its local, potentially atheist form must not be accepted. Herschel strove to separate his father's allegedly solid observational model of nebular cosmology from the Laplacian story of this planet's origins. Nichol reacted angrily. The ban on any inference from separate but serial nebular types to a real law of progression was just like contemporary conservative attacks on transmutation of animal species. Nichol deliberately posed as the Geoffroy Saint-Hilaire of astronomy. "It is full time that such speculations cease to be confounded with 'Atheism'—at least by our *Learned!*"[47]

Nichol's literary technology reinforced his political and moral claims. He appealed beyond the closed circles of the conservative corporations to a literate, radical audience of the reviews and the schoolrooms. In 1838 Nichol told John Herschel that "good introductions to astronomical science are abundant, but it often occurred to me that the *Moral results* of Astronomy might be made quietly apparent to many minds, and quietly introduced among their modes of thought, by treatises, not without the show of science, but without its form—having in that something of the form of Romance."[48] His language was an object of contemporary comment. Nichol's friend, the Dundee minister George Gilfillan, wrote that "we have been amazed and delighted to witness the impression he contrives to make upon the humblest minds, by the joint effect of his subject—his gorgeous style—his gigantic diagrams, and the enthusiasm which speaks through his pallid visage and large grey eyes."[49] Nichol's techniques attached the claim that the stars were the source of life to the claim that he was the stars' representative. Vitality was an attribute of his celebrated prose style, his account of the heavens, and the stars themselves. His verbal rendition of Herschel's picture of the Orion nebula became a topos of Victorian popular radical lecturing (see Fig. 10.3). Nichol con-

Figure 10.3. John Herschel's drawing of the Orion nebula, published in the works of John Pringle Nichol in the 1840s. Reproduced from John Pringle Nichol, *Thoughts on Some Important Points Relating to the System of the World*, 1st American ed. (Boston: Munroe, 1848), opposite p. 88.

nected Orion to the contemporary representations of fossil life on Earth. The punctuation conveyed his breathless message of organic fulfillment: "What is the intention of such a mass? Is it to abide for ever in that chaotic condition? Void, formless and diffuse in the midst of order and organisation—or is it the germ of more exalted being—the rudiments of something only yet being arranged?"[50]

Nichol's lectures and handbooks were a means of disseminating a radical interpretation of nebular astronomy. It was adopted by John Stuart Mill, Nichol's close ally and editor, as the emblem of progress; it was incorporated by the Edinburgh journalist Robert Chambers, Nichol's friend, into his anonymous classic of scientific naturalism, *Vestiges of the Natural History of Creation* (1844); and it temporarily became an article of faith among evangelical writers such as David Brewster, who had backed Nichol's campaign for a redesigned observatory at Glasgow.[51] These uses brought into the public domain the esoteric niceties of nebular astronomy. The fragility of astronomical language and representation was dramatized. Mill was forced to excise most references to the nebular hypothesis from later editions of his *System of Logic* under the hostility of orthodox savants such as Herschel and Whewell. Chambers clarified and revised the exact model of nebular condensation which he held supported the rule of natural law in organic development. Brewster abandoned his endorsement of Nichol's claims for true nebulosity, not because of the work of the Earl of Rosse, but simply because of the naturalistic deployment of the nebular hypothesis in *Vestiges* for what Brewster saw as "mischievous speculation" and atheist purposes.[52]

Nebular stories had to be carefully treated. John Herschel tried this task at the British Association for the Advancement of Science in 1845. He told his Cambridge audience that "every astronomical observatory which publishes its observations becomes a nucleus for the formation around it of a school of exact practice." Inexact speculation was a vice in business and journalism. It was ruled out by the manners of scientific and financial commerce. Herschel damned both *Vestiges* and *System of Logic* as works that had broken these rules. He instructed his audience on the difference between "my father's general views of the construction of the heavens," which would "still subsist as a matter of rational and philosophical speculation," and more materialist and dangerous versions of the nebular hypothesis.[53] But the key problem remained: the savants needed the public forum, yet they found that this forum was unstable when presented with exciting astronomical representations. The link between literary and social technique was close. Herschel spoke of nebulae "which exhibit no regularity of outline, no systematic gradation of brightness. . . . The wildest imagination can conceive nothing more capricious than their forms, which in many cases seem totally devoid of plan." Instability was a feature both of the natural spaces of the astronomical heavens and of the social spaces of astronomical representations.[54]

As in many other fields of early Victorian culture, the astronomers' public was a large and literate community. To attract this public, astronomers treated their pictures of the heavens as immediate representations of the heavens and themselves as uniquely skilled representatives. Thus they simultaneously drew their audience's attention toward and away from the skilled labor involved in picturing. When Airy spoke at the Astronomical Society's award ceremony for John Herschel, he emphasized the virtue of Herschel's engravings of the nebulae incorporated in the great survey of the heavens. The pictures, he alleged, were the only means of giving accessible expression to otherwise incommunicable aspects of the nebulae. Conventional astronomical techniques of polar coordinate mapping simply failed here. "The peculiarities which they represent cannot be described by words or by numerical expressions. It would be absurd to define the place of every point of a nebula, and the intensity of light there, by co-ordinates of any kind. These maps contain that which is conspicuous and distinctive to the eye, and that which will enable the eyes of future observers to examine whether secular variation is perceptible." This point was important—true nebulosity was to be recognized through changes of shape, so pictures acquired an unusually high status. But in the same breath Airy pointed out that such pictures demanded rather uncommon skill, thus questioning the degree to which intersubjective comparison might be possible: "few observers possess the delicacy of hand of Sir JOHN HERSCHEL; yet it were to be wished that his example might be imitated by many, and that careful drawings, the best that circumstances admit of, might frequently be made of the same nebula."[55]

In order to achieve this end, it was important to disseminate representational skill. The puzzle was to work out a means to check the competence of any astronomical draftsman, to check whether this skill had been reliably transmitted. On the one hand, this competence could only be checked against the virtues of the observer's telescope and the characterization given of a known object. But, on the other hand, astronomical drawings were supposed to be used to check whether telescopes were reliable and whether the object surveyed had changed its shape and ap-

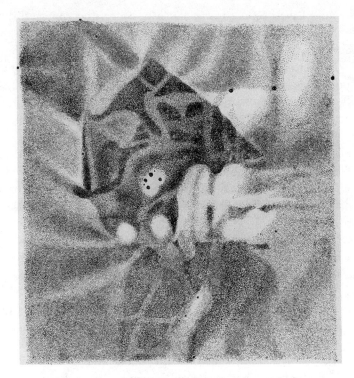

Figure 10.4. The Parsonstown picture of a region of the Orion nebula, completed and published in 1867. Reproduced from Parsons, p. 190.

pearance. This set up a potentially vicious circle, to be closed through the deployment of the culturally powerful conventions of contemporary astronomy. Who was to be trusted? Criteria of competence and trustworthiness could change. In the 1830s whatever John Herschel observed provided the tribunal before which others' observations were judged; after the construction of the Leviathan of Parsonstown, Herschel had to explain away his own drawings as illusory. The division of labor at observatories also mattered. Rosse himself took great care to stress the skill of the draftsmen he employed (see Fig. 10.4). For pictures of the great nebula in Orion, he employed Bindon Stoney, "a highly educated civil engineer, well accustomed to use his pencil," whose "drawing was made with great care, and he was engaged upon it the whole season." In the 1860s he used Hunter, "an accomplished artist." Rosse asked Herschel for comments on his Orion pictures, and the Earl then reported that "the engraving is on the whole very accurate; a little more softening off in the faint outlying parts would have been desirable, but Mr. Basire [the engraver] did not think that it would be practicable consistent [*sic*] with the reasonable durability of the plate; the forms, however, are correct." Nichol also consulted Herschel for validation of his own incendiary interpretations of the appearances of Orion: "Of your approbation of my attempts to convey the idea of nebula by white on a black ground, I am happy in being assured. The worst of Lithography is that the impressions are most unequal—at least of any lithography I have been able to command."[56] Support staff is invisible when messages are deemed successful, and highly visible when representation is problematic. Correctness was not self-evident and it was not easy to blame printers and engravers for all of astronomy's representational troubles. It needed careful technical and social management to construct and then propagate such credibility.

Representational technologies, both literary and pictorial, that astronomers and other commentators deployed in making meaning out of the nebulae show how flexible and how troublesome was the search for authority. Nichol often discovered this. When he lectured to the Newcastle Literary and Philosophical Society he earned 100 guineas, but was assailed by the irascible sage William Martin, self-styled Christian Philosopher, who judged that "the book of his, the sooner it is consigned to the flames the better, and not to be imposing upon the ignorant by charging half a guinea for it."[57] The decades of Reform saw an explosion of popular and radical literature. The very term "journalism" was a French import deployed by conservatives to decry the "plebification" of knowledge. Chambers, Nichol, and Mill plied their trades in this world of the steam-press and the vagaries of public opinion. Great charts and dramatic rhetoric were the inevitable accompaniment of campaigns to make the heavens serve terrestrial purposes.[58] Journals taught their readers how to interpret the observatories' messages. In 1845 the *Pictorial Times* explained the difference between the ravings of *Vestiges*, which had referred to the evolution of this world from a primordial fire mist, and the secure deliverances of nebular astronomy. "Our provincial correspondents must not confound the 'fire mists' and 'fire mist vortices' of recent speculations with the nebulous matter and nebular phenomena of astronomic observers. The former are fancies unsupported by evidence, the latter are facts confirmed by everyday experience." The *Pictorial Times* picked out the appearances of the Orion nebula as safe astronomy: "These are phenomena which may be watched by every observer, and the present warm nights are favourable for the purpose."[59]

It was, however, not obvious how to distinguish commonplace phenomena from speculative images. Perhaps this is a general predicament of astronomical picturing. In their study of contemporary work in this field, Lynch and Edgerton argue persuasively that working astronomers use the term "aesthetics" to denigrate work designed for popular consumption or work by rival representatives. The suggestion is that "this dissociation is not so much between different professional cultures," such as between "science" and "art," but rather "between aesthetic standards associated with different audiences."[60] Astronomers, journalists, natural philosophers, priests, and radicals were overlapping communities aiming at a complex of different audiences. These representational differences cannot be located simply in terms of high and low science, between esoteric and exoteric realms. The same terms might occur in a wide range of forums and a historical account of such language and imagery needs an account of techniques for image-making and of the audiences who interpreted these images.

The Leviathan and Its Uses

The Leviathan of Parsonstown was represented as a technological triumph. Many discussed the casting, polishing, mounting, and staging of the telescopes. Rosse was described in the London press as "a practical mechanic and chemist on a gigantic scale." The Leviathan was made comparable to Brunel's rail systems or Nasmyth's steam hammers. It was also widely seen as a moral victory. Many also referred to the moral and cultural significance of its promoter's social place, a noble member of the Protestant Ascendancy in an Ireland wracked by famine and revolt. In autumn 1843 the *Illustrated London News* claimed that Rosse "seems to love science for his own sake, and, untempted by any desire for applause, he has been working silently and for himself, until the magnitude of the results have forced themselves on the notice of the world." Both the new Earl of Rosse—he inherited the title in 1841—and his closest adjutant, the suave and prodigious Thomas Romney Robinson, director of the Armagh Observatory, were pillars of the Church of Ireland and the Ascendancy. Whereas Rosse supported the Whigs and favored moderate Catholic emancipation, Robinson preached in Dublin against Papism, materialism, unbelief, and evolution theory. It was not surprising that Robinson found the views of *Vestiges* objectionable. As early as spring 1841 he already told Rosse of his loathing for Nichol's *Architecture of the Heavens*, which he described as a catchpenny book designed for mass consumption.[61] It was clear from the late 1830s that were Robinson in charge of the great telescopes, they would be turned at once against subversion and the science of progress.

When the new reflecting telescope at Parsonstown was built in the 1840s, Rosse and Robinson needed some quick results, and their attack on Nichol and the Orion nebula was an important task. To undermine the status of previous work on that object would be to call into question the stability of the telescopic astronomy produced by the great reflectors and especially the judgments of proficient observers of the nebulae. This was a real trouble, because the public standing of astronomical doctrine was not guaranteed. Throughout the 1840s, some of the Parsonstown team set out to make Orion's nebulosity just as illusory as *Vestiges'* fire mists. There were not many reliable means by which the Victorian public could safely judge what was a genuinely everyday experience. In establishing criteria for such judgments at Parsonstown, the emphasis on Rosse's *individual* endowments was vital. The telescope became his personal property, its results his responsibility.

Rosse supervised the process of casting, grinding, and erection at Parsonstown. He began in the late 1820s, after training at Trinity College Dublin and at Oxford, and after convincing himself that Fraunhofer's success with achromatic lenses did not rule out further progress in great reflector technology, even though this technology had hitherto proved incommunicable. Robinson recalled that "the construction of a large reflector is still as much as ever a perilous adventure in which each individual must grope his way."[62] Between 1827 and 1839, Rosse assembled a skilled team at Parsonstown. A local smith helped train workmen in the construction of grinding and polishing machines. They worked out means of casting specula in separate pieces and then fusing them. By 1839 a 36-inch mirror had been made with a perfected parabolic shape. In particular, Rosse set out to mechanize the entire polishing process. A two-horsepower steam engine was used to drive a cast iron tool against a mirror rotating in a constant-temperature water cistern. Rosse also challenged the Herschel style in another significant way. He rejected the Slough pattern of suspending the telescope with the observer sitting at the mouth of the tube, as this would cause image disturbance because of air heating. The Newtonian suspension he used required a large team to move the telescope. Further workmen were hired to shift it. In September 1839, Rosse was in command of a workable instrument, but his three-foot mirror had been cast in sixteen separate pieces. By summer 1840, he had replaced this with a mirror made in one piece. When the Parsonstown team was satisfied by the 36-inch mirror's performance, they immediately projected a telescope of twice the light-grasp. Between 1840 and spring 1842 this formed much of Rosse's concern at the observatory. He commissioned a huge foundry with three furnaces, stepped up the copper content of the speculum metal, and, after five separate trials, cast a perfect six-foot mirror. The drama of this moment was an irresistible occasion for Robinson's oratory. Speaking at the Royal Irish Academy two weeks after the casting, Robinson emphasized the contrast between the sublime, precisely celestial, milieu of the foundry and the cool managerial control exercised by the Earl. "The workmen executed their orders with a silent and unerring obedience worthy of the calm and provident self-possession in which they were given."[63]

Self-possession and providence were the virtues of the Parsonstown Leviathan. The descriptions of its construction on the public platforms of Victorian culture made its prestige. In a later account of the Parsonstown specula, Robinson explained that polishing the mirror "requires the most scrupulous attention *in every part*, that a very slight omission will make it fail, and that even the most experienced operator is not *secure* from disappointment." Publicity and security were the keys, and Rosse was widely lauded as the first to render secure and then publish his methods (see Fig. 10.5).[64] Mechanization at Parsonstown was a symbol of Rosse's power and the subservience of his trained artisans to genteel management. "All these gigantic constructions . . . have been executed in Lord Rosse's workshops," Robinson commented, "by persons taken from the surrounding peasantry, who, under his teaching and training, have become accomplished workmen, combining with high skill and intelligence the yet more important requisites of steady habits and good conduct." Technology helped moralize the workforce. The *Dublin Review* dutifully noted that where Herschel's techniques remained secret, the Irish earl's "triumph consists in the invention of a means, by which, with perfect ease and security, the figure can be imparted to specula of *all* dimensions, from one inch to six feet in aperture; and by which what before was always precarious, and at best a labour of days, and even of weeks, may now be accomplished with infallible accuracy, and under the superintendence of a common workman, in a few hours." The Royal Irish Academy, the Royal Society, the British Association, and the Royal Astronomical Society were all told of the telescope's technical specifications and the way it had been

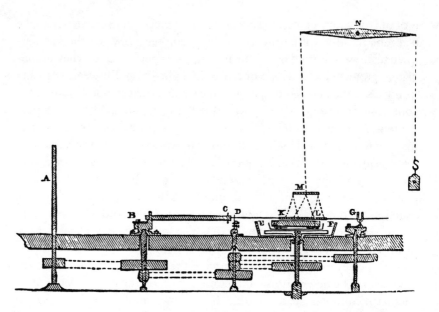

Figure 10.5. Rosse's machine for polishing telescope specula, as described by Thomas Romney Robinson in 1857. Reproduced from John Pringle Nichol, *Cyclopaedia of Physical Sciences* (Glasgow: Griffin, 1868), p. 782.

designed and built. Airy celebrated the fact that Rosse's grinding machine "imitates very closely, but with that superior degree of regularity which is given by machinery, the operation of polishing by hand." Whereas the *Illustrated London News* lauded Rosse's silence and solitude, Airy in contrast noted that Rosse "had shewn it was possible, without any important manual labour, to produce with certainty, by means of machinery, mirrors of a size never before attained . . . and he had, by publication and private communication, made these methods accessible to the world."[65]

Display of technology was crucial. The drama of the structure was a warrant for the telescope's performance. The telescope mounting was remarkable. Rosse commissioned two piers of masonry 72 feet long and 56 feet high, of gothic magnificence. The six-foot, unlike its smaller predecessor, could only move in the meridian; it simply could not make repeated, regular observations of one object. Its architecture was built purposely for heroic individual resolutions. At least four workmen were needed to make it move and function each night, and in west (and wet) Irish weather, the number of working nights was severely limited. Thus the technology of the Leviathan helped define who would now count as a proficient astronomer. This demarcation became clear as Parsonstown was transformed into a site of pilgrimage. Visitors were encouraged. A guidebook, *The Monster Telescopes Erected by the Earl of Rosse* (1844), was produced and widely distributed. The gentlemen of science paid their respects. Robinson's political and intellectual ally, William Rowan Hamilton, was there as early as February 1835, composed sonnets in its honor, and was the first to point the three-foot to a celestial object.[66] After the meeting of the British Association at Cork in summer 1843, an entire deputation visited Rosse's observatory. In summer 1848, when Airy visited, he learned that "it is easy to see that the vision with a reflecting telescope may be much more perfect than with a refractor," though he also discovered the "astigmatism" of the telescope's field away from the meridian and at low elevations.[67] But these visits also reinforced the telescope's status as a uniquely "technical" object. This was just the term Airy used when dealing with an inquisitive lady who asked whether she could look through the instrument. "The appropriation of the telescope on a fine night to any body but a technical astronomer is a misapplication of an enormous capital of money and intellect which is invested in this unique instrument." In other words, tourists were useful so long as they stayed just outside a carefully bounded zone where Rosse, Robinson, and their closest colleagues debated and interpreted the work they made the instrument perform, and where these "technical astronomers" began to package its results for public consumption.[68]

The technologies embodied at Parsonstown made for an asymmetric distribution of power. Visitors to the great telescope were impressed. At Rosse's observatory, the authority of the Earl, of Robinson and of their team, was commanding. But elsewhere, in the learned societies, in the public prints and in the popular handbooks, their command over the lessons of the telescope was weaker. Complex deals were struck between the authors of the nebular reports and the managers and commentators in other, more public, forums. Part depended on a significant loss of prestige and authority by the Parsonstown team. This was especially relevant because that team itself was not homogeneous and did not have a single message to convey; and because the technologies of print and of picturing were highly malleable at just this period. The resolution of the Orion nebula into stars provides an example of the way these compromises and debates worked.[69]

Rosse's first report on the performance of his three-foot mirror was a paper sent to the Royal Society in late 1839. He gave full details of his heroic technology, but was hesitant about the criteria to be met for a nebular resolution: "in describing the appearance of these bodies, I am anxious to guard myself from being supposed to consider it certain that they are actually resolvable, in the absence of that complete resolution which leaves no room for error; nothing but the concurring opinion of several observers could in any degree impart to an inference the character of an astronomical fact." This caveat was carried in a footnote—the main body of the paper was far less cautious, promising near-resolutions of several significant nebulae.[70] But elsewhere the Parsonstown team treated Herschel's authority with less respect. Rosse told his friend, the Bradford priest and science lecturer William Scoresby, that "the nebulae were very diligently observed by Sir W. Herschel, who had ample funds, and the resources of London at command, that there has been no improvement in machinery, discovery in chemistry or optics since his time, which was applicable in any way to the improvement of the telescope. There is nothing in our machinery which could not have been as well made [then]." A history of telescope engineering helped Rosse undermine past achievements. John Herschel, for one, was disturbed by this public skepticism of the virtues of his family's reflectors. In a report to the Royal Society, Herschel pointed out that his father had used a very reliable polishing mechanism, had not relied on manual feel, and "had obtained by such instruments a complete, separate and independent command of the force" exerted on the mirror.[71]

This retort prefigured conflicts with the Parsonstown team. Rosse made it evident that a collective act of observation at Parsonstown was necessary, but it was also sufficient to justify his own "command." When he set up the one-piece three-foot mirror in autumn 1840, Rosse invited Robinson and the celebrated London astronomer James South to use the new instrument. The three men had very different interests: Rosse wanted to assure the status of his instrument; South remained a leading member of the metropolitan astronomical community; Robinson had the nebular hypothesis firmly in his sights. Robinson and South reached Parsonstown on October 29 and stayed one week. The very first night Robinson noted that this was a better machine than Herschel's. In every case, he compared what he saw with the published drawings, always finding in favor of the Parsonstown telescope; and in every case Robinson made resolvability the clear aim of observation. Of the Crab nebula he noted that "I feel confident that either a finer night or a few more inches aperture would actually resolve it beyond dispute," but South thought this might simply be due to "optical illusions produced by defective adjustment." This was another calibration problem. Robinson was prepared to use his own antinebular observations as a standard. South stayed with the Herschel drawings. "Perfection of figure of a telescope," he later noted in a widely distributed letter to *The Times*, "must be tested not by nebulae but by its performance on a star of the first magnitude." Even Robinson, however, agreed that Orion resisted these powers: "the nebula of Orion is very fine; the flocculent distribution of the light more beautiful but not the least suspicion of resolvability."[72] On Guy Fawkes night, November 5, South and Robinson acquired a set of South's fine eyepieces, "with whose quality I was familiar," and again proved the instrument superior to the Herschels' telescopes. Robinson was waspish on the theme. William Herschel's "almost absolute refusal to let anyone look through it [is proof] that it was indifferent in definition." He immediately continued with the implication of this former failure and Rosse's future success. Were Rosse to build a larger mirror he would succeed in "rectifying received opinions." That is, Robinson was already "doubtful of the existence of that assumed nebulous matter which is the basis of much speculation respecting the formation of stars and planets."[73] This speculation was Nichol's, and the Glaswegian astronomer and his radical allies became the immediate target of Robinson's program for Rosse's machine.

An opportunity for delivering a well-aimed blow against the nebular hypothesis arrived immediately at the Royal Irish Academy. Literary technology had to be managed at least as carefully as visual representation. A promise was extracted that Robinson "would not make an extempore harangue but that he would bridle his warm heart's feelings with having before him a written sermon." The promise was broken. With William Rowan Hamilton in the chair, Robinson attacked the Herschels for the incompetence of their instruments and their woeful drawings of the Orion nebula, he praised Rosse immeasurably, and despite vain efforts to dissuade him, told a cheering audience that "all nebular observations must be reformed." The Academy agreed to print a carefully revised version of this extraordinary speech, and suggested some studious revisions of the style, including the insertion of such qualifiers "as 'probably,' 'I think,' 'having the appearance of.' " But even in the watered-down version that was printed in the *Proceedings*, Robinson repeated the charge that Herschel's mirrors were polished by *feel*, and the claim that the nebular hypothesis "has, in some instances, been carried to an unwarrantable extent."[74]

The events in Ireland in November 1840 signaled the trouble in any transfer of observations and interpretations from the telescope to the public. South's hints at the right mores of literary style and Rosse's careful discrimination between private enthusiasm and public caution were swept aside by Robinson. In 1845, when the Leviathan was ready, Orion became its target. Robinson soon recorded in the observing book, and told the public, that "no REAL nebula seemed to exist among so many of these objects chosen without bias: all appeared to be clusters of stars."[75] South and Herschel also told their publics of the triumphs of the Leviathan, but neither astronomer was prepared to make the inference that this destroyed the nebular hypothesis. South denied that all nebulae had been resolved, while Herschel denied that the genuine nebular hypothesis was that peddled by Mill and the author of Vestiges. He sent Rosse new drawings of Orion, pointed out how dreadful was the Irish sky, and told the British Association, for example, that it was "a general law" that resolvability was limited to spherical nebulae.[76] The climax of Robinson's campaign was reached after Christmas 1845, when Orion was observable from Parsonstown, and Nichol himself visited Rosse, at least partly to arrange the transfer of some of Glasgow's best telescopes. He was the first to point the mirror at the Orion nebula: "I can recognise much of what I saw then, in the drawing by Sir John Herschel."[77] On March 19, 1846, however, Rosse wrote to Nichol that "there can be little, if any, doubt as to the resolvability of the nebula." As Nichol put it in print later that year, the Orion nebula now seemed little more than a "SAND HEAP of stars."[78]

This announcement, and Nichol's publication of it within months, did not and could not determine the response of Rosse's audience. The news that the Leviathan of Parsonstown had resolved the Orion nebula was interpreted in terms of the local purposes and immediate interests of each commentator. The impact of this news dramatizes the malleability and instability of early Victorian public debate. The linguistic technologies of pulpit oratory, carefully judged gentlemanly comment, and learned exegesis each offered different accounts of the presentations of Rosse's team. Nichol accepted the resolution, but denied Robinson's inference. Nichol drew his audience's attention toward Vestiges, a work fresh off the press, whose author many judged Nichol to be. While Chambers worked very hard in Vestiges and in its sequel, Explanations, to distinguish between the different types of evidence for the nebular hypothesis from zodiacal lights, from the coplanarity of the planets, and from Saturn's rings, Nichol worked equally hard to distance his version of the progressive cosmology from the unpleasant associations of these dangerous books.[79] The resolution of Orion had not affected its epistemic status. The law of progress was a necessary principle of the physical and moral universe. The hypothesis that this solar system was generated from a condensing cloud had "as much light as we may expect any object to be clothed with, when seen from so vast a distance." When Rosse began to report spiral nebulae, as he did in 1845, Nichol treated these new pictures as evidence of some kind of fluid present in interstellar space. The lessons of the Leviathan were reconcilable with Nichol's "science of progress" and its nebular creed.[80]

Just the same was true of the rival publicists David Brewster and William Whewell. Brewster treated the resolutions of nebulae reported from Ireland with extreme skepticism, until after the appearance of *Vestiges*, which he loathed. Then he used what Rosse and Robinson provided as fatal weapons against materialist atheism. Only now did Brewster announce, in contradiction to all his previous views, that the nebular hypothesis was "improbable in its very nature and gratuitous in all its assumptions."[81] In similar terms, Whewell, the author who coined the very term "nebular hypothesis," found Rosse's *spiral* nebulae indispensable evidence for the existence of nonstellar, and therefore nonplanetary, material in space. He wished to show, against Brewster, that this Earth was the only inhabitable world. So he asked Rosse for powerful "diagrams" of the spirals. Brewster then used the Parsonstown "resolutions" against Whewell's deployment of the Parsonstown "spirals."[82]

The responses of Chambers, Nichol, Whewell, and Brewster help show that no Victorian commentator was at a loss for evidence for or against true nebulosity on the basis of the Parsonstown work. Authority could not be secured simply by gesturing at the Leviathan or by citing its astonishing achievements. The credibility of the great telescope was the result rather than the cause of these rival representations. Robinson made the Leviathan divine. When it was completed in early 1845, like his friend Hamilton he composed verses in its honor: "Welcome to thy new existence, / Child of Intellect and Might! / Welcome to thy wide Dominions, Lord of Ether and of Light! / Thou shalt lead us on in triumphs, / Yet to mortal power unknown; / Realms which Angels only visit, / Shall yield hommage at thy throne." Brewster also made such instruments godly. In late 1844 he told his evangelical readers that "the telescope was never invented. It was a divine gift which God gave to man, in the last era of his cycle, to place before him, and beside him, new worlds and systems of worlds—to foreshew the future sovereignties of his vast empire—the bright abodes of disembodied spirits—and the final dwellings of saints that have suffered, and of sages that have been truly wise."[83] Nichol instead secularized the great machine. In a lecture in 1849, he compared the great but hidden purpose of the telescope with the great but hidden purpose of the division of labor in the British social order. At a conjuncture of major proletarian unrest, this was a valuable lesson.[84] Nichol helped reinforce the image of the astronomers' complex machine by linking it to Benthamite commonplaces about the fate of labor. He was acutely aware of the powers of the press: he recognized that by the mid-1840s nebular hypotheses "are discoursed of in almost every popular periodical, as having passed within the domain of common knowledge."[85]

This common knowledge was not consensual. Papers such as the *Times* and the *Athenaeum* carried stories about Rosse's telescope, the nebular hypothesis and the resolution of the Orion nebula, and the implications of *Vestiges* and its doctrines. The *British Quarterly* found evidence for an overall condensation in the resistance of the interplanetary ether, the medium through which light traveled and which affected cometary and planetary motion. The journal told its readers that "the sensible agents employed in effecting changes so majestic upon systems so immense" need not be "complex, incomprehensible and unmeasured"; and it warned that the best evidence of providence and divine Will was provided by "the commonplace, familiar, unobtrusive things beneath their feet." Stellar religion was the resort of atheists; natural history provided the basis of true faith.[86] The *Dublin Review* expressed the same careful skepticism toward the news from Parsonstown. It printed Rosse's drawings of two of Herschel's nebulae, noted that these showed how different these objects appeared in different telescopes, and concluded that "the observations hitherto made upon nebulae are too imperfect to form a safe foundation for any hypothesis." The pictures Rosse distributed were judged as evidence of the conjectural and pictorially variant representations of the heavens.[87]

In the later 1840s, Rosse's visitors made these representations serve their own purposes. Among these visitors was William Scoresby, the Arctic explorer, evangelical naturalist, and successful preacher. He needed Rosse's help to evict Methodists from his Yorkshire parish. He traveled to Parsonstown on several occasions, and was "permitted to have free access to, and examination of, all the observatory records and drawings." Against skeptics of the telescope's performance, Scoresby lectured throughout Britain on its triumphs of space-penetrating power, and helped lobby the government for support for Rosse's proposal to erect a Leviathan in South Africa. Scoresby reported that the Earl "will yield up his laboratory, machinery, and men, to the service of Government, and is willing, moreover, to give the direction and guidance of his master-mind."[88] Each commentator located the mastery of the telescopes in a broader field of rival authorities. The Leviathan could be seen as the representative of British power in Ireland. It could be seen as the representative of a celestial power much greater than itself. It could be seen as a puny attempt to challenge the overarching law of progress, or of providential development, visible in the cosmos. Each reading needed resources drawn from elsewhere in Victorian culture. No reading of the telescope was totally compelling. The public forum was the site at which the power of the Leviathan was made.

Resolutions and Conclusions

The means the public used to make the Parsonstown telescopes authoritative mattered because that authority was never secure within the astronomical community itself. It is not now believed that the Orion nebula is stellar. Its resolution at Parsonstown may have been due to the interposition of many small telescopic stars.[89] In the 1840s and 1850s, the resolution of the Orion nebula was not a compelling result. Instead, where it served local interests, it was acceptable as a means of calibrating other instruments; where it did not, it was queried or rejected. The great German savant Alexander von Humboldt, who had long conjectured on the condensation of stars from nebular fluid, wrote in summer 1848 that he would ignore the fact that "the Irish dissolver and our friend Brewster deny the existence of nebulosity, including that of Orion." A year later, he was even more explicit. Rosse had admitted that nebulosity remained in the Orion nebula, and, according to Humboldt, "even stronger telescopes, after having resolved what remains to us at present of nebulosity, will create nebulosity anew because they will penetrate further stellar layers which hitherto have escaped the observer."[90]

Others sought to match and, if possible, improve on the Leviathan's powers. Eminent were James Nasmyth, the great Manchester industrialist, who put his foundry expertise to work during the 1840s in casting mirrors, and the grand amateur William Lassell at Liverpool, whose new polishing machines and clock-drives allowed remarkable observations of Orion, which nevertheless seemed to resist all attempts at complete resolution.[91] Refracting telescopes apparently as strong were also available. In September 1847 the Harvard astronomer W. C. Bond installed a fine fifteen-inch refractor. He used news from Ireland to make his refractor worthy. To calibrate its performance he turned it on Orion: "the whole appearance of this nebula was altogether different from the representations given in Books—it is much more extensive, appears . . . sprinkled with stars." Bond recorded this in his observing diary, but in a letter of the same night to the university's president, Bond was much less reticent. He now claimed the nebula had been resolved, and pointed out to his patron that neither the Herschels nor, as far as Bond knew, Rosse had been able to achieve this. Bond had not heard, yet, of the March 1846 result in Ireland. So he used his own refractor's resolving power to make it credible: "I feel deeply sensible of the odiousness of the comparison, but innumerable applications have been made to me of evidence of the excellence of the instrument, and I see no other way in which the public are to be made acquainted with its merits."[92] The ambition of this claim was further clarified by later entries in Bond's observing book. In early October he recorded that although the Orion nebula looked different from Herschel's picture, yet "in some parts there are no indications of resolvability." Later in the month Herschel sent him the volume of Cape observations, and by December Bond had convinced himself that in Orion stars "are lost and won frequently, this sort of interrupted vision is common even in good nights." In public, Bond's telescope needed the authority it gained by beating the Leviathan of Parsonstown. Within the Observatory, however, the "winning" of the nebula was never so clear.[93]

Nor was the resolution of the Orion nebula announced from Parsonstown so sure. In summer 1851, the Astronomer Royal told the British Association that he denied the inference that "all nebulae are resolvable."[94] Many astronomers, including John Herschel and the astronomers at Pulkovo in Russia, led by Otto Struve, persisted in claiming observations of changes in Orion's shape. Struve told Rosse in 1853 that he reckoned that "the alleged miracles of resolution are nothing but illusions."[95] In the last major publishing venture of his life, a wide-ranging *Cyclopaedia of the Physical Sciences* (1857) distributed by his University's printer, Nichol repeated his opinion that "the additional power that resolves some nebulae, discovers others having the aspect of milky lights, and to the great telescope of Parsonstown, which can penetrate so profoundly into space, there appear a great many more unresolved nebulae, than can be seen with any other." As for Orion, though he much admired Bond's Harvard drawings, he could judiciously report that "we have not as yet obtained a full figure of this nebula from Parsonstown." In any case, he told his many readers, the Nebular Hypothesis of planetary origin was still well evidenced by solar atmospheres, zodiacal lights, and interplanetary ether.[96]

Such arguments for true nebulosity were given much point by the work of publicists of the 1850s, especially of the team at the *Westminster Review*. In 1857–58 the former railway engineer and hardworking journalist Herbert Spencer published there a series of articles on what he now called the "development hypothesis," using Laplacian nebular planetogeny to back up his evolutionary picture. "The nebular hypothesis works out beautifully," he told his father. Spencer later recalled that his article on "Recent Astronomy and the Nebular Hypothesis" was designed "to show the illegitimacy of the inferences drawn from Lord Rosse's disclosures." Spencer reckoned that even if the Parsonstown resolutions were accepted, this would not damage the nebular hypothesis. Like Nichol and Humboldt, he noted that stronger telescopes would reveal more unresolved objects. Alternatively, it could well be that all primordial fluid might now be condensed, and the telescope itself could not be viewed as arbitrator: "What are we to think when we find that the same instrument which decomposes hosts of nebulae into stars, *fails* to resolve completely our own Milky Way?" Spencer reinforced these reflections by consulting Herschel, Airy, and the Cambridge expert on the mathematics of spinning rings, James Clerk Maxwell, all of whom backed his skepticism toward the Parsonstown reports, even while reserving judgment on the radical implications Spencer drew from the truth of the nebular hypothesis.[97]

One convert to Spencer's evolutionist creed was the astronomer William Huggins, who made the spectroscopy of the 1860s into a vital technological weapon against the picture of nebulae as purely stellar. Huggins, a wealthy amateur with a fine observatory at Tulse Hill and allies among the London chemists, was entirely hostile to the picture of nebulae as "sand heaps," in Nichol's evocative phrase.[98] In 1864 he showed that some nebulae exhibited gaseous line spectra rather than the continuous spectra to be expected from a congeries of stars. "The riddle of the nebulae was solved. The answer which had come to us in the light itself, read: Not an aggregation of stars, but a luminous gas."[99] It was as hard for Huggins to make astrospectroscopy legitimate as it had been for Rosse to secure the great reflectors two decades earlier. Both sides tried to make their instruments representative. In May 1865 Huggins gave a dramatic lecture at the Royal Institution, insisting on the absence of any previous *observational* knowledge of the constitution of stars and nebulae, and indicating that even the Orion nebula revealed three bright lines in his spectrometer. So a new interpretation of Rosse's and Bond's observations was required. If their observations were to be accepted, Huggins indicated, "these nebulae must be regarded not as simple masses of gas, but as systems formed by the aggregation of gaseous masses."[100] The key term in Huggins's discourse was "gaseous." His spectroscope was represented as the only reliable means of calibrating gases. Resolutions and changes of shape were to be displaced by line spectra. In a key paper of the same year devoted to the Orion nebula, Huggins staked his claim: "the detection in a nebula of minute closely associated points of light which has hitherto been considered as a certain indication of a stellar constitution, can no longer be accepted as a trustworthy proof that the object consists of true stars."[101]

During the next three years Huggins, Rosse, and the London scientists debated the rival merits of the Leviathan and the spectroscope. There was no immediate closure on this issue. In June 1868 Rosse sent the Royal Society a long paper, composed by his son, summarizing observations and drawings of Orion produced at Parsonstown in the previous two decades. The paper included a new drawing of the nebula made by Stoney, whose credibility Rosse was careful to stress. He also tried to answer Struve, whose denial of the resolutions was now clear. He listed new stars and transcribed the Parsonstown observing notes. "There are several places where we have reason to suspect that a change of form may have taken place in the nebulosity since our observations commenced."[102] After a survey of Herschel's and Bond's stories about the nebula, Rosse printed the remarks of his draftsman, Hunter, claiming that the main region of the nebula was "clearly resolved," and providing his readers with instructions for seeing Hunter's drawing as evidence of this great fact. "This resolvable appearance can be seen on good nights only, and with a very good speculum." The implication was that only one observatory, Parsonstown, would ever capture this appearance. The astronomers there had acquired a spectroscope; Rosse's son and his tutor Robert Ball, a future astronomer of great celebrity, worked the instrument. They confirmed that the Orion nebula gave the line spectrum Huggins had already reported. But they had resources for explaining away this fact. They alleged that because the brightness of a spectral line fell with increasing width, a continuous spectrum, if present, would be very much fainter than a few, bright, gaseous lines. "We can hardly expect that the remaining light could produce any but the feeblest continuous spectrum." In other words, no spectroscope could calibrate the clear deliverances of Hunter, Rosse, Ball, and their colleagues.[103]

Rosse sent Struve and Huggins copies of this paper. Huggins continued, eventually successfully, to insist on the power of the spectroscope and its ability to pick up a continuous spectrum. Struve was sterner. He gave Rosse a lesson in the meaning of the word "resolvability":

In my opinion if a nebula is resolvable it will offer the same appearance on any occasion when the images are sufficiently favourable. Thus admitted, your own observations show that, with regard to the central part of the nebula of Orion, this term ought not to be applied, for in different nights you see traces of resolvability in different parts of the nebula. . . . I should say instead of resolvability, there is a tendency of the nebulous matter to form itself in separate knots, sometimes in this, sometimes in an other direction.[104]

By 1880, all nebular astronomy had ceased at Parsonstown. Initiative and authority had already shifted elsewhere, to new disciplines, such as astrophysics, and to new institutions, such as the observatories at Kew and South Kensington. The Leviathan's career in the mid-nineteenth century dramatizes the tenuous resources such institutions could command. There was no moment when consensus reigned amid astronomical communities and their audiences. The picture of science as an esoteric zone, where facts are made, and an exoteric zone, where they are consumed, is wrong. Following Ludwik Fleck, we can say that facts and representations are formed in the spaces between these zones. Facts become robust by drawing both on esoteric private work, produced in labs, observatories, and technical institutions, and on the general strictures of popular culture. Fleck persuasively pinpoints the significance of the *vividness* of this more exoteric realm.[105]

This gives some point to our initial comments on the restricted scope of the model of literary technology. Every literary form of fact-making is linked with local complexes of technical and social practice. Representations of nebulae varied immensely between these locales. The representative quality of the Leviathan varied too. To stabilize any story about the nebulae it was important to define the role of the telescope, the observers, political and moral cosmologies, and well-characterized audiences. Stories about nebulae were potent and mutable parts of the business of Victorian Britain. The Leviathan of Parsonstown can thus be juxtaposed with other contemporary technological enterprises to engineer the universe, such as Charles Babbage's calculating engines in the 1830s and 1840s. Like the great reflector, these machines were the costly and spectacular concerns of a wide range of communities—artisans, machinists, entrepreneurs, journalists, and gentlemen of science. Well aware of contests between the heroic author-inventor, Babbage, and the precision engineers whose "rules of trade" made it unclear whose property the calculating engines were, Rosse himself nevertheless tried to mobilize the Royal Society and other institutions behind Babbage's campaign for the mechanization of intelligence. In both cases, the credibility and authority of the engine relied explicitly on the workings of publicity, marketing, and the workshops.[106] Careers of such devices as the telescopes and the calculating engines in this period show how to situate technologies in historical settings. Like other devices, the Leviathan's authority was not straightforwardly dictated by a noble engineer and his visions, but was produced with the various literary, material, and social technologies of early Victorian Britain.

Technology, Aesthetics, and the Development of Astrophotography at the Lick Observatory

According to articles and books published at the end of the nineteenth century, the introduction of photography to astronomy was one of the most notable events in the discipline during a century not lacking for notable events.[1] Edward Holden, director of the Lick Observatory between 1887 and 1898, was rapturous about the promise of photography: it would simplify observation, increase the reliability of data, and produce permanent records of the heavens untainted by distraction, ill discipline, or bias. This would happen, he argued, because of the mechanical virtues of the camera:

It does not tire, as the eye does, and refuse to pay attention for more than a small fraction of a second, but it will faithfully record every ray of light that falls upon it even for hours and finally it will produce its automatic register . . . [that] can be measured, if necessary, again and again. The permanence of the records is of the greatest importance, and so far as we know it is complete. . . . We can hand down to our successors a picture of the sky, locked in a box.[2]

Likewise, an anonymous writer in the *Edinburgh Review* predicted that photography would bring about "a new birth of knowledge regarding the structures of the universe." Central to this new astronomy was the relationship between astronomers and stars mediated by photography: in this new order, "stars should henceforth register themselves," an unforgettable phrase that suggested both stellar complicity and photographic neutrality. It goes without saying that aesthetics would not even exist in this new order. "Their pictorial beauty," the *Review* said of astronomical photographs, "is the least of their merits," and comparisons of photographs would be made without considering whether one was more beautiful or striking than another.[3]

Embedded in these accounts was a view that the changes photography brought to astronomy were at once revolutionary and evolutionary. On the one hand, it opened up new career and research paths, forced changes in the design of instruments and the architecture of observatories, contributed to the development of astrophysics as a discipline, and readjusted the moral economy of astronomical practice.[4] On the other hand, photography realized nineteenth-century astronomers' dream of producing visual records that would let them describe the heavens with mathematical precision, and detect changes over decades or centuries in the appearance of any celestial object.[5] Because both aimed to produce realistic representations of the heavens, drawings and photographs could be evaluated according to the same standards, and comparisons quickly revealed that photographs were far superior to anything produced by astronomers working with telescopes and sketch pads. The camera's light-gathering power gave it the ability to resolve objects that the human eye never could; the photograph's permanence assured that it would be useful decades or even centuries after it was taken; and photography's mechanical nature made it a disinterested, objective recorder of the heavens. More broadly, the development of astrophotography can be seen as part of a larger eclipse of human observation by mechanized technologies of imaging described by Lorraine Daston and Peter Galison. They argue that a central characteristic of late nineteenth-century science was a search for mechanical substitutes to human observation, and a growing distrust of forms of visual representation dependent on human judgment. Big instruments and rapid photographic plates not only revealed more than the sharpest eye and best telescope; they offered the possibility of building a completely mechanical and automated system for producing pictures, a closed system that would transmit pictures from the observatory to the pages of the journal without ever being touched, interpreted, or otherwise corrupted by human frailties.[6]

But as historians of science know, although public accounts of scientific triumph are interesting as literary performances and political declarations, they can rarely be trusted to tell the whole story. The unpublished correspondence of astronomers involved in the development of astrophotography tells a very different tale from what appeared in the contemporary literature. Making and reproducing pictures of nebulae, star clusters, eclipses, comets, and galaxies was not easy. There were countless technical problems to be solved, new practices to be learned, and connections to be managed with instrument-makers, chemical suppliers, and printers. Relations with printers were particularly important, for photographs made at the telescope could not move from the observatory into the wider world without passing through their hands. Letters between them and astronomers are especially interesting, for they reveal a world in which pictures are oddly plastic, images are born and die in baths of acid or under a retoucher's tool, the real and the unreal are sometimes indistinguishable, and the words used to describe photographs are unfamiliar. Read as a whole, they show that although photographs were intended to be realistic pictures of natural objects untouched by style or bias, technical difficulties prevented them from being rendered entirely by machine. In truth, the reliability of astronomical photographs depended on the successful use of aesthetic standards as guides in their production and evaluation, and the judicious use of experienced artisans to correct the errors of mechanical processes. The aim was to craft pictures that signaled their objectivity by conforming to the proper aesthetic standards and by being so skillfully worked that they betrayed no evidence of human intervention.

Normally these processes of image-making and -reviewing remain secret, and the words used to describe photographs, when heard after almost a century, remain the stuff of confusion and conjecture. But on a spring afternoon in 1992 in the archives of the Lick Observatory in Santa Cruz, California, I discovered a window that let me watch the processes and decipher the words. Gathered into a file with the unpromising title "Miscellaneous—P, 1906" was a cache of letters between the director of the Lick Observatory and the Photogravure and Color Company of New York City. They documented the Company's attempts to produce plates for an atlas of star clusters and nebulae photographed by the Observatory's late director, James Edward Keeler. The papers had lain untouched and unread for some twenty years. Other files with drab titles like "Photography 1904–1911" and "Miscellaneous Bin–Biz 1874–1904" yielded more letters about the Observatory's dealings with photographers and engravers, documenting their interactions and illuminating the problems of making and reproducing astronomical photographs. They show how astronomers talked to one another about, and made choices between, different instruments and printing options. They reveal how difficult it was for printers to reproduce astronomical images, and how important human intervention was in maintaining the stability of mechanized printing processes. Finally, they show how astronomers looked at photographs, and the role aesthetics played in shaping responses to images.

This essay uses this material to show how astronomers at the Lick Observatory made and made sense of photographs between the 1880s and 1920s. It is organized as follows. First, I consider some of the issues involved in the study of scientific representations and argue for the importance to historians of studying the ways they are produced and reproduced. Next, I describe the major printing technologies of the day and explain how astronomers chose between them. In particular, I look at the technical problems in photoengraving that kept them from being simple copies of original photographs, and which created a need for management and intervention by engravers. I then turn to the process of producing reproductions of photographs. I describe the various problems encountered in the process, and show how aesthetic standards defined the quality of pictures and helped assure the stability of the medium. In particular, I show how in a period in which human judgment and skill were looked upon with increasing suspicion, aesthetic standards were still needed to define what good pictures were. In fact, it turns out that a central irony of the desire for intervention-free pictures was that aesthetic standards and craft knowledge were necessary to assure the reliability of mechanical reproductions. The challenge was not to banish aesthetics and craft skill, but to define and use them in a way that made them self-effacing.

Studying Scientific Representations

The history of scientific images stands uneasily in a borderland between the history of ideas, the history of technology, and the history of art. Until recently, that meant that it was ignored by most historians of science; Lynn White's comment that "it is strange that historians of science have paid so little attention to the development of the visual arts," still held true 40 years after it was made in 1947.[7] The same cannot be said today: scholars are now busy examining images and representational practices in fields as diverse as Renaissance natural history, early modern astronomy, Victorian geology, and modern biology. That much of this work is aimed at examining "images" or "representations" rather than works of art or even "pictures" tells us something about how it has developed. The very malleability of the term "representation" has attracted some—after all, what cannot be considered a representation?—as has the promise of opening up a new field for the application of literary theory or cultural studies. Fruitful though this work has been, I believe that it falls short in an important respect. Images are interesting for the information they contain, and can be "read" in some ways like books, but they never exist apart from the technologies and media that create them; yet to date very little attention has been given to that fact, to considering its historiographic consequences, or to exploring the material life of pictures.[8] The application of literary criticism to the study of pictures must be undertaken with particular care, for choices about printing processes and materials affect the content of images in critical ways, and relationships exist between images and media that can hardly be said to exist with text. Images of nature do not effortlessly move from the sketch pad or photograph to the engraver's steel plate to the printed page. Every change of state requires a reconstruction of the image, and those moments of reconstruction are full of dangers: a photograph taken in the field can be overexposed, an artist might unconsciously alter some critical detail while copying an original, an image accidentally dissolves in an etcher's acid bath. To write a fully informed history of scientific representation, we must enter the world of pictures and picture-making, and examine the interaction of materials, practices, and interests during the creation of original pictures in the laboratory or observatory, the copying of pictures in the artist's studio, and the mass production of illustrations for journals, atlases, and books. This is just one of several opportunities that the history of scientific illustrations offers for developing a history of science that reincorporates aesthetics, artistic talent, and other things that are often given little attention in current scholarship.

To some, pointing out the materiality of pictures will seem a demonstration of the self-evident and unexceptional. But recognizing that images are *things* as well as ideas allows us to see that as much can be discovered from looking at their material lives as from analyzing their formal qualities and their resemblance to their predecessors. It encourages us to ask questions about what technologies were used to make pictures, who copied and printed them, and how things like artistic training, artisanal skill, and theoretical commitments shaped their content. It forces us to distinguish between different stages of an image's life cycle, and to be alert to changes in its appearance in each stage. It may also encourage a greater self-consciousness about the difference between "images," which can be thought of as visual messages or information, and "pictures," which are images embedded in materials. To date, we have written histories of images. I propose to write a history of pictures.

The proximity of the study of visual representations to the history of art also recommends that we look for opportunities to import analytical and critical tools from that discipline into our studies. The works of Bernard Smith and Samuel Edgerton are elegant, excellent models of how the art historian's eye can be applied to scientifically important pictures.[9] More generally, we should create a place for aesthetics in judgments about scientific illustrations, and even photographs. We tend not to think of photographs, particularly photographs of things like galaxies, as being produced according to aesthetic standards, or to consider such standards part of their scientific value. However, in an interesting study that combined anthropology, art history, and science studies, Edgerton and Michael Lynch have argued that even contemporary astronomical images—digitized, abstract, and infinitely more malleable than photographs—reflect the aesthetic standards of their makers. Their study reveals that researchers and technicians draw on their knowledge of a detector's flaws and eccentricities, their software's abilities, and their own experience with the "characteristic profiles . . . of astronomical entities," to build a practical aesthetics of image-processing. This aesthetics is somewhat different from that of the arts, as they explain, but its elaboration and use deserves attention:

What aesthetics means [in astronomy] is not a domain of beauty or expression which is *detached* from representational realism. Instead, it is the very *fabric* of realism: the work of composing visible coherences, discriminating differences, consolidating entities, and establishing evident relations. . . . This hands-on process of "interpretation" can be treated as an art situation within the performance of scientific practice.[10]

I will argue that the development of a similar aesthetic helped stabilize astrophotography in the late nineteenth and early twentieth centuries, and became a resource for turning natural images into natural facts.

At the same time, I do not want to argue that because these pictures were shaped by social and aesthetic factors they reveal nothing about nature, and can be read as political monuments that neither refer to nor contain information about their natural subjects. This is certainly an argument that one can make, but rather than make the unmasking of claims of objectivity and naturalism the end point of my analysis, I want to show how my subjects worked through the uncertainties and problems inherent in scientific research to produce pictures that could be used as representatives of nature. The assumption that because scientific practices are social they are nothing *but* social misses what is really interesting: not how practices are "truly" irrational, but how they come to produce things that are reasonably trustworthy. Astronomers spent years negotiating practices that might not stand between nature and the observer, methods of intervention that would erase intervention, collaborative strategies that would turn malleable visual records into solid visual certainties. There is an obvious irony in this; but it worked, and produced pictures that are recognizable and draw admiration from scientists 90 years later.

Photoengraving Technology, Craft Skill, and the Problem of Intervention

Original photographs were generated in the observatory, at the telescope and in the darkroom. To be circulated among more than a handful of people, however, a photograph had to be published, and this required that it be engraved—move from the observatory to the photoengraver's studio and press, the image transferred from collodion on glass, to the steel-faced copper plate, to the printed page. Before discussing how astronomers made decisions about reproducing pictures, therefore, we should examine the technologies they had at their disposal, understand how they worked, and investigate the relationship between original photographs and photoengravings. Most astronomical illustrations were printed as halftones or photogravures, so we will begin by reviewing these two technologies.

Halftone printing began with the photographer making a special negative of the image to be printed. Negatives were made on wet plates, mixed by the photographer according to his own personal (and often secret) formula. Dry plates were faster to use, but were not "as rich in contrast, transparency, and sharpness."[11] The image was shot through a halftone screen, and the camera was equipped with a diaphragm. The shape and size of the diaphragm determined the negative's sharpness, intensity of contrasts, and depth of shadows. More than one diaphragm could be used to achieve different results in bright and dark areas, though one writer warned that "skill in manipulation is necessary to achieve satisfactory results."[12] Contrasts were determined by the size of the half-tone dots, and the degree of their connection; as long as you did not look too closely, you would not see the dots, but fields of grays or blacks. The screen broke up the image into a field of pixels, and the amount of detail in a halftone depended in large part on the density of the screens.[13] While the photographer shot and developed the negative, an assistant polished a copper plate and coated it with enamel or a solution of albumen and bichromate. After the negative was developed, it was stripped off the glass plate and put back on reversed; meanwhile, the copper plate was warmed over a Bunsen burner until the enamel turned brown. The negative and copper plate were then clamped together in a wooden frame, and exposed to a light source for several minutes. The parts of the enamel exposed to the light through the negative would harden. Because the copper plate was underneath a negative, the lights—now a field of halftone dots—corresponded to dark spaces in the original photograph. When it was finished, the copper plate was removed from the frame and the soft unexposed gelatine (the lights in the original) was washed away. The plate was then covered with ink and a dark red powder called dragon's blood, and heated. The plate was now ready for etching: the dragon's blood melted and fused to the hard gelatine, forming an acid-resistant layer that slowed the etching process and made it easier to control.[14] The plate was immersed in a tray containing a strong solution of perchloride of iron. The acid bit first into the most exposed areas of the plate, while the areas under the gelatine and dragon's blood were protected. As the etching progressed the acid turned red, and the plate was brushed occasionally to dislodge any sediment. The acid's progress was monitored by scratching off a small area on the plate's margin and feeling the depth of the bite with a fingernail. It would be shallow compared to a relief line engraving or woodcut, otherwise the dots would flatten or collapse when squeezed by the press.[15] When the acid had done its work, the plate was removed from the bath, rinsed and cleaned, and the asphalt taken off the back. The result at this stage was referred to by printers as a "flat etching." At this stage, the first proof from the plate was made and examined for flaws. If one area was too light, for example, the acceptable areas were covered with asphalt, and the exposed section deepened—and hence made darker—with a routing tool or more acid.[16] After the finisher made corrections, another proof was drawn and examined, and further alterations were made if necessary. The finished plate was coated with a thin layer of steel to prevent degradation during printing.

The other major reproductive technique used in printing astronomical photographs was the photogravure. Unlike halftone printing, which had a number of excellent firms, work in this area was dominated by a single company, the Photogravure and Color Company.[17] Its founder, Ernest Edwards (1837–1903), was one of the most famous nineteenth-century printers. The son of a headmaster of King's College, London, Edwards served in the Crimea and studied at Cambridge before turning to photography and invention. He emigrated to the United States in 1871, living first in Boston, then moving to New York in the 1890s. His reputation in America was made in high-quality art reproductions and popular book illustrations: he printed a series of Dürer, Rembrandt, and Marc Antonio engravings using the heliotype process, a process Edwards invented himself.[18] By the time he began work for the Lick Observatory, Edwards also had considerable experience in scientific illustration, having worked with Alexander Agassiz, director of the Museum of Comparative Zoology at Harvard, and with Eadweard Muybridge on the illustrations of *Animal Locomotion*.[19] Photogravure was widely regarded among nineteenth-century collectors as the finest and most demanding printing method ever created. It was praised for its "softness, richness of ink impression, [and] lack of visible screen," and its "enormous capacity . . . for smoothly graduated chiaroscuro."[20] Edwards described the process in a talk before a workingman's institute in 1887. He began by making a glass positive from a negative of the image. The positive was made on "collodion, on a gelatine dry plate, or on carbon . . . according to the subject and the quality of the original negative."[21] Collodion plates were probably used for the most finely detailed pictures. A second glass plate, coated with light-sensitive gelatine, was attached to the positive, and an exposure was made. The gelatine turned hard and water-resistant in exact proportion to the intensity of the exposure, and thus the picture was fixed in the gelatine. It was then peeled off the glass and affixed to a sheet of copper with an aquatint ground, a fine random granulation. This copper plate was the medium from which the prints would eventually be taken. Its back was coated with asphalt varnish to protect it from the etching acid, and it was set in a tray containing a solution of ferric chloride. "The action of the bath commences first . . . [on] the shadows or darks of the picture . . . where the gelatine has not been affected by light," Edwards explained. "The lights are protected from the action of the chloride, the gelatine covering them having been water-proofed by light, whilst the half-tones being partly water-proofed are partly protected."[22] After the acid etching was complete, the plate was dried, the varnish removed from the back, and the plate was steel-faced by "depositing an infinitely thin layer of steel all over the surface of the plate, thick enough to preserve the copper from injury, but thin enough to retain the very finest gradations."[23] By steel-facing the plate, the original imprint was protected and its life span extended almost indefinitely. The plate was now ready to be printed.

   Even though these were mechanical processes, they could not function without human oversight and correction. In both halftone and photogravure, engravers worked with such hardy materials as copper plates, asphalt, acids, routing tools, and burnishers, but they needed sharp eyes and steady hands. Careless handling or unfamiliarity with a process's eccentricities could ruin a plate. Each process also had its own special problems. Dragon's blood was difficult to spread evenly, and the dusted plate had to be warmed slowly for the powder to melt right, otherwise it would etch unevenly and produce artificial patches of light and darkness. Unfinished halftone plates were always handled by the edges, for the "slightest touch of grease from the skin will disturb the evenness of the etching process."[24] Experienced photographers cleaned their halftone screens more often during the winter, because water vapor condensed on them more quickly and made negatives foggy and indistinct.[25] Further, at every stage of the process engravers confronted technical choices that would affect the quality of the finished image. Negatives were retouched by artists who drew in lines, sharpened contrasts, and deepened or lightened shadows; this work was important and legitimate enough to have specialists, with their own tools, studio space, and manuals comparing methods used in different countries.[26] Newspaper illustrations were made with screens of 100–130 lines per inch, halftones for magazines with 150–75 line screens. Screens with more than 300 lines per inch required "exceptional facilities and skill in printing," the finest glossy art paper, and stiff ink. Astronomical halftones were made with screens as fine as 400 lines per inch.[27] Making proofs of flat etching was another technical step that was more difficult than at first sight. It was critical that this proof be well made, for it would serve as the basis for all further reworking of the plate. Flat etchings were thus hand-drawn by a pressman who specialized in this work, and who often also learned something of hand-drawing and retouching techniques to improve collaboration with finishers.[28] A large number of variables—the thickness of the ink, the dampness of the rollers, the quality of the paper, the pressure from the press—affected the quality of the proof, and experience and skill were needed to make one that "accurately" represented the plate. "Here is a large scope," one engraver commented, "for the artistic perception of the machine-minder."[29] Having a pressman and finisher who worked well together helped guarantee good results. Finally, like negatives, flat etchings were retouched by specialists, who worked with their own tools and techniques.

Likewise, photogravure printing involved careful work, selecting among different technical options, and manipulating an image that was highly plastic. The image could be altered, intentionally or unintentionally, at nearly every stage of the process, and the quality of those alterations depended largely on the artist's skill. The glass positive could be retouched "to any desired extent" before being copied. The etching process was especially tricky. "Almost everything depends on the skill and judgment of the operator at this stage of the process," Edwards declared of the etching. Experience and a sharp eye were needed to determine the strength of the acid bath and duration of the etching, for "a mistake made now cannot be rectified subsequently, and a mistake now is very easily and very quickly made." Finally, the intaglio plate could be altered by hand before steel-facing, but the finisher who did this work, Edwards said, "must always be an artist. A great deal may at this stage be done to the plate in the way of correction and alteration, and a very great deal may be done to assist in giving the plate the proper values of the original." Consequently, this was highly specialized work: even an etcher or engraver accustomed to retouching line engravings "makes a poor show when attempting to finish a photogravure plate. It is only an adept who knows how, to dare to alter the marvelous etching of the sun." Finally, the steel-facing had to be done by a "skilled printer who is also an artist."[30]

Even the actual printing could not be entrusted to a machine working alone. E. E. Barnard's experience printing photographs of Comet Morehouse (destined for the December 1908 *Astrophysical Journal*) illustrates the difficulties. "The proofs were excellent," he told Lick Observatory director Wallace Campbell, but the prints themselves were "a great disappointment." A trip to the printer revealed the problem. The printing was done with a steam-press, so the slight adjustments in inking and pressure that could assure quality in a hand-pressed run could not be made. In the interests of getting better results with another picture that would appear in the January issue, "I spent an hour or so at the press with the printer, trying to get the proper depth of printing, etc., but as they could not print them singly . . . it was a matter of give and take."[31]

Danger and opportunity thus coexisted in the moments of etching, proofing, retouching, and printing. At those moments the image became unstable and malleable, and it could either slip out of control, dissolving in acid or under the artist's burnisher, or stabilize and be secured for the next stage of its journey. This was the central problem in astronomical printing, and one of the critical issues in scientific representation: mechanical processes could not succeed without skilled intervention, and one had to *define* what kinds of intervention constituted improvement versus alteration. Halftone photographers had to choose different screens, diaphragms, light levels, plate formula and exposure times for each subject; proof printers had to calculate which inks, papers, and presses would give the best representation of the flat etching; retouchers had to judge how much a certain detail should be enhanced or contrasts redrawn. Even during the "chemical manipulation" in which an image was etched onto a plate—the only method Campbell authorized for engraving the main subjects of a plate—engravers varied acid strengths and immersion times, and made decisions about how far to let acid bite into different parts of a plate. The etching process operated without attention to aesthetics or visual languages, but it was not free of human judgment. Quite the contrary: human judgment was required to make it work properly. More generally, all of the decisions made in the engraver's studio were directed to the purpose not of changing the appearance of images, but of maintaining it, making sure that the images did not change as they were inscribed and reinscribed on one plate and another. To have removed humans from the process would have been to guarantee their failure. Cronenberg pointed to the irony of the importance of human experience and judgment in making halftones, which above all were advertised as "mechanical" products, when he warned that engravers must at all times "avoid working like a machine" for the process to succeed.[32] Likewise, American engraver Carl Nemethy argued,

A good plate cannot be produced by accident, nor is it a simple mechanical result. There is much skill, taste, knowledge and talent required to make a good negative as there is to any other kind of work; and a good etcher is fully entitled to be called an artist. . . . The machines do not work alone and the fine outfits and choice chemicals do not turn out perfect plates. Skill, practice and a thorough knowledge is required to have a first class result; without these all else is useless.[33]

It might be argued that astronomers might not have known all this, that they were content to keep printing black-boxed and unproblematic. How engravers and printers handled printing problems and made choices between methods, after all, was their business so long as the pictures had good tonal contrasts, lots of detail, dark backgrounds, and plenty of snap. But astronomers *were* aware that graphic skill and experience were essential for issuing epistemological warrants covering inscriptions and re-inscriptions; that is why they requested certain experienced workmen be assigned to their projects. "I wish you would give your *personal* attention to each" picture, Holden told a printer making plates of the lunar surface, "to make it *the best possible*."[34] Positives of Halley's Comet were sent to a Chicago company with a note saying, "the subjects are pretty difficult, and I hope that we may have the personal interest of the engraver."[35] Fifteen years later, another set of photographs "of considerable scientific importance" were sent to the same firm with instructions to "have one of your experts" choose the positives best suited for reproduction.[36] On a large project like an atlas, it paid for printers or shop owners to visit observatories to see firsthand "what was really wanted."[37] When his company went into receivership in the middle of a project, Ernest Edwards took pains to assure the Observatory that "the expert artists and workmen who have worked with me for so long will remain" in a newly formed company, and no novice engravers would be assigned to the project. Edwards's letter suggests that the same etchers and printers were often given all the astronomical work that came into a shop, and became expert in the special challenges of astronomical printing. In fact, some engravers were so well known that astronomers asked after them by name. When ordering plates of the Andromeda nebula, Campbell asked the C. L. Wright and Company if "Mr. Schmidt," the printer already "acquainted with the requirements of reproduction in astronomical work," would be available to oversee the work.[38] (When that company also went bankrupt and was reorganized under the name Royal Engraving, Campbell wrote to ask whether "Mr. Schmidt, who did the former nebular work for us," and the fine-screen equipment were both with Royal; Wright reassured him that "the fine screen and Mr. Schmidt and I are all in one place.")[39] The Chicago Photogravure Company printed a number of astronomical subjects for the Lick and Yerkes observatories because its manager, A. B. Brunk, acquired a reputation as a skilled and careful printer.[40] Campbell made a point of asking that "Mr. Brunk . . . see what he can do" with a set of eclipse positives sent to the company in 1915.[41] E. E. Barnard, who worked closely with Brunk when the Company printed his Milky Way photographs, wrote that the printing succeeded because "Mr. Brunk has given his personal attention and a remarkable devotion" to the project.[42]

The character of photoengraving processes meant that good halftones and photogravures could not be completely free of human intervention. But there was a final complication in the relationship between a block and an original negative: photograph and illustration existed in a symbiotic relationship. Engravers were supposed to produce illustrations that were faithful copies of photographs, but photographs had to be altered to make them easier to copy. In particular, printers asked astronomers to vary the strength and density of glass positives to suit the demands of engravers or the eccentricities of different processes.[43] One company expressed a "very great objection . . . to dense positives," explaining that their "shadows are so very dense that we are unable to get any detail" out of them.[44] Barnard thought the Lick Observatory's glass positives "too thin and weak to give the best results."[45] He went on to explain that there was "quite a difference" between the two qualities: thinness or density referred to the level of contrast and brightness in a positive, weakness or strength to the amount of detail the positive retained. Thus strong positives could be thin, but still have "sharp" and "definite" details.[46] Astronomers and printers agreed that strong positives were best, but printers' preferences about thinness varied. One asked for positives so gray they "astonished" Campbell.[47] When working with paper positives, some firms also preferred to work with enlarged images, because "in the reduction we strengthen up the picture somewhat and eliminate . . . irregularities and roughness."[48] Another company that worked for both Lick and Yerkes felt that "we can get a more faithful reproduction from . . . a glossy velox print . . . [than] positives on glass. . . . The reason is that our engravers working from the velox print have copy which serves as a guide for tone, color, variation, etc. In the case of the glass positive it is of course necessary to observe this through transmitted light and this is deceptive."[49] Campbell, however, replied that "all of our best results in the past, in the way of half-tone cuts, have been made from positives on glass rather than the prints."[50] Likewise, astronomers might tailor positives to suit their own purposes, or to accentuate the success of capturing a difficult subject or virtuosity of a new piece of technique. Glass positives of a little-known subject, for example, might be "developed with a view to obtaining the best general representation of the nebula" rather than showing the full extent of its outer edges.[51]

Choosing a Method and Managing the Process

Astronomers traded information with one another about their experiences with different printers and printing methods. An especially good photograph might prompt a query about who printed it; similarly, a bad experience would serve as a warning to the circle that a printer could no longer be trusted.[52] From these exchanges we can reconstruct how they decided to reproduce pictures generated in the observatory. Conveniently, the fortunes of a single photograph allow us to see how astronomers evaluated these different processes. In 1899 James Keeler made a photograph of the Trifid Nebula, one of the best-studied nebulae in the Northern sky. It was an exceptional picture; George Ellery Hale reported from the Yerkes Observatory that "everyone . . . considers it far superior to anything . . . they have ever seen."[53] He sent a glass negative of his Trifid Nebula plate to the Photogravure and Color Company in early 1899, for publication in the *Astrophysical Journal* and *Publications of the Astronomical Society of the Pacific*.[54] In early March, Keeler received a proof, which he thought "quite satisfactory, though of course not equal to the glass positive."[55] The actual prints were not quite as good. The contrasts and tone were muted somewhat, and Hale found them "unnecessarily harsh." The image had literally degenerated on the printing press. "This was precisely the trouble that led me to give [them] up," Hale explained. "The contrast tends to increase with the number of plates printed. I believe that by means of the very best halftone methods decidedly better results could be obtained."[56] As a result of this and similar experiences, Hale favored the relief halftone: they did not have as much detail as photogravures, but after years of editing the *Astrophysical Journal*, he had come to the conclusion that a photogravure was fine in proof, but the delicate shadings and details could not survive a run of a thousand plates. Proofs of relief halftones, in contrast, "fairly represent the average result to be obtained . . . [and there is] no special change in a halftone block during the printing of a few thousand plates."[57] Keeler agreed that the quality of the plates was uneven, but "I was not unprepared for this," he replied. "The specimen copies . . . are always good; but the trouble is to get a good lot. Moreover, a nebula is a tough subject."[58] In his view, the delicacy and detail the best photogravures provided made them worth the trouble and uncertainty. He told Edwards he was "pleased with the photogravures you furnished, and hope some time to send you a larger order."[59]

The Trifid Nebula photograph was used again when Campbell selected a printer for volume 8 of the *Lick Observatory Publications*, a series of nebular photographs taken by James Keeler with the Crossley reflector. This was important work for several reasons. Keeler had rebuilt the Crossley reflector for photographic use, and the pictures, which were on the cutting edge of astrophotography, were demonstrations of his engineering and astronomical talents. Finally, Keeler had died during the project (in 1900), and been beloved at the Lick: the quality of the work and its incompleteness gave the volume a certain poignancy, and made it a memorial to his memory and talents. In February 1902 Campbell wrote to engravers in Chicago, New York, and Boston, asking for samples of their work and estimates for a volume containing 60 or 70 prints. All had either worked for the Lick before or been recommended by friends.[60] Several responded that spring, proposing to use either fine halftone, heliogravure, or photogravure.[61] By the fall of that year, the leading contender was C. L. Wright, a New York firm that had worked for the American Museum of Natural History. Wright himself promised engravings of "the very highest [quality] ever seen in any scientific work," and even agreed to have an Observatory member work in his shop, but the work was too expensive for the Observatory's budget.[62] (On the strength of their proofs, a second company received orders to engrave several subjects in the next two years.)[63] Campbell resumed his campaign in 1904. After securing funding from private subscribers and the state, he returned to the search for a printer.[64] He sent copies of the Trifid Nebula photogravure from the *Astrophysical Journal* and glass positives from the original negative to several companies, along with instructions on how it could be improved. "The bright parts of the nebulae" were represented well in the photogravure, he thought, and generally the picture had a "bright and snappy appearance," but "all the faint masses of nebulosity have been lost." In their test plates, "all the faint details shown on the glass positive should be in the reproduction."[65] All got mixed reviews—two were discolored but captured "the extent of the nebulosity beautifully" and had a "snappy character," another was a good black and white but showed limited nebulosity—but the contract finally went to the Photogravure & Color Company.[66]

After choosing a method, there were still plenty of mundane problems and details that needed tending. First, negatives had to be copied and prepared for shipment. Original negatives, which would have yielded the most perfect copies, almost never left the Observatory and were therefore inaccessible to printers. Campbell made only one offer to an engraver to work from originals, and then only if an assistant from the Observatory went with them. The negatives would be "at all times completely in his charge," and would be given to the engravers one at a time.[67] Another company refused to accept originals because of the responsibility that came with them.[68] Copies of original negatives were themselves quite valuable and demanded careful treatment. Campbell issued detailed instructions about how plates should be packed for shipment: each was wrapped in "clean plain unruled white paper," and a dozen plates were packed in a cardboard box, which was then sealed in a padded wooden box. Finally, he ordered, "mark each package with value at the rate of $10 per negative. This does not pretend to be their value," he added, "but it will make the Express Company careful."[69] Ernest Edwards promised that both the glass plates sent by the Observatory and the photogravure plates were to be kept in a fireproof safe.[70]

The printing process usually took several rounds. The positive was sent to the printer, who made a copper plate or relief halftone of it. Proofs were struck from the plate and sent to the astronomer for review and correction. The proof was then compared to the positive from which it was made. Most followed the practice of Yerkes astronomer Edwin Frost, who never judged the quality of a halftone "without having the original positive or print on hand for comparison."[71] One proof would be returned to the printer with notations written in the margins of the photograph or on a separate sheet of paper, and the other would be kept for future reference. The printers would then make changes to the plate, strike a second set of proofs from the corrected plate, and send the new proofs off for approval. Proofs were to printers what positives were to astronomers, and were treated by astronomers as a direct measure of the quality of the original. Wallace Campbell once ordered a new plate on the basis of a badly made proof, and later discovered that the corrections had been unnecessary; he did not let the printers forget the experience.[72] Normally no more than one round of corrections had to be made, but especially difficult subjects could require three or four rounds. Only on rare occasions was this cycle of printing and reviewing abandoned. When work needed to be done quickly and there was "no time for corrections," one had to take what one could get.[73] The proofs were usually thrown away once the plate was approved.

It seemed to astronomers that any number of things could slow down printers. Complaints over delays were legion, and phrases like "unsatisfactory progress" are sprinkled throughout astronomers' letters to printers.[74] Certain delays were seasonal: work tended to slow down in the summer due to the effects of heat on the printing process, and companies had to complete rush orders before they closed for Christmas and New Year's.[75] (Others had even more baroque causes. On two different occasions work ground to a halt when the printers went bankrupt and had to reorganize their companies.)[76] Still, astronomer Charles Perrine reminded his colleagues dealing with slow engravers of "the advisability of not forcing them do rush work on fine subjects:" it was better to endure the delays than force printers into carelessness.[77] Many of the technical problems that held up printing could be solved by a visit to the shop. When Photogravure & Color began work on the Keeler memorial volume in 1904, their first proofs were covered with light spots.[78] Charles Perrine, who was going East for a conference, was sent to the shop to sort things out.[79] "So far as I can judge from this morning's work and conversation, it ought to be possible to get them 'trained' in a week or two," he reported to Campbell after a few days in the shop. "The trouble is in the etching and the only way is to go through every step in the process with them." However, Perrine seemed optimistic: "they seem anxious to get it just right and I think we can make a go of it."[80] Karl Arvidson, manager of the Photogravure & Color Co., and his men made two new plates with Perrine standing beside them, commenting on their appearance as they were engraved. The first "shows faithfully every detail in the original positive," and he expected that "with a little care" proofs of this quality could be made from most of the plates. The second was made from "the most difficult [positive] of the entire lot, one in which the contrast is very weak, and the background necessarily thin." It took a couple of days, but a usable plate was finally made. "There is little doubt that we will be able to get, eventually, what we want," Perrine concluded.[81] E. E. Barnard's experience mirrored Perrine's. Working with A. W. Elson on his remarkable plates of the Milky Way, he recalled that "their results were very disappointing" at first, but after a visit to the shop and some experience "they succeeded very well."[82] But for only a time: after he left, "the Company became careless, and the pictures became poor and frequently defective."[83]

Aesthetics, Improvement vs. Alteration, and the Stabilization of Photoengraving

Of course, we have no records of what astronomers told printers and engravers as they bent over an acid bath or stood together examining a proof of a fresh halftone; but most of the communication between them took place by mail. Astronomers' letters to printers, and their comments on proofs of newly engraved and unfinished plates, give us a view of how they thought about and looked at photographs and engravings as they took shape. They record the passage of photographs and prints as they move across the threshold of the observatory, and astronomers' attempts to extend and maintain dominion over their pictures. From these, we can see what was most important about pictures, what qualities made a print good or bad, and what features astronomers tried hardest to maintain as the image moved from glass to metal plate. We can also see how astronomers divided engravers' labors on a plate into "improvement" or "alteration," thus providing one crucial measure of whether a photoengraving was acceptable. Finally, we can discern the role aesthetic standards played in providing common frames of reference for both astronomers and printers when judging engravings.

Briefly, a good picture was a "faithful reproduction" of a photograph that maintained a good contrast or "intensity values," preserved both the bright areas and all the small and faint details, and rested on a perfectly uniform and dark sky.[84] English photographer Alfred Brothers described a "technically perfect" photograph as one in which "the stars are clear round dots, and the background is quite black."[85] Campbell summarized the desires of the entire community when he told an engraver what he wanted in a picture of the Andromeda Nebula. "We are anxious to reproduce all the features of the nebula, as nearly as possible, with the contrasts shown on the positive [glass plate]," he said, "and at the same time [to] preserve a background as dark as possible. In particular, we hope to preserve the details in the denser part of the nebula and bring out the faintest outlying mass of nebulosity visible on the plate."[86]

As Campbell's instructions indicate, details were always important. Engravers were constantly admonished to "take particular pains in securing as accurate and complete a reproduction of the detail," or to "be very careful to bring out the details."[87] New or special phenomena were the focus of the astronomer's, engraver's, and reader's attention alike, and had to be accurately reproduced. A 1928 coronal photograph was returned because a set of dark arches over one prominence were not quite visible. "We want to show those arches whatever else fails to show," the printers were told.[88] Astronomers were also concerned to maintain proper contrasts between the brightest and faintest parts of an image.[89] Engravers were urged to "respect . . . the faithful reproduction of intensity values and . . . details."[90] Edward Holden's main concern when printing his lunar photographs was to maintain "a good contrast between the full limb and the surrounding black" sky, and he was full of praise when the engraver succeeded in "preserving (but not exaggerating) the contrasts" on the lunar surface.[91] Still, printers often returned proofs that were judged "harsh," with sharp contrasts gained at the expense of fine detail. Spectrograms were simpler in many respects than regular photographs, but still things could go wrong. "Many of the faintest lines on the original were lost in the reproduction," Campbell complained of a spectrogram of Halley's Comet. "It looks as if your photographer had over-exposed the subject, thus burning out these faint lines."[92] Proofs of Mars and Jupiter photographs were condemned as "rough" and "too harsh," lacking any "gradation of shade" that defined a good picture.[93] The aim always was to keep contrasts "as nearly like the prints furnished."[94] Campbell usually suspected that the steel plate had been left in the etching acid too long, producing a plate that was too deeply cut.[95] A high-contrast plate was more dramatic, and engravers accustomed to the coarser standards of newspaper and magazine work may not have even noticed the finer points of an astronomical photograph at first. Repeated orders to avoid "greater contrast with the consequent blotting out of the finer markings" in favor of flatter but more detailed pictures suggest that printers and astronomers looked at these pictures with different eyes.[96] Astronomers assumed that readers had the visual training necessary to detect fine features, and did not need to have contrasts exaggerated or pictures interpreted for them.

Detail and contrast existed in a delicate balance. To judge by the number of times Campbell and Aitken urged engravers to attend to both, astronomers feared that one would come at the expense of the other. By far the most difficult images to reproduce were those that had wide variations in brightness and details in both the brightest and faintest areas. Photographs of the solar corona were considered almost impossible to print well, because a corona's outermost regions were a thousand times fainter than its inner, and it was shot throughout with filaments and tiny structures.[97] Comet photographs were difficult to reproduce for similar reasons.[98] One printer was reminded to "secure as great a contrast as can be done" between the head and the surrounding sky without sacrificing *either* the details in the head or the faint tail.[99] Spiral and irregular nebulae, star clusters, and sections of the Milky Way were also hard to reproduce. Star cluster photographs had to show as many individual stars in the center as possible, and clusters were often surrounded by faint nebular wisps.

The color or depth of the background sky was also watched closely. Backgrounds were supposed to be uniformly dark, featureless backdrops that did nothing to call attention to themselves. Proofs with backgrounds that were "not of uniform density" or not "uniformly dark," were "muddy," or showed "variation" or "various irregularities," were rejected.[100] Campbell asked for backgrounds with a "uniform tint" or "uniform shading free from blurs or mottled markings."[101] Alfred Brothers had similarly argued for the importance of uniform black backgrounds, and singled out the mottling of backgrounds as the chief problem of overexposed plates.[102] However, printers found uniform backgrounds as hard to make as singers find sustaining a single note over several measures. The source of the problem could be the positive, the plate, the paper or ink, or the printing press. Printers at Barnes-Crosby in Chicago explained that "the difficulty with solid black backgrounds is to get the proper kind of ink so as to prevent mottling and still show the delicate vignette" of the central image.[103] Ink might not adhere properly to the paper, it might gather "in the center of the plate, causing a wavy effect or ridges all over it," or it might not be the necessary deep black.[104] "Blurs" and "mottled markings" in the background were by definition artifactual, products not of nature but of the printers' plate or flaws in the negative.[105]

There were also more elusive, subjective visual qualities whose properties were difficult to articulate. One of them was "snap," which meant something more than contrast or sharpness. The term is first used by Campbell in 1904, when reviewing proofs of nebula photographs.[106] The term comes up again and again in the following decades. Campbell instructed a company to "aim for a snappy, contrasty block" of a nebula in 1918.[107] A few months later he apologized for some positives he sent, saying "we hope that you can use [them] . . . though they are not very snappy."[108] On another occasion, Campbell judged several halftone proofs "to be very deficient in 'snap' and to be too flat."[109] Its use with other terms makes clear that it is not synonymous with either good contrast or detail: a photogravure of the Trifid Nebula is praised for its "bright and snappy appearance . . . but all the faint details shown on the glass positive" disappear during the engraving and printing.[110] It is possible that "snap" was affected by color: one engraver is asked if a "more snappy effect" of a print could be produced by remixing the ink.[111] Other pictures were described as being "lifeless." Some pictures made by Barnes-Crosby in 1918 were sent back because they "lack life and look faded,"[112] while in another case good nebulae images were set in "lifeless" backgrounds.[113]

When proofs came back from the engravers' and were examined by astronomers, they were judged according to how well they conformed to the standards described above. They were also studied with an eye to how they would be scrutinized by readers, and to how they would be used.

Photographs of well-known subjects like the Orion or Andromeda nebulae would be examined by people familiar with half a dozen classic illustrations in old atlases or articles, and this placed an extra pressure on them to be especially brilliant. Good prints of familiar subjects were often published as demonstrations of the power of a new instrument, and serve as images of technical virtuosity. James Keeler published four articles discussing well-known nebulae photographed with the Crossley reflector in 1899 and 1900; each begin with reviews of antecedent drawings by figures like the Earl of Rosse and William Lassell, followed by descriptions of their shortcomings as revealed by the new photographs.[114] Prints of the most familiar objects could not be left to printers' own aesthetic judgment. Edward Holden even explained how to judge prints of the *Moon*. He told the Photogravure & Color Co., "9 out of 10 judge our maps by their sharpness and by the sharp divisions between the light colored Moon and the dark sky." Thus, preserving contrast and sharpness were essential for winning over both popular and scientific audiences. But, he continued, "the 10th person, who knows, first decides whether that sharpness is sufficient, and then next whether the contrasts are well kept—the details in the lights for example."[115] Good contrasts were harder to maintain than sharpness, and a print that combined high contrasts with fine detail was a true masterpiece. Holden is familiar to historians of astronomy as a notoriously difficult person, but these instructions cannot be explained away as just more domineering and micromanagement. James Keeler, who had a far gentler personality, also described how different pictures should be studied. His review of Holden's lunar prints (the very ones Holden describes in the sentences above) gives us another perspective on these photographs, and is all the more valuable because they are compared with the Paris Observatory's new lunar atlas. "If we regard the plates in these two atlases as pictures," he wrote, "the advantage is altogether with the Paris heliogravures; they are larger, more brilliant, more impressive. But," he continued,

pictorial effect is evidently no just criterion of scientific value, and if we regard the atlases from the latter standpoint, we see that each has certain advantages of its own. In the Paris photographs the enlargement has, perhaps, been pushed beyond the limit of usefulness, and it would seem that everything which appears on the plates would be shown equally well if the scale were only half as great. If this is so, the impressive appearance above referred to has been gained at the expense of handiness. Further, an examination of the Lick Observatory plate shows that brilliancy of effect has been sacrificed to secure other and more solid advantages.

The printing has been carried so far that details appear in even the highest light, with the result that, while much is shown that otherwise would have been lost in the process of reproduction scarcely any pure white is found in the picture, and a general flatness of effect is produced. Each atlas has, therefore, its own special value. The Paris atlas will be eminently useful for consultation in its place on the library table; the Lick Observatory atlas will find its chief use in the hand of the observer at the telescope.[116]

Thus context, use, and audience had to be taken into account when deciding how much brilliance, contrast, sharpness, flatness, snap, and life a picture should have.

But what astronomers looked for most closely was evidence that engravers and printers understood the boundary between improvement or correction of plates and doctoring, and had worked accordingly. Astronomers tried to be as clear as possible about the difference. "In scientific subjects of this nature," one typical explication went, "no local intensification, tooling, or 'fudging' can be allowed."[117] Changes in the appearance of a nebula or comet could be made only "by the method used for making the cut," that is, through chemical means. "No shading of any part, retouching, local reduction or the like can be permitted on these scientific subjects."[118] The line between improvement or correction and alteration was often a literal boundary, for astronomers saw pictures as a combination of areas that could be touched up by hand without rendering the entire image "scientifically inexact," and areas that could only be dealt with chemically. A sky that was not perfectly dark, that was mottled or spotted or uneven in tone, could be altered by whatever methods the engraver deemed appropriate, including making alterations in the plate by hand. As long as interventions were confined to the *sky* they were acceptable, and counted as corrections rather than dangerous alterations.[119] The principal subject of the plate—the comet or nebula or corona—was always forbidden to engravers' tools. Let us consider again the problem that developed early in the printing of the Keeler memorial volume, when the proofs came out covered in spots. Karl Arvidson wrote that the spots that "do not come too close to the objects to be reproduced without any mechanical touching," that is, the nebula or star cluster, could be easily touched out and the others "can be gotten rid of in the etching." But, he fretted, "the thing I am most afraid about is that we might take out some little spot that should be left in." The major problem was that some of the spots were obviously artifacts, but some of them looked like stars, and no one was willing to risk accidentally removing the latter along with the former. The engraver could touch out the more obvious mistakes, but it would be too easy to cross that literal boundary between space that could be altered and a subject that could not.[120]

The lines between improvement and manipulating grew clearest when astronomers found effects in pictures that had been artificially induced. Astronomers were quick to point out such transgressions and remind engravers of the rules.[121] After discovering several subjects that were "faked somewhat in reproduction" in 1916, Campbell reminded the printers of what constituted a permissible alteration of the image. By "faked," he wrote, "I mean that the attempt has been made to bring out certain features unduly. . . . Careful comparison of it with the glass positive will show that certain streamers appear to have been accentuated." Any change in a plate that went beyond what a positive contained did not qualify as an improvement, and even if it proved impossible to record "some of the finest and faintest details of the original" through mechanical and chemical processes, engravers could not enhance specific features manually. The "finest and faintest details" might be put in by hand, but they would be illegitimate, for as a rule, "any attempt to accentuate these faint details . . . gives us a representation which is scientifically inexact." He concluded, "Any manipulation in the process of making . . . [prints of] scientific subjects can not be tolerated. Our aim is to get a block which shall represent the original negative as exactly as possible."[122] In another case, Campbell caught alterations on a proof of a comet. "If you will examine the proof of this in connection with the glass positive," he told the printers, "you will see at once that the streaks in the comet's tail are brought out far more prominently in the proof than they are on the positive, and the block looks as though these had been put in artificially."[123]

One of the most dramatic examples of unacceptable alteration came in 1908 when engravers at the Photogravure & Color Co. tried to highlight the stars in a plate of the Orion Nebula. In November, while examining final prints of the Orion Nebula, Perrine noticed "faint dark rings around some of the stars in the denser parts of the nebulosity" that "give rise to a suspicion that the effect has been enhanced by the engraver."[124] This was an especially bad mistake, for the Orion Nebula was perhaps the best-known astronomical subject and had been included in every major atlas of drawings and photographs, and because this plate was going to be the volume's frontispiece. The engravers were, in fact, trying to communicate to viewers that the stars were supposed to be brighter than the surrounding nebula; adding dark space to signal the presence of a contrast (if not to trick the eye into seeing it) had been part of the visual language of engraving for decades.[125] Campbell was furious that the print had been doctored. "You will remember that our first instructions were that there must be no hand-work (tool-work) in the nebulae themselves," he reminded Arvidson, and "we have rejected all such work when discovered."[126] It was too late to make new plates, so an erratum slip explaining the engravers' "corrections" was hastily inserted.[127] Barnard spotted them immediately, and his praise for the entire volume was somewhat qualified. "In general . . . the results are excellent," he told Campbell, but also noted "there are some cases however where it looks as if the Photogravure people had done some doctoring."[128]

For their part, engravers did their best to reassure their customers that they understood the difference between correction and deception. Printers assured their employers that they understood that "fidelity to the subject is very essential," as one company put it.[129] Engravers, for their part, wanted to be on the right side of the line just as much as the astronomers. In negotiating with Edward Holden for the lunar photographs, Ernest Edwards promised to "touch out all the pinholes and scratches in your original negatives where this can be done without changing the aspect of any lunar feature."[130] In explaining the mistake with the Orion Nebula, Campbell told Barnard, "Photogravure & Color Co. understood thoroughly that they were to do no re-touching," and "claimed that their over-zealous workmen had done this against their orders."[131] Barnard complimented one printer when he said, "he thoroughly understands that no retouching must be done in any astronomical picture."[132]

The division of plates into zones that could be worked by different methods, and the definition of chemical means as the only acceptable method of producing an image in the most important areas of a picture, served two purposes. They were an attempt to shape work in the way that guaranteed the best possible outcome, by protecting images from the appearance of artifactuality; they were also a means of establishing boundaries of authority and control between the observatory and the print shop. Astronomers were well aware that both the etcher's acid and the engraver's knife altered images, and engravers' skill was required to operate both properly; but because the engraver's prejudices and ideas could be seen to operate more indirectly in the former, it was chosen to operate on the subjects and centers of pictures.

Advocates of mechanical reproduction declared that the new systems offered pictures that were exact duplicates of originals untainted by human error. Lorraine Daston and Peter Galison have argued that human intervention and judgment became suspect in this period, which led to the development of self-consciously unaltered images. This desire was one of the impulses behind the adoption of photography in the observatory. The ideal of self-registration in the observatory was matched by a vision of reproduction in which photographic plates would register themselves on halftone plates or photogravures, thus to be impressed on the pages of scientific journals. But the realities of photomechanical reproduction were quite different from the ideal. They show that printing technologies were such that throughout the period under discussion, skilled workers and skilled work were necessary to monitor and correct the flaws and errors of mechanical production: the intervention of master craftsmen was essential if engravings of astronomical photographs were to be error-free. Printing processes had several steps, each of which was an opportunity for introducing inaccuracies and mistakes in images, or for correcting and restoring their information-content. Within the observatory the rules that assigned credit for plates was commonly known and respected, but as pictures moved outside, issues of authority and authorship reemerged and had to be renegotiated. At the most basic level, astronomers and engravers had to come to agreement about what constituted "intervention" versus "alteration," and how to judge what constituted a good picture. Ultimately, skill and aesthetics were still required to produce pictures that could be accepted as accurate and useful images of nature.

Standards and Semiotics

❖❖ In the whole history of science there is perhaps no more
fascinating chapter than the rise of the "new science" of
linguistics. In its importance it may very well be compared
to the new science of Galileo which, in the seventeenth
century, changed our whole concept of the physical
world.
—Ernst Cassirer, "Structuralism in Modern Linguistics"

It is a commonplace of modern intellectual history that all disciplinary
knowledge is bound up in the web of language, either as the bearer of
tradition, memory, and tacit and unexpressed habits of thought, or as the
grammatical a priori of all possible expression.[1] Surveying some rather
different versions of this claim—those of hermeneutics, structuralism, and
analytic philosophy—Michel Foucault diagnosed the conditions for its
possibility in the early nineteenth-century turn to make language an
object of scientific inquiry.[2] Foucault was certainly right to identify the
science of language as a crucial nineteenth-century discipline, "reverberating with politics" and interacting at key points with other disciplines. Yet historians of science, even those advocating a "linguistic turn"
in the discipline, have curiously neglected the nineteenth-century sciences of language.[3] In trying to make our way through the web of language, it may be useful to reconstruct how it was woven in the first place.[4]

Like so many other modern institutions, the web of language was at
least partially spun in the laboratory, with the aid of precision measurement and media technologies.[5] From 1875, when the French Société de
linguistique joined forces with the laboratory of Etienne-Jules Marey,
important networks of collaboration between physiologists and linguists
were developed to investigate speech acts with graphic inscription devices. When this collaboration resulted in the founding of a permanent
laboratory for experimental phonetics at the Collège de France in 1897,
the *maître* of French linguistics and one of the principal sponsors of the
new laboratory proclaimed that the new laboratory ensured the demise of

German speculative philology and the consolidation of a science of language. According to Michel Bréal:

linguistics will finally be in a position to record the facts instead of asserting a priori principles. There will be no more phonetics *in vacuo*, aided by technical terms that, although no doubt extremely learned, convey only misleading or vague ideas. . . . Instead of speculating on hypotheses such as the "primitive velars" or "the Indo-European j" we will see what a given individual's articulations are at the moment when they are produced in his mouth. Thanks to the instruments of Edison and Marey, we will be able to write sounds—or rather they will write themselves, so that what the ear perceives in a necessarily confused and fleeting way will be able to be examined minutely and at length by sight.[6]

Bréal's remarks call attention to a remarkable conjuncture: the coterminal construction of the modern laboratory as a "system of literary inscriptions" and the new studies of forms of signification.[7] Both developments have been the focus of much study, but their points of interaction have been almost entirely overlooked. Historians of linguistics, for example, agree that a sea change occurred in the science of language during the late nineteenth century, but they have signally missed the role of the laboratory in that transformation.[8] Historians of the scientific laboratory, on the other hand, have described inscription devices and graphic methods in some detail without having mentioned the late nineteenth-century battles over the status of the sign. This essay seeks to describe the interaction between the two disciplines in and around the laboratory of the physiologist Etienne-Jules Marey. The principal aim shall be to show how the phonetics laboratory was strategically conceived as part of the emerging disciplinary identity of French linguistics and how over the course of the last third of the nineteenth century, experimental phonetics altered the discourse of Indo-European philology and shaped the new modern science of linguistics.

This essay is a contribution to a burgeoning material history of linguistics. Through the work of Friedrich Kittler we have begun to understand that a complex nexus of phonographs, typewriters, psychological instruments, and a host of other devices decisively interposed themselves everywhere language was at play in the decades around 1900.[9] In a work of social history also informed by poststructuralism, Timothy Mitchell has described how as a technical system and a matrix of interests, the telegraph network in French colonial Egypt transformed both Arabic philology and French notions of language in the same period.[10] Mitchell stresses some of the crucial calibrations of both technical standards and human beings necessary to make colonial communication systems work.[11] One need not have looked to the colonies for the expanding networks of control from Paris, however; as Eugen Weber has shown, in the fin-de-siècle the same procedures advanced apace in the French provinces: the imposition of the metric system of measures, standardized time, and the rule of the French language.[12] Historians have documented the salient role of Paris laboratories in the former initiatives; this essay will describe their role in the latter.

The historiography of French linguistics points ineluctably to its canonical formulation in Saussure's *Cours de linguistique générale*. Although not renouncing entirely such Whiggery, the present essay aims to fill out some of the largely neglected context of the Genevan linguist's Paris sojourn. For, as Hans Aarsleff has reminded us, "Saussure did not arrive in Paris from Leipzig and Berlin with the ideas that generated the *Cours de linguistique générale*, and he did not leave Paris without them."[13] Aarsleff insists on the importance of Saussure's work with Michel Bréal, whom the linguist called his "excellent teacher" (*maître excellent*), and the Société de linguistique de Paris, reminding us of just how much of Saussure's thought was inherited from his French colleagues. The present essay takes this claim a step further, arguing that the French linguistic work tacitly inscribed in Saussure's *Cours* was itself shot through with the political and social interests of the French state, which in key respects were mediated by the phonetics laboratory.

At the heart of these interests lay the familiar notion of "linguistic value," which formed the central notion of Saussure's problematic.[14] Aarsleff has pointed out that Saussure inherited this term from his *maître*, Bréal, who introduced it into French linguistics in 1879. Aarsleff failed to notice, however, the many valences of the term itself, among them the laboratory measurement of phonetic values to which Bréal's Collège de France colleague Marey had devoted a chapter the year before in his treatise, *La méthode graphique en sciences expérimentales*.[15]

Marey's treatise summarized more than three decades of work on techniques of scientific representation across a range of fields, with special emphasis on its employment in his own discipline of physiology, to which he had made significant contributions.[16] Since the late 1850s, Marey had become the foremost French representative of graphic inscription techniques, which had thrived in Germany in the laboratories of "organic physicists" such as Ludwig, Helmholtz, Vierordt, and Brücke since the late 1840s.[17] The appearance of *La méthode graphique* galvanized calls on both sides of the Rhine for a special discipline devoted entirely to the techniques of graphic representation in science.[18]

Graphic methods played a crucial role in the constitution of the science of linguistics in the late nineteenth century. With the help of inscriptive apparatus, linguists rendered the fleeting and unseen phenomena of speech as a materialized and visible object. The *image vocale* or *image acoustique*, to use the terms of Bréal and Saussure respectively, served to codify the concept of the *phoneme*, what Sylvain Auroux has called "the key notion for the constitution of the science of linguistics."[19] The material signifier or acoustic image (Saussure habitually spoke of these as one) became transformed into a circuit model of communication in which words or verbal messages are exchanged, put across, got over, sent, passed on, received, and taken in.[20] Articulate sounds, elementary phonemes or entire words, thereby became analogies to the dots and dashes transmitted by cable throughout the French Empire. "Words . . . are like telegraphic signals," wrote Bréal, with preassigned values to be transacted.[21]

The transmission of telegraph signals was, of course, dependent upon the hard work of calibrating electrotechnical standards.[22] In Saussure's work the calibration of linguistic values is also taken to be an accomplishment, not of specialists, but of the community of speakers, who shape not only the significations of words but their manner of articulation in the human body. In language as in telegraphy strenuous calibration of standards enables the signifier not only to be arbitrary, but to be more robust because of it. Insofar as the linguistics of Bréal and Saussure became extended far beyond the specific phenomena of language to the gamut of communication signs, the role of laboratory-based precision measurement would seem to have affected a crucial mediation of modern values.[23]

The Politics of Language

Throughout the first half of the nineteenth century France could boast nothing comparable to the thriving philology industry of Germany.[24] By the 1860s the study of language became yet another plank for Franco-German rivalry, when Ernest Renan declared the massive Prussian victory over the Austrians at Sadowa a victory for German science.[25] Victor Duruy, the Minister of Public Instruction, targeted philology as one of the disciplines to be reconstructed in France, along with physiology and mathematics. The reform-minded Duruy anointed Michel Bréal, a young Alsatian philologist, as champion of the new French science of language, in parallel with Claude Bernard's role as discipline-builder for physiology. With Duruy's patronage, Bréal secured a chair in Comparative Grammar at the Collège de France in 1866 and took up the position of perpetual secretary for the newly formed Société de linguistique de Paris. Firmly placed at the head of his "troops," as he referred to his colleagues in the Société, Bréal spearheaded a campaign to define a French science of language comparable with, but different from, the kind practiced across the Rhine (*outre-Rhin*).

The first battle concerned the proper name of the discipline. "Linguistics" (*la linguistique*) marked off the French preference for the study of the living language, as opposed to the fixed textual entity of German philology.[26] In this and many other senses, Bréal's *apologia pro scientia sua* marched in striking parallel to the efforts of his Collège de France colleague Claude Bernard for French physiology. Like Bernard, Bréal set out to distance his discipline from its historical roots, which in this case meant the rejection of a strongly historicist approach to language. On several crucial points Bréal drew explicit analogy to Bernard's program, arguing for the study of language in vivo, as opposed to the anatomic dissection of dead historical language entities favored by the German tradition.[27] But as with Bernard's critique of medical anatomy, the definition of disciplinary identity required a complex process of both affirmation and negation.[28]

Bréal's careful pattern of appropriation and rejection can be discerned clearly in his treatment of German philology in the 1860s. He was uniquely qualified for this crucial yet delicate aspect of French discipline-building. Born to French Jewish parents in a part of Rhineland that since the Congress of Vienna had been a region of Bavaria, Bréal spoke German as a first language in his early childhood, before his widowed mother moved the family to Wissembourg in Alsace. After French schooling and education at the École Normale, Bréal had obtained permission from Duruy to go to Berlin to study Sanskrit and comparative grammar with Franz Bopp.[29] Upon returning to France in 1862, he translated Bopp's magnum opus into French.[30] He followed up his translation of Bopp with renderings of key works of August Schleicher, the standard-bearer of German philology in the generation after Bopp.[31]

Although more kindly disposed to Bopp (who in addition to being his teacher had been educated in Napoleonic France), Bréal diagnosed a dangerous pathology in the entire tradition of Indo-Germanic philology. In his Introduction to Bopp's *Vergleichende Grammatik* he relentlessly attacked German philology at its source: the Sanskrit studies of Friedrich Schlegel, especially the inspired work *Über die Sprache und Weisheit der Indier* (1808), which prompted a generation of Germans to embark on an intellectual quest for the South Asian subcontinent.[32] At the bottom of Schlegel's conception lay notions of language as an organism informing the very spirit of a people. Sanskrit represented the original perfection of the languages that Schlegel and his countrymen termed Indo-Germanic or Indo-European, which were marked by similar vocalic roots and analogous systems of grammar and internal structure. For Schlegel and most subsequent German philologists, these resemblances explained not only a linguistic but also a spiritual and racial kinship system, set apart from the other languages of the world. To study Sanskrit was to partake of "a mysterious education that the genius of mankind or at least a privileged portion of the human family might have received in its infancy."[33]

Bréal attacked Schlegel's claim that the first speakers of proto-Sanskrit, or Indo-European, had through divine grace grasped the precise relation between sound and concept with an intuition—*une faculté créatrice*—brought on through the ecstasies of phonetic autoexperiment.[34] Bréal ridiculed on methodological grounds all Romantic claims to derive the alphabet of Nature (*Buchstaben der Natur*) in acoustic experiment or through self-experimentation on vital organs, including those of phonetic articulation.[35] Even more pernicious was the assumption of privilege ascribed to speakers of Indo-European and their ancestors, which not only linked race to language, but denied any such grace to speakers of Semitic, Chinese, or American Indian languages.[36]

Bréal credited Bopp with having resisted the more mystified impulses of this tradition, such as those of the Schlegel brothers, Windischmann, Creuzer, and Goerres. Bopp, he wrote, "did more than anyone to dispel the mystery in which these lofty minds—lofty but friends of the twilight—were pleased to envelop the first productions of human thought."[37] He rejected aesthetics and metaphysics in favor of a desacralizing analytical decomposition and decontextualization of the Sanskrit texts into paradigmatic lists in which the morphology of declinations and conjugations served as the basis for syntax and lexicology. But Bopp's analytical morphology could only make sense of languages whose orthography or transcription-system could plausibly conform to their pronunciation, because it only dealt with letters, not sounds. His method of phonetic comparison began by compiling word-pairs whose members were clearly identical; it was thus easy to recognize Latin *frater* in Sanskrit *bhrâtar* ("brother"); Greek *menos* in *manas* ("mind"); Latin and Greek *jugum* and *zugon* in Sanskrit *yugam* ("yoke"). From there the philologist proceeded "as in a game of solitaire, where we begin by arranging the obvious cards, these then helping us to find the rest."[38] Out of the regularity of the correspondences between the appearance of one letter in one language and another given letter in a second language laws of the correspondence of vowel and consonant sounds were derived.

Bréal charged that without a credible system of transcription,

Bopp's great edifice remained unstable. Any claims about the status of phonetic laws based solely on the regularities of letters could not be deemed reliable. Bopp's approach assumed that the very being of language resided in written texts, outside of any human agency, individual or collective. Bréal countered that language had to be seen as a social fact, as a set of conventions determined by whim, calculation, and not least, fashion. He had little difficulty convincing Parisians of the latter claim. Citizens of the French capital, Bréal reminded his listeners, quite regularly took up linguistic affectations, such as the eighteenth-century vogue for pronouncing *r* like *s* and *s* like *r*: *Paris* became *Pazis*, *mari* became *mazi*, and conversely *un oiseau* became *un oireau*. Some affectations, such as this one, were duly discarded, but others could become incorporated into language.[39]

Bréal and his allies reserved their harshest attacks for second-generation German philologists, especially August Schleicher and his school, who had discarded Bopp's metaphor of laws of gravity for the potent synecdoches of Darwinian or Haeckelian evolution.[40] "Languages," wrote Schleicher, "are natural organisms external to human will which behave in accordance with fixed laws; they are born, they grow, develop, grow old and die; they therefore manifest that range of phenomena called life."[41] The French objections, however, focused less upon Schleicher's Darwinism than upon the organicist and racialist tradition to which it had been assimilated.[42] Bréal argued that Schleicher had made language "a fourth realm of nature," a charge echoed by his staunch ally Gaston Paris, a professor at the Sorbonne: "The development of language does not have its cause in itself, but in man, in the physiological and psychological laws of human nature; in this way it differs essentially from the development of species, which is the exclusive result of the encounter of the essential conditions of the species with the exterior environmental conditions."[43]

Militant denials of the autonomy of the language had been heard in the Société de linguistique since its inception, and for reasons that went beyond direct quarrels with *outre-Rhin* philology. Part of the motivation was political: one year before the founding of the Société, Duruy had renewed the campaign to extend the use of the French language throughout the country and to suppress the various patois still widely spoken within French borders.[44] Duruy's 1864 initiative, which ordered a survey of the linguistic habits of French citizens, harked back to investigations conducted by the Abbé Gregoire on behalf of the Jacobin authorities.[45] French linguists seized upon the opportunity to pick up where the Abbé Gregoire had left off, charting the linguistic map of France, and furthering the civilizing mission associated with the advance of the national language.

In the last years of the Second Empire, Duruy and the linguists strengthened their collaboration on several fronts. In 1867 the Minister supported the creation of a section on philological sciences at the newly formed École pratique des Hautes Études (E.P.H.E.), whose six positions he filled with German-trained, but relatively anti-German philologists, including Bréal and Gaston Paris.

During the next decades the E.P.H.E. became a backbone of the linguists' alliance with the French state, as the principal site for training the teachers who would bring the French language, and just as importantly, the self-identity of French, to the provinces. From 1881 to 1891 Saussure taught at the E.P.H.E. and served as one of the principal cadres of its civilizing mission.[46]

This political alignment hardened further after the bitter Franco-Prussian war, thereby facilitating the extension of the Duruy model to the whole of France. It was an ironic development, because German notions of language and culture had been cultivated in opposition to Napoleonic domination of Central Europe, which promoted not only the French language but French theories of language and epistemology.[47] Now the repressed had returned to haunt the French with ferocity. Numerous linguists—including Bréal, Gaston Paris, and Ernest Renan—joined the ranks of the Société de l'Enseignment supérieur, an organization of leading scientific intellectuals committed to avenging their humiliation and suffering by strengthening French educational institutions.[48] With Renan's treatise *La forme intellectuelle et morale de la France* (1871) as their constant reference, the French savants lashed back. The historian Fustel de Coulanges charged that: "All science serves [the Germans] as an armament against France. They invent the insupportable theory of the Latin races to give their dynastic ambitions the false veneer of a quarrel of races. They use philology and ethnology to demonstrate that our most French provinces are their legitimate property."[49] Renan made a similar argument, even couching it in terms of a threat to his German interlocutor:

The individuality of each nation is surely constituted by race, language, history, religion, but also by something much more tangible, by actual consent, by the will of those of different provinces to live together. . . . The overly sharp division of humanity into races not only rests on a scientific error, but because very few countries are of a truly pure race, it can only lead to wars of extermination. . . . You have raised in the world the flag of ethnographic and archeological politics in place of liberal politics; those politics will be fatal to you.[50]

The French linguists rallied around conceptions of language that severed any essential links with race. Gaston Paris argued that the liberal conception of politics was also true of the nature of human communities to language, because a people could, and often do, cast off the regional dialect (*Volkssprache*) by learning a new language.[51] Paris argued that the German notion of Latin or Roman peoples failed to grasp this central fact of linguistic voluntarism in the history of France and its neighbors:

When one speaks of the Latin races one uses an expression that has absolutely no justification: there is no Latin race. The Roman language and civilization were adopted, more or less voluntarily, by the most diverse races: Ligurians, Iberians, Celts, Illuriens, etc. It is thus in the sacrifice of their own original nationality that the unity of the Roman peoples rests; this basis is thus entirely different from the principle that constitutes Germanic or Slavic unity. . . . With those peoples nationality is exclusively the product of blood; Romania is by contrast an entirely historical product. Its role would thus seem, in the face of societies who are nothing but aggrandized tribes, to represent the fusion of races by civilization. Against the principle that rests solely on a physiological basis, one happily opposes that which founds the existence and the independence of peoples on history, on the community of interests and the participation in the common culture.[52]

Animated by patriotism, the politicians of the French language would carry out their struggle in the name of natural science. Against the organicism and natural history of German philology they would enlist the tactical support of the physiology laboratory, invoking the moral authority attached to self-registering instrumentation.[53]

Visible Speech

Within the Société de linguistique, interest in traditional French theories of language grew during the 1870s, especially those connected with the

Idéologues and the attendant tradition of study of deaf-mutism, which took inspiration from Diderot's *Lettre sur les Sourds et Muets* and from the work of J. M. de Gerando and Jean-Marc-Gaspard Itard on the Wild Child of Aveyron.[54] Although the famous case of training the *enfant sauvage* had been a manifest failure, it provided important lessons for speech education, not just for the mentally and physically handicapped, but also for those other *sauvages*—the provincials—whose linguistic disability could be readily attributed to geography.[55]

In 1874 the Société elected the expert on deaf-mutism Leon Vaïsse as its president. Vaïsse, a former colleague of Maria Montessori and author of *De la parole considérée au double point de vue de la physiologie et de la grammaire* (1853), took his election as a "testimony of [the linguists'] interest in my lifelong professional work" because in the education of deaf-mutes "some of the most interesting questions of the science of language arise and are sometimes resolved."[56] The approach of the French experts on deaf-mutism, Vaïsse assured, offered "a sureness of method that the philologists of another time could not have suspected." French linguists, he claimed, possessed the intellectual maturity to "abandon the path of rash speculations" favored by German scholars: "To the studied patience of the erudites from across the Rhine, you know, French linguists, how to unite the practical sense from across the Channel, and among you delicate critique is never a foreign importation."[57]

Vaïsse's critique focused above all on foreign notions of overarching sound-laws derived from the written, textual history of languages. As a specialist in deaf-mutism, Vaïsse used an approach to language that favored the individual speaking subject and its volitional capacity to alter its manner of articulation.[58] Vaïsse's Montessorian program thus dovetailed with Bréal's attack on the German notion of language as a mental abstraction (*abstrait imaginaire*): there must be physical and psychological determinants of the act of speaking. In his presidential address to the Société, Vaïsse unveiled a new physicalist research program that would unite the diverse interests of the organization:

Language, considered in its physical elements, is what forms the raw material of our common research. A bit of air is released from the lungs and strikes the ear, after having been agitated, broken, and deflected in various ways against the walls and the surfaces of the mouth and by the diverse dispositions of the tongue—this is the matter out of which articulated language is composed. . . . But in its workings, speech [*la parole*], the last and most important manifestation of the human soul, transports thought on these molecules of air, the infinitely small is reconciled with the infinitely large, the atom carries the world![59]

Vaïsse maintained that the new doctrines of energy conservation enabled linguistics to treat the act of speaking as a matter of force and movement.[60] To this end he led a delegation of members from the Société that approached Etienne-Jules Marey, one of the foremost French proponents of the thermodynamics of life, about the possibility of collaboration on studies of speech phenomena using the kinds of inscriptive apparatus for which the physiologist had become famous in his work in cardiology, respiration, and other bodily functions. Vaïsse thought that Marey's graphic methods could be used to provide a system of transcription similar to that used by the Scottish professor of articulation Alexander Melville Bell, father of the inventor of the telephone. In his system of "visible speech," Vaïsse reported, "our colleague Bell traces on a tableau the characters of another phonographic (or rather glossographic) alphabet that he has himself used for a number of years in his lessons of articulation to pupils deaf since birth. Reduced to their rudimentary lineaments, the characters of this alphabet offer the figure of the essential organs of speech, following the median line, with the disposition that they affect in the production of our different phonic elements."[61]

Bell's work alerted Vaïsse to other possible graphic representations of speech. Only a few years earlier, Marey had stirred the Paris medical community by using graphic methods to show the succession, duration, and intensity of the movements of the inner cavities of the heart.[62] After this initial success, Marey committed himself to extending techniques of graphic inscription to as many physiological functions as possible, and advocated nothing less than a transformation of the signifying scene of the experimental laboratory. Not surprisingly, he jumped at the proposal to bring linguistic phenomena under the purview of graphic methods, with the intention of replacing the "flowers of language" in science with the precision of machines, the "language of the phenomena themselves."[63]

For the linguists, tailored investigatory methods in a well-tooled laboratory would confirm the autonomy of their discipline.[64] So would a unique category of phenomena: speech. Although the laboratory did not enable linguistics to sever completely its ties to traditional philology, it definitively called into question the traditional primacy of the written word in the sciences of language. In the guise of speech, language once more became a source of communication, a theme almost entirely absent from philology. But the communicative turn in linguistics, which would eventually become part of Durkheimian sociological investigations of the social bond, rested on a firmly physiological or physical basis in the scenario of articulation and audition.

The initial project of the new phonetics research group, consisting of Marey, the linguist Louis Havet (secretary of the Société), and the physician and deaf-mute expert Charles Rosapelly, endeavored to capture in simultaneous interaction physiological functions connected with the acts of speech: the thoracic cage, the larynx, the lips, and the air pressure within the nasal passages. Although problems of deaf-mutism remained among their concerns, Havet directed their arsenal of physiological instruments at one of the canonical texts of German Sanskrit philology, the *Pratisakhya*, a Vedic text reckoned to be more than two thousand years old that had been a focus of Sanskrit philology for more than two decades.[65]

The *Pratisakhya* occupied a special place among the Vedas, because it furnished an exhaustive description of the vicissitudes of Sanskrit pronunciation: enunciation, tone, duration, pitch, evenness, and compounding. Brahmanic tradition held that the Vedas were a mere transcription of an oral tradition practiced by the rishis in heightened states of consciousness (*Dhyaana*), so that the phonetics of the mantras found within them could not be easily reproduced on the basis of conventional orthography. Moreover, defective pronunciation of sacred texts was regarded as a profanation.[66] To ensure a more perfect state of pronunciation, then, the *Pratisakhya* contained descriptions of the parts of the body from which certain syllable sounds originate and through what kind of effort they are brought forth. Many of the sounds described were without representation in writing. Some sounds, for example, were intercalated in pronunciation, within couplings such as *kn, km, tn, tm, pn, pm*, and so on. The ancient Hindus called this pairing of a "mute" consonant with a nasal consonant a *yama*, or twin.[67] Despite exhaustive attempts at self-experimentation, however, the Hindu sages proved unable to decide which sound was the "twin" of the other. During the formation of the mute consonant that commenced the pair, the soft palate remained closed. It opened, however, at the moment of the formation of the nasal consonant.

The French savants set out to first resolve this dilemma of Vedic linguistics. But the stakes were higher: if successful, they would show the limits of phonetic self-experiment in general and thereby render invalid the greater part of the conclusions of German philology. Charles Rosapelly indicated the intention of the physiologist's intervention to over-come "organic analysis" in matters philological.[68] "Modern phoneticists have preciously assimilated these rather minute, but entirely exact analyses [in the *Pratisakhya*], of the different inflexions of language; physiological experimentation sheds still more light on the different acts of the lips, the tongue, and the soft palate in the formation of consonants."[69] Rosapelly's experimental methods were guided by the "principle of least action," according to which "all human acts tend to be executed with the least possible effort."[70] German physiologists had used the principle with great success to describe the way persons learn to calibrate the muscular move-ments of the eye when learning to see.[71] Marey and Rosapelly invoked the principle to ascertain the analogous processes involved in learning to articulate the sounds of language.

The physiologists' critique of "organic" phonetic self-experiment centered on the limits of human hearing capacities. Experimental phys-iology claimed advantage over the unaided auditory sense by virtue of its ability to render phonetic phenomena to other modalities of sense, espe-cially to vision. "Our aim in the experiments," wrote Rosapelly, "was to replace the auditory sensation with an objective expression of the acts of phonation."[72] By "objective" Rosapelly meant ocular, more specifi-cally the well-known "acoustics of the eyes" developed by the physicists Rudolf Koenig and Jules Lissajoux, who had produced different graphic representations of sound phenomena. The aim of these techniques, like those of Helmholtz described by Timothy Lenoir, was to present a visual analog of the exact form of the acoustic phenomena.[73] The primacy of the visual sense derived from the capacity to fix it in an immobile im-age, which could be examined and perhaps subjected to mathematical treatment. This inscriptive apparatus, then, rendered the hitherto fleet-ing phenomena of speech into materialized scientific objects: the pho-nemes. Although the term first appeared innocuously in a lecture by A. Dufriche-Desgenettes to the Société de linguistique, it soon be-came directly associated with the laboratory experiments of Havet, who also introduced the term into the international vocabulary through the Association phonétique internationale. Thus the phoneme, which, in Sylvain Auroux's words, became the "key notion for the constitution of the science of linguistics," grew as a direct product of the signifying apparatus of the laboratory.[74] Much of the rest of this essay will chart its biography, from Marey's laboratory through its role in French linguistic surveys to its deployment as *image vocale* or *image acoustique* in the systems of Bréal and Saussure, respectively.

Here is how it was done. Under Marey's direction, Rosapelly and Havet set out to test Hindu phonetic wisdom by constructing an apparatus that would simultaneously inscribe the movements of pressure within the nasal passages, the vibrations of the larynx, and the movements of the lips (see Fig. 12.1). The first apparatus consisted of an inscriptive manometer that traced a horizontal line when there was no nasal emission, and elevations in the curve in the cases where articulated phonemes exerted pressure. The second instrument consisted of an electromagnetic stylus (modeled after a Deprez galvanometer), which made a trembling line tracing the vibrations of the larynx and a straight line when there were no reverberations. The third inscription device traced the movements of the lips, dropping when they parted and making a straight line when they were together. Finally, a Marey pneumograph was used to inscribe the rise and fall of the chest during respiration.

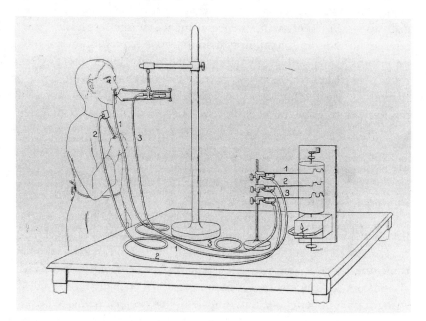

Figure 12.1. Marey-Rosapelly vocal polygraph. From Charles Rosapelly, "Inscription des Mouvements Phonétiques," *Travaux du Laboratoire de M. Marey* (Paris: G. Masson, 1875). Courtesy of the Syndics of Cambridge University Library.

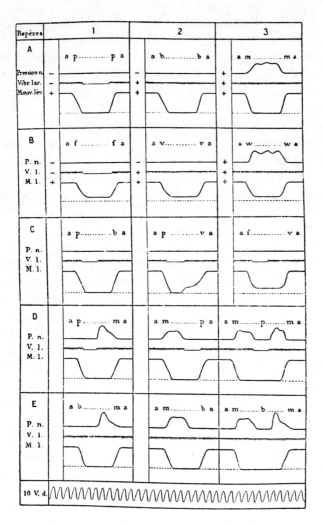

Figure 12.2. Vocal polygraph inscriptions of Sanskrit *yamas*. From Rosapelly, "Inscription des Mouvements Phonétiques." Courtesy of the Syndics of Cambridge University Library.

For Marey and Rosapelly the results signaled the triumph of physiological methods over organic self-experiment. (See Fig. 12.2.) The separate bodily indications of the phonemes revealed that the sound was a twin of the mute consonant and not, as the Hindus believed, of the nasal sound, because the simultaneous traces showed (most clearly for the word *apma*) that the soft palate opened before the labial act that signaled the emission of the consonant *m*.[75] For Havet, on the other hand, the studies suggested the importance for comparative linguistics of sounds not represented in writing. Traditional philology was shown to be too exclusively focused on the sounds of written consonants, rather than on the nuances of living speech.[76]

Focus on the unwritten aspects of speech led to studies of vowels, which even traditional philology regarded as problematic, because some phonetic scripts such as Arabic and Hebrew often neglected to indicate them at all.[77] Reverberating vowel sounds had been at the center of several attempts to graphically inscribe the human voice in the 1850s and 1860s.[78] One of the most notable had been the handiwork of an ambitious young stenographer named Leon Scott de Martinville. Scott designed his *phonautographe* as an attempt to elevate stenography (also called "phonography" until the end of the century) to an automatic and universal pasigraphy, by forcing "nature to constitute by herself a general written language of all her sounds."[79] (See Fig. 12.3.) The pure voice of Mother Nature would supersede the myriad systems of notation based in mere convention that Scott had recounted in his *Histoire de la sténographie*.[80] In order to accomplish the direct inscription of the voice, Scott's subject spoke into the wide end of a bell. A mobile membrane attached to the bell's interior registered the vibrations, which were inscribed by a stylus in a revolving drum covered with smoked paper. The resulting curves were highly irregular and without much use.

Despite its manifest failure, Scott's instrument inspired further attempts to perfect vocal inscription techniques, most notably by Franziskus Donders in Holland, and soon thereafter, by Rudolph Koenig in France.[81] The Dutch physiologist tested the theory of Helmholtz that the sound of vowels is comparable to the timbre of different musical instruments, which the ear can recognize even when the instruments all play the same note. Donders argued that the timbre (*Klangfarbe*) of different vowels was formed in the cavity of the mouth after originating in the larynx. The best evidence for this was provided by the vowels uttered in a whisper, without any involvement of the larynx. This theory suggested that vocalization could be studied as a purely acoustic phenomenon, without any involvement of the organs of articulation.

Koenig responded with a splendid instrument that would display sound to the eye, the so-called *flammes manométriques*—which used gas flames to produce the dogtoothed patterns of graphic inscriptions.[82] (See Fig. 12.4.) The apparatus was as ingenious as it was dramatic. At the

Figure 12.3. Scott phonautograph. From Jules Marey, "Inscriptions des phénomènes phonétiques. Part 1. Méthodes directes," *Revue générale des sciences pures
et appliquées* 9 (1898): 449. Courtesy of the Syndics of Cambridge University
Library.

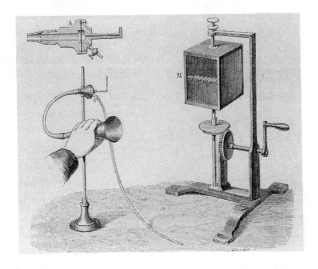

Figure 12.4. Koenig manometric flame apparatus. From Marey, "Inscriptions des
phénomènes phonétiques," 452. Courtesy of the Syndics of Cambridge University Library.

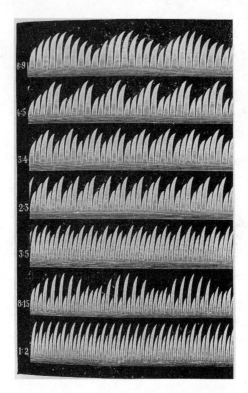

Figure 12.5. Photograph of manometric flames. From Marey, "Inscriptions des phénomènes phonétiques," 453. Courtesy of the Syndics of Cambridge University Library.

extremity of a gas pipe Koenig placed a small metal reservoir carrying a thin burner to light the gas. A thin rubber membrane formed an inside wall of this reservoir. When a series of vibrations were transmitted to this membrane, the gas of the small reservoir became alternately compressed and dilated, thereby submitting the flame to rapid rises and falls too quick for the eye to directly perceive. With a four-sided mirror turned by a crank, however, each flame gave the image of a long luminous ribbon with a pattern of dogtoothed indentations. These images remained too fleeting for analysis, however, until Koenig managed to devise a method for recording them photographically. (See Fig. 12.5.) The temporal appearance of the flame-images was gauged with a simultaneous flame vibrating not to the sound of the voice, but to the action of a tuning fork with known periodicity.

Koenig's manometric flames proved incapable of resolving Helm-holtz's claim that the timbre of vowels depended upon the relative inten-sities of different harmonics. That awaited the invention of the phono-graph, which physiologists seized upon during the first Edison shows. When the instrument was first displayed in Paris on the Boulevard des Capucines before pressing crowds, Donders came to the French capital to visit the show together with Marey. Between demonstrations the two physiologists asked to use the instrument for a "scientific experiment of great interest," in which Donders sang the five vowels. He then asked the demonstrator to change the speed of the reproducer before playing back the recording. A newly assembled audience then heard the vowels played at altered speeds. Each of the vowels came out distorted, *a* became *o*, *e* became *ou*, and so on, showing that the timbre could not be the result of the relative harmonic intensity, because that relation would be preserved at different speeds and the character of the timbre preserved.

Over the next twenty years the phonograph found constant use in deaf-mute instruction, particularly in the work of H. Marichelle, who showed, in Marey's words, "that the phonographic trace is the only ex-pression that perfectly defines the nature of a vowel." Marichelle used the phonograph not in the customary way, but as a graphic inscription device, whose traces he examined through a corneal microscope and measured with an ocular micrometer. He also became adept at the procedure per-fected by Ludimar Hermann in 1889 of photographically reproducing the phonograph records and analyzing the sinuous curves whenever possible with the harmonic principles of Fourier's theorem and Ohm's law.[83] Mar-ichelle used this method to produce a complete series of drawings of the phonographic grooves for French vowels.

An additional advantage of this approach over customary physiologi-cal methods, Marichelle claimed, was that a single phonographic instru-ment could be used to capture most of the multiple inscriptions taken in earlier studies by different devices.[84] The deaf-mute instructor reckoned that all phonemes, vowels and consonants alike, take their characteristics from the region in the vocal apparatus from which they emanate. They became vowels or consonants only by virtue of a further modification, the degree of openness or closure in the "generative orifice." Marichelle insisted that it was useless to teach vocal sounds in isolation, not as they are used within words, where minute differentiations occur depending on the articulative conjunction.

After 1891 Marichelle worked together with Marey's assistant (*pré-*

parateur), Georges Demeny, to produce an "optical equivalent of the phonograph" to monitor speech.[85] Demeny produced numerous close-up chronophotographs of a subject, usually himself, producing short phrases at rates of 15 to 24 images per second. Demeny synthesized the images with an instrument of his design, the "photophone," a peep show device that combined a zootrope or a Plateau phenakistoscope with a light source to produce a moving image of vocalization. (See Fig. 12.6.) Images of Demeny mouthing "Vive la France" and "Je vous aime" quickly spread far and wide, as an improved version of the instrument, the phonoscope, competed with Edison's kinetoscope for popularity in parlors and exhibitions around the world.[86]

Contemporaries exclaimed that Marichelle's use of media technologies—phonograph and chronophotograph—had finally wrenched phonetics "away from the acousticians" and "profoundly changed the theory of speech."[87] These uses in turn gave rise to a veritable explosion of technologies designed to produce transparent and efficient communica-

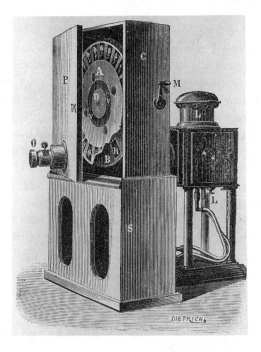

Figure 12.6. Demeny photophone. From Marey, "Inscriptions des phénomènes phonétiques," 449. Courtesy of the Syndics of Cambridge University Library.

The Edison Commercial System

Conducted with the Business Phonograph

FROM BRAIN

TRADE MARK
Thomas A. Edison

TO TYPE

Appliances Manufactured and installed by the
National Phonograph Company, Commercial Dep't
Main Office Orange, N.J.U.S.A.

Figure 12.7. Edison business phonograph. Advertising circular, National Phonograph Company, 1907. Courtesy of the Hagley Museum and Library, Wilmington, Del.

tion using phonetic inscriptions and algebras. Instruments such as Edison's Graphophone (Fig. 12.7) flooded the burgeoning market for technologies of stenography and internal business communications.[88] At the macroscopic level, the Danish linguist Otto Jespersen worked together with an international committee that included Wilhelm Ostwald and the French logician Louis Couturat to develop an "algebra for speech sounds" as the basis for an international auxiliary language for use in commerce, science, and diplomacy.[89]

The Mother('s) Tongue

During the 1880s French linguists turned fervently away from Paris to the provinces, implanting the phonetics laboratory in the byways of the countryside as a means of making visible the elusive and prolific patois that thrived there.[90] "The present state of linguistics," announced one linguist, "imperiously demands a more profound study of the patois that have succeeded in surviving the ever more dangerous attacks of the written language."[91] "Imperious" referred to the politics of the situation where, a century after the revolutionary surveys, government-sponsored education had begun to succeed in eradicating some patois, while other dialects remained stubbornly intact. Continuity of purpose, if not of method, reigned between the original surveys of the Abbé Gregoire and those of the 1880s. To the latter studies, although still devoted to the implantation of French and the eradication of patois, was added the curatorial function of detailed description and when possible, of phonographic archiving of vanishing tongues.[92] The linguistic anthropologist Lucien Adam, for example, requested that his correspondents gather folktales and provide a careful phonetic transcription as well as grammatical characteristics and lists of words.[93]

By the late 1880s field studies of patois had become dominated by the extramural implantation of the phonetics laboratory. Previewing the "Triple Alliance" between physics, physiology, and philology that would define the linguistics of the future Eduard Koschwitz remarked:

The study of patois is the alpha and omega of all historical grammar. To study patois properly, one has to be a true phonetician, that is a naturalist phonetician, physicist, and physiologist. Just as historical grammar, which cannot do without the study of patois, forms an integral part of philology, it will be not only grammar but all of modern philology that becomes a natural science. . . . One has for too long forgotten . . . that languages are composed of sounds, which by their acoustic effect belong to physics, by their formation to physiology, and that the letters of the alphabet are nothing but the very imperfect signs of living sounds of the present and the past. The study of the real value of these past or present letters can only be determined by a scientist who knows how to recognize the emissions of the voice hidden under the letters.[94]

These remarks served as a homily to the work of the Abbé (Jean-Pierre) Rousselot, the first person to take the work of the phonetics laboratory out to the French countryside where he was born and bred. An unusual homecoming, it must be said—but in keeping with the character of the Third Republic—to return with an instrumental arsenal to study the mother tongue, and then to dedicate the work to his mother! After coming to the French capital to study with Gaston Paris and Bréal amid the rising tide of patois studies, Rousselot worked in the laboratory of Marey and Rosapelly and together with France's foremost linguistic geographer, Jules Gilliéron, cofounded a journal devoted to the dialects of France, *La revue des patois galloromans*. This twofold training effectively bridged the gap between the laboratory and field linguistics, disciplines that had hitherto exchanged work but not really merged. "What was still

lacking around 1890 and what one found in Rousselot," stated an obitu-
ary notice after his death in 1924, "was a linguist doubling as an experimenter, a savant knowledgeable about problems of phonetics and capable
of resolving them by perfecting the procedures of the graphic method."[95]

Rousselot's dual interests grew from a life marked by the rapid modernization of rural France in the Third Republic. Born in the village of
Saint Claud, Charente, he spoke patois at home and French in school. He
reflected this divide in his career, vacillating constantly between geographical and genealogical studies of patois and efforts to promote standard French pronunciation. This pairing formed the core of his brilliant
dissertation, a study of the village patois spoken by his own family in
Charente, which he rendered in the cosmopolitan idiom of the graphic
method using instruments partially of his own design. He published the
work in two monographs, one devoted to his findings, the other to the
technical arsenal used in his investigation.[96]

Through an ingenious combination of methods Rousselot depicted
the development and character of his native patois, in contrast to his
acquired standard French. Beginning with a preamble entitled "Why I
Studied my Patois and How I Studied It," Rousselot remarked that he had
been impressed by the recent work of linguistic geographers, especially
Charles, Baron de Tourtoulon's *L'étude géographique sur la limite de la langue
d'oc et de la langue d'oïl*. But he found himself "shocked" to see that the
work "was concerned with the letters rather than the sounds of which the
letter is the symbol."[97] For that reason Rousselot decided that "instead of
studying the dead letter" he would study "living speech." De Tourtoulon's emphasis on geographic boundaries led Rousselot to see the importance of his own native dialect, which marked the limits of the idioms of
the North of France and those of the Midi. He began with a general
reconnaissance of the region, traveling with the help of local priests who
were native to it.

Rousselot's first move sought to enframe his mother('s) tongue within
the compass of larger geographical patterns of dialect, from Charente to
the Allier and the Loire, then he turned to a more detailed study of the
phonetics, morphology, syntax, and lexicology of the patois of Cellefrouin, then to a study of the language spoken within his own family, and
finally to a detailed study of his own mother's speech, which he compared
with his own. Arguing that "speech is composed of a multitude of sounds
and noises of which only the principal ones are represented in alphabets,"

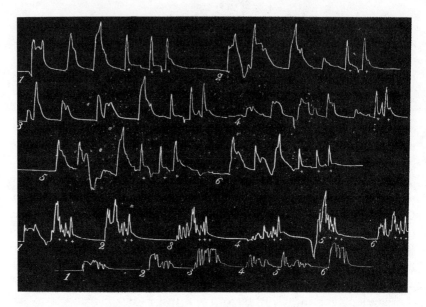

Figure 12.8. Graphic recordings of patois. From Rousselot, *Les modifications phonétiques du langage étudiée dans le patois d'une famille de Cellefrouin* (Paris: H. Welter, 1891). Courtesy of the Syndics of Cambridge University Library.

Rousselot based his study on field samples taken with graphic inscription devices.[98] (See Fig. 12.8.) Variations of sonority, spirographic measures of the breath expended in speech, and the duration and musical pitch of the sounds comprised the principal phenomena under investigation.

Vowels served as a focus of Rousselot's study, as they had in the original surveys of the Abbé Gregoire. In fact, the association of vowels with dialects and patois was an idea of ancient provenance, often reiterated in metropolitan studies of provincial speech. In contrast to the stable, form-giving and -maintaining function of consonants—the only true letters of the written alphabet—vowels were taken to define the mutability of oral culture and thus became virtually synonymous with patois itself.[99] "Patois is variation, and the vowel designs its nature": these two convictions went virtually unchallenged in the traditional conception.[100] To fix the nature of the vowel would thus be to capture the very essence of patois, and in a sense, language in its most natural state. Largely unfettered by literature, teachers, dictionaries, grammars, and state policies, patois, "on the contrary, is uniquely transmitted by oral tradition. Almost nothing disrupts its natural evolution." Hence Rousselot's scientific credo: "With regard to cultivated languages, I would frankly side with the preference the botanist accords the plants of the field over the plants of our gardens."

Rousselot thus configured patois as an element of the natural landscape, ripe for conquest, civilizing, and policing through technologies of writing. To Cellefrouin he brought along an astonishing arsenal of linguistic surveillance equipment. The mobile laboratory contained about a dozen apparatus, all built by the renowned Paris instrument maker Charles Verdin and controlled by Marey and Rosapelly.[101] The core of the inscriptive apparatus consisted of a recording cylinder with a Foucault regulator, a Desprez electric signal, and a Marey *tambour à levier*. The latter device consisted of a rubber membrane with a lever inscriptor that could be easily affixed to diverse organs of the body—the lips and tongue, thorax, larynx, teeth, or nasal passages—with the help of several supplementary appliances. Rousselot's kit contained several simple inscriptive devices of his own invention, such as the "larynx explorer" and the "speech inscriptor." (See Fig. 12.9.) The latter was an attempt to synthe-

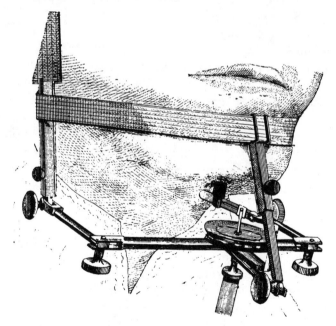

Figure 12.9. Rousselot Speech Inscriptor. From Rousselot, *La méthode graphique appliquée à la phonétique* (Paris: Macon, 1890). Courtesy of the Syndics of Cambridge University Library.

size the principal advantages of Koenig's manometric flames and the Edison graphophone and phonograph. Rousselot used a Verdin microphone with an embouchure borrowed from the Edison graphophone, which enabled one to speak without having to touch the lips to the material. The signal-transmitting apparatus featured strong electromagnets and a powerful battery to overcome resistance and to give all phases of the current without interruptions.[102]

Rousselot controlled the traces of speech with those that rendered the specific bodily organs of articulation. The aim was to determine the exact locations of the palate, larynx, thorax, and so on, from which sprang the articulation of phonetic utterances. These anatomical "regions of articulation" were then arranged spatially on the map of Cellefrouin, showing the parallel distributions of phonetic variation across the entire region.[103] Rousselot's linguistic geography paralleled a common practice among anthropometers, ethnographers, and demographers of producing maps charting geographical distributions of characteristics ranging from stature, cephalic index, hair- or eye-color, to occurrences of epilepsy and color-blindness.[104] (See Fig. 12.10.) Rousselot never published such a map, but he delineated the very same portrait of Cellefrouin in prose. He supplemented his measures with a historical ethnography of the region drawn from municipal and parish records, oral history, and his own family history. The aim of this exercise was to give as complete as possible a diachronic picture of the evolution of the dialects in the region, with emphasis upon the "outside influences"—natural or human—that may have affected the development of dialectological differences.[105] With this movement between linguistic anthropometry and cultural history Rousselot distinguished "the laws that have their raison d'être in our physical nature and those that have their raison d'être in the generative faculty of our spirit."[106]

At the very heart of Rousselot's cartography lay his mother, bearer of his mother tongue and the very embodiment of Mother Nature in the linguistic landscape. In Rousselot's linguistic geography the writing technologies for the surveillance of patois, marshaled as an aspect of the conquest of nature by civilization, converged with what Gillian Rose has described as the masculinist thrust of geographic fieldwork, which makes a fetish of the landscape in images of Woman and Mother Nature.[107] Luce Irigary has described how the equation of woman with space or place constitutes a central aporia of Aristotelianism and its derivatives, especially in Roman Catholic theology: "If, traditionally, in the role of

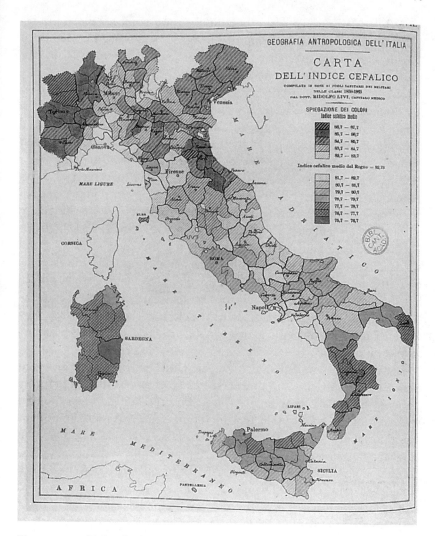

Figure 12.10. Livi anthropometric cartography. Distribution of cephalic index in the Italian peninsula. From Rudolfo Livi, *Anthropometria Militare* (Rome, 1896). Courtesy of the Syndics of Cambridge University Library.

mother, woman represents a sense of *place* for man, such a limit means that she has become a *thing*, undergoing certain optional changes from one historical period to another. She finds herself defined as a thing. Moreover, the mother woman is also used as a kind of envelope by man in order to help him set limits to things."[108]

While Madame Rousselot became the focal point for the entire region of Cellefrouin, the attempts to invigilate and probe her degree of "civilization" were generalized to the entire area under study. Rousselot made efforts to gauge the effects of the French state's attempts to subdue patois and instill standard French in the region. The results were surprising. Standard French had utterly failed to influence the vocalism of *la France profonde*. When it came to relating the vowels of Cellefrouin and common French, Rousselot wrote that "French has still not succeeded in imposing its vocalic system. In adopting the language of Paris, the province has been able to recast its consonants . . . but its vowels have not been affected one bit."[109] Once again patois had remained impervious to the advances of the civilizing mission.[110] Rousselot diagnosed this failure as the constant tendency to privilege the written over the oral language: "It is not spoken French that is spreading; it is the French of books; and this is still more accommodating. Each person only reads the vowels in his own manner of speaking. Education is coming to rectify certain points, but these are few, and our indulgence on these matters is large. In addition, the instructors frequently share the defects of the region where they teach, if they are not even worse. More than one barbaric sound can invoke his paternity."[111]

By the 1880s standard French had fared little better in the region than the metric system, standard time, and a range of other ordering practices imposed from Paris. For linguists, however, the gulf between the oral traditions of the countryside and the language of Paris raised important questions about the possible relations of spoken (*la parole*) and written language (*la langue*), and the associated question of synchronic and diachronic linguistic structures. For many commentators Rousselot's work heralded a new era for linguistics. Young German linguists extolled Rousselot's approach as a means to break the philological hegemony of the Young Grammarians (*Junggramatiker*) by enrolling the strength of German physiology laboratories. Reviews claimed repeatedly that Rousselot's methods of self-registration sounded the death knell of the ironclad sound-laws postulated by historical philology, which rested on the autonomy of language and denied the agency of individual speakers. The following remarks, though anonymous, are typical:

All linguistic research moves in the broad continuum between the individual and humanity: it registers the faintest alterations in the same individual and the most fleeting differences between two persons to the subtle alterations that grow into sharp differences. In the attempt to understand it goes forward from the psychophysical experiment and ends with the consideration of natural influences through which humanity differentiates itself. The real, internal progress of linguistic science seems to me to lie in this anthropological-ethnological direction, and only in it.[112]

In 1893 Rousselot spent the summer semester at the University of Greifswald, where he lectured and demonstrated his apparatus, and sent well-equipped Germans into the bush to study regional accents (*Volksmundarten*) from a fresh standpoint. In 1897 the Leipzig linguist Hermann Breymann published a survey of the phonetic literature that effectively took Rousselot as its hero for abandoning the "subjective" methods of the philologists and "placing phonetics on a purely experimental basis" by using "methods of self-registration long known in other sciences."[113] Many Germans argued that Rousselot's ethnographic phonetic would transform the traditional approach to historical syntax. "Without a more profound study of the different patois," wrote Koschwitz, "there will be no historical grammar and, as a consequence, no comparative grammar."

The French Société d'anthropologie echoed this view when on May 3, 1900, they moved to support the creation of a *Musée glossophonographique des langues, dialectes, et patois* to order and classify the full range of speech patterns not only in France, but throughout the world.[114] The museum, explained its promoters, Paul Azoulay and Julien Vinson, would create a phonic parallel to recent attempts to found a national photographic museum, on the grounds that phonography is to hearing what photography is to sight, and attempts to found photographic museums were underway throughout the world. Azoulay argued that both inscriptive media should be taken as forms of writing. He quoted the promoter of the photography museum, the astronomer Pierre Janssen, to underscore this point: "Photography registers the chain of phenomena during time, just as writing registers the thoughts of men during the ages. Photography is to sight what writing is to thought. If there is any difference, it is to the advantage of photography. Writing is subject to conventionalities from which photography is free."[115] The Société d'anthropologie enthusiastically endorsed the proposal on the grounds that such a glossophonographic archive would extend the important work of Rousselot. One member suggested that its compilers might follow the example of the anthropometric surveys conducted on military conscripts. Fresh recruits would provide a full and accurate picture of the "extreme diver-

N°   Musée phonographique   *N°*

PHONOGRAMME N⁰   à écouter à l'oreille; à distance.

1⁰ LANGUE

    Dialecte
    Patois
    Région géographique
    Non écrite; écrite en caractères

          Phonétique, vocabulaire, phrases usuelles
          Conversation avec   sur
Sujet Enregistré.   Récit (Folklore) sur
          Déclamation sur
          Lecture de

2⁰ CHANT (Nature   ) sur
3⁰ MUSIQUE (solo, concert) nature
    Instruments à : corde   , vent   , percussion
    nommés :

Nom et prénoms du phonographié,
homme ou femme
âgé de   ans solaires ou lunaire
né à
pays ou tribu :   continent :

(Père né à   , Mère née à   )
(habitant à   , habitant à
    habitant : ville; (port); faubourg; village; plaine; montagne; à popu-
    lation dense, rare (combien?   ` ;) depuis quand?
    sédentaire; ayant voyagé; où ?
Lettré ou illettré
Parlant autres langues
Profession

Conditions de l'enregistrement.

ENREGISTRÉ : à voix, ou son bas, ordinaire, fort, très fort,
au cornet : nature   dimension
         à la distance de
au tube parleur
sur phonographe   ; calibre du mandrin
avec diaphragme   pesant
    dont membrane   épaissse de $\overline{100}$ m/m
à la vitesse de   tours (mètres) à la minute
à la température de
à   le
par (nom, prénoms, profession)

Texte imprimé, écrit : transcription, traduction.
N⁰ˢ des photographies : corps   face

         OBSERVATIONS : ¹

Figure 12.11. Musée phonographique survey form. From *Bulletin et Mémoires de la Société d'Anthropologie* 1 (1900). Courtesy of the Syndics of Cambridge University Library.

sity, in one epoch, of the pronounciation and the intonation of a single language according to the social classes and the regions of the country."[116] Echoing Rousselot's simultaneous attempt to inscribe and eradicate his mother('s) tongue, Azoulay remarked without any sense of irony or perversity that mechanical reproduction could preserve for civilization what it had itself exterminated. "Thus," he concluded, "one could save from death the voice of peoples who are themselves dead."[117] The Universal Exposition of 1900 in Paris, then newly underway, would provide an excellent opportunity to begin such a collection, just as anthropological study had profited from previous international exhibitions.[118] Within weeks Azoulay had secured the assistance of Sir Robert Hart, customs officer for the Chinese colonies, and produced a series of phonographic recordings of the sixteen principal dialects of China.[119]

International interest in the project mounted immediately after the French voted to support the museum. The Vienna Academy of Sciences passed a resolution to support a similar institution within weeks of the French vote.[120] In early June, Azoulay called upon the Société de'anthropologie to move quickly on the matter, so that they could derive as much advantage as possible from the Universal Exposition. Azoulay proposed that an international commission be convened immediately to establish standards for the apparatus, accessories, and the methods of glossophonographic collecting and reproduction.[121] It was highly important to proscribe attempts of instrument makers, especially the Americans who, "jealous of being commercially supplanted," introduced unbelievable variability in the market. In addition to research that would "encourage all inventions capable of perfecting still more the present phonograph, or even replace it with another apparatus for recording and reproducing the sounds," it was important to establish the methods of classifying and retrieving the images.[122] Azoulay and Vinson proposed for consideration an "identity card," reminiscent of Alphonse Bertillon's *portrait parlé* used in the judicial identification service of the Paris police department, to be filled out and filed for each sample.[123] (See Fig. 12.11.) In addition to information pertaining directly to the vocal recording, the identity card would also classify personal, geographic, ethnographic, and familial information about the speaker. Through the most sophisticated methods of French bureaucracy, the mother tongues of all the world would be technically reproduced and stored for posterity in a Parisian archive. And this at the very moment when Rousselot's apparatus was being exhibited in the "methods of colonization" section of the Paris Universal Exposition, in an exhibition of the Alliance Française's techniques of teaching French to colonial peoples.[124]

Saussurean Values

Disciplinary tradition in linguistics has taken the work of Ferdinand de Saussure as its theoretical canon and point of departure, and elevated the Genevan to the status of patriarch. As often happens in science, sanctification eradicates context; the "clean" and theoretical tone of Saussure's *Cours de linguistique générale* makes scant reference to the people and work recounted above.[125] Yet conceptual fissures and the late nineteenth-century politics of language can still be read throughout the text, above all in the trademark notions of langue/parole, synchrony/diachrony, and linguistic values. The rest of this paper shall be devoted to delineating the role, tacit and acknowledged, of laboratory phonetics in Saussure's *Cours*.[126]

Without the physiological study of speech in vivo, Saussure acknowledged, linguistics could have never overcome the bias of traditional philology toward the written word. The linguist needed

a natural substitute for the artificial aid. But that is impossible unless he has studied the sounds of the language. For without its orthographic sign a sound is something very vague. We find ourselves at a loss without a system of writing, even if its assistance is misleading. That is why the first linguists, who knew nothing about the physiology of articulated sound, constantly fell into these pitfalls. For them, letting go of the letter meant losing their footing. For us, it means taking a step toward the truth. For the study of sounds will provide us with the help we need. In recent times linguists have learned that lesson. Turning to their own account the investigation begun by others (including physiologists . . .), they have provided linguistics with an auxiliary science which sets it free from the written word.[127]

Saussure regarded speech as fundamental to language, inasmuch as it provided a fundamental "acoustic image" of a given linguistic phenomenon—a term he took from Bréal's "vocal image"—which referred to the sound pattern imprinted upon our memory and which we could reproduce more or less at will. The notion of the acoustic image came directly from the graphic phonetic studies described above, and referred directly to the materialities of communication that gave precise metrological values to speech. As indicated by two of his most famous epigrams, "in the language itself, there are only differences" (*dans la langue il n'y a que des différences*)[128] and "the language itself is a form, not a substance" (*la langue est une forme et non une substance*),[129] Saussure envisioned a living language as a vast array of relative and contrasting acoustic images, distinguished from one another by measurable degrees of difference. In principle, these could be referred to the differences in the graphs produced in phonetics laboratories. But for the linguist, this graphic image could remain for the most part a reminder of the way language was calibrated in the ear and brain.[130] Every speaker of a language had to be psychophysically equipped to properly encode and decode all semantic, grammatical, and phonological signs within the same margin of accuracy as other speakers. The resulting circuit, called by Saussure a *chaîne parlée* or a *chaîne acoustique*, resembles an acoustic translation of the Morse code, where dashes and dots become beeps and bips.[131] In their fully standardized form, where individual speakers produce sound-waves identical enough to be recognized without error, these sound-configurations and their attendant semantic and grammatical relations, become what Saussure called "linguistic values."[132] These are, in the first instance, differential: a linguistic sign's "most exact characteristic is to be what the others are not." That is what enables an individual speaker to connect the right sounds with the right meanings: "The sound of a word is not in itself important, but the phonetic contrasts which allow us to distinguish that word from any other. That is what carries the meaning."[133]

Linguistic values, Saussure argued, are fundamentally collective institutions, although they can only be made manifest in individual speakers of the language. They exist, like the *langue* itself, "perfectly only in the mass" and the individual "is powerless either to create it or modify it."[134] Instead the individual speaker bears the language like a rough-and-ready glossophonographic museum in the brain, calibrated psychophysically and anthropo-geographically, capable of being reproduced within a margin of accuracy at any time. "The individual's receptive and coordinating faculties build up a stock of imprints," Saussure claimed, "which turn out to be for all practical purposes the same as the next person's."[135]

The Saussurean individual speaker thus occupies a relation to the language community much like the relation between the Durkheimian individual to the collective. "If we could collect the totality of word patterns stored in all those individuals, we should have the social bond which constitutes their language. It is a fund accumulated by the members of the community through the practice of speech, a grammatical system existing potentially in every brain, or more exactly in the brains of a group of individuals; for the language is never complete in any single individual, but exists only in the collectivity."[136] Saussure's argument for the nature of the social bond directly parallels that of Durkheim. In the *Rules of Sociological Method*, the sociologist wrote similarly of a collective fund, from which the coinage of communication is drawn: "The system of signs I use to express my thought, the system of currency I use to pay my debts, the instruments of credit I utilize in my commercial relations, the practices followed in my profession, etc., function independently of my own use of them. And these statements can be repeated for each member of society. Here, then, are ways of acting, thinking and feeling that present the noteworthy property of existing outside the individual consciousness."[137] Saussure assumed that the social body must remain stable and cohesive enough to enable an unerring recognition of acoustic images in the circuit of transmission between speakers. The work of standardizing the language-system, therefore, involves most aspects of the maintenance of the social bonds themselves, in addition to those pertaining to strictly linguistic concerns.

In a carefully chosen move that underscored the nexus of metrology and sociology, Saussure drew his principal illustrative metaphor for linguistic value from economics, specifically from monetary theory.[138] Like units of currency that express the worth of the whole monetary system in delimited parts, linguistic values become comprehensible to thought by an analogous process of segmentation or articulation. Equating phonic substance with linguistic units, Saussure asserted that sound "is not a mold into which thought must of necessity fit its forms, but rather a plastic material which in turn divides itself into distinct parts in order to furnish thought the signifiers it requires."[139] The significance of the single unit of value is thus sustained by the structure. In the formulation of Bréal, from whom Saussure borrowed the notion of value, speakers of a language "imitate bankers who, when they handle values, treat them as if they were the coin itself, because they know that at a given moment they could change them for the coin."[140] Saussure took the metaphor a step further, arguing that coins arbitrate not only the value of bread, but also the worth of other coins, including perhaps those of another system, such as the relation of a franc to a dollar.[141] Value, he argued, always involves (1) something *dissimilar* that can be exchanged for the item whose value is under consideration, and (2) *similar* things that can be *compared* with the item whose value is under consideration.[142]

Saussure remained at pains to ensure that the substance of which the coin was made had no bearing on its value. "It is not the metal in a coin which determines its value," he asserted.[143] Thus there could be no gold standard in language any more than a Walrasian economist could support one in monetary policy. Coinage, monetary or verbal, must remain in all cases arbitrary. Value can only be determined by a given unit's relational placement within the system, and not by reference to an external measure.[144] Words, then, are "digitalized," to use Harry Collins's term for the process of rendering signs arbitrary and constant.[145] They are already always calibrated, and therefore remain more robust than they would be if they simply bore the values of their constituent materials.

Saussure thus insisted on denying the importance of the signifier's material embodiment: it should matter no more that words are borne by sound waves, than it does that coins are made of metal. Similarly, linguistics in its ultimate form—semiology—must be shorn of all traces of the productive scene of speech, all vestiges of bodily labor and the material channels of communication (textual or phonic), in order to treat language formalistically, as "a system of pure values." "The vocal organs," Saussure argued, "are as external to the language system as the electrical apparatus which is used to tap out the Morse code is external to that code."[146] Saussure therefore replaced the primacy of the vocal apparatus with the notion of "articulation":

One definition of *articulated speech* might confirm that conclusion. In Latin, *articulus* means a member, part, or subdivision of a sequence; applied to speech, articulation designates either the subdivision of a spoken chain into syllables or the subdivision of the chain of meanings into significant units; *gegliederte Sprache* is used in the second sense in German. Using the second definition, we can say that what is natural to mankind is not oral speech but the faculty of constructing a language, i.e., a system of distinct signs corresponding to distinct ideas.[147]

With the notion of articulation Saussure resolved several antinomies at once. In the term itself he bridged the chasm between speech and language, and by making sign-production "what is natural to mankind," he dissolved the seemingly homologous opposition between nature and civilization.[148] But most crucially he deferred the ontology of the language-system from the individual speaker to the language-community. Although the relations between signifiers and signifieds remained arbitrary, the viability of the system as a whole vouchsafed its naturalness.

In the case of the French language in the late nineteenth century, the viability of the language derived in large measure from the willingness, if not zeal, of the French state to arbitrate and guarantee its standards. In this sense language policy corresponded closely with initiatives to promote norms and standards in other spheres of practice—this was a raison d'être of the Third Republic. The spatial imperative of long-distance control through telegraph systems gave way to the necessity of sustaining society in time, less by imparting influence than by creating, representing, and maintaining ritually shared beliefs. Linguistic values, like other values, had to be arbitrary in order to remain robust, but natural in order to remain valid. In this sense standards were semiotic and semiotics were standards.

Experimental Systems, Graphematic Spaces

❖

In recent years, it has become fashionable among historians and philosophers of science to turn their attention to the field of experimentation. We are witnessing, after the Kuhnian move from continuity and verity of scientific knowledge to discontinuity and relativity, another turn—from the Kuhnian predilection for science as theory to post-Kuhnian engagement with science as experiment. A philosophical landmark in this move has been Ian Hacking's *Representing and Intervening*. Hacking reminded philosophers of science that experiments have a "life of their own."[1] Historian of science Peter Galison[2] has taken up this challenge and argued for "a history of experimentation that accords that activity the same depth of structure, quirks, breaks, continuities, and traditions that we have come to expect from theory."[3]

A growing history and philosophy of science industry today is carrying on that move and amalgamating it with what has come to be labeled "Science as Practice and Culture."[4] "Social construction of science" has become a shibboleth for those wishing to be members of the club. Actors, interests, politics, power, and authority have acquired the status of key terms in a "strong program" to treat science on a par with any other cultural activity whatsoever. That Thomas Kuhn is "among those who have found the claims of the strong program absurd: an example of deconstruction gone mad,"[5] might not surprise, and might perhaps be put aside as a matter of taste. From a fundamental epistemological point of view, it is Bruno Latour who most explicitly and most radically has called attention to an impasse of "science and society" studies from which there seems to be no easy escape.[6] To put it crudely: What do we gain by substituting "social conditions" for what has been taken as "natural conditions" of scientific activity?[7] If, in the perspective of social construction, we have lost the illusion of an ultimate reference called "nature," what do we gain by trying to compensate for this loss with the mirror image of "society" as a new and insurmountable reference? From where do we hope to derive its epistemic legitimacy? With the tetragonic opposition of theory and practice, nature and society, we remain, despite all rotation of competences, within the confines of a conceptual framework that Jacques Derrida without doubt would qualify as the logocentric legacy of occidental metaphysics.[8]

This essay does not claim to transcend these confines with an encompassing gesture. Its purpose is more humble and modest. It sticks to the Derridean program of reworking such oppositions *from within*, of trying to render their limits different/deferent. It starts, therefore, from the more narrowly conceived Hacking and Galison move from theory to experiment within the realm of scientific activity and develops a framework in which experimentation takes on meaning as a set of specific kinds of *epistemic practices*. Nevertheless, this essay is ambitious: it tries to characterize those structures as hybrids that are recalcitrant to classification in either realm, the natural or the social, the theoretical or the practical.

Experimental reasoning, then? Even this expression can easily be misunderstood. Its grammatical structure presupposes reasoning as the *genus proximum*, whose specific difference is to be guided by experiment. What is at stake, however, is just the opposite. It is a kind of movement oriented and reoriented by generating its own boundary conditions, *in* which reasoning is swept off by tracing, a game of material entities. Gaston Bachelard has spoken of instruments as "theories materialized"[9] and has concluded: "Contemporary science thinks with/in its apparatuses."[10] And he has spoken about a "scientific real (*le réel scientifique*) whose noumenal contexture it is to be able to orient the axes of the experimental movement."[11] In analogy to Wittgenstein's well-known expression, we could call this a *tracing-game*. Wittgenstein says: "I shall also call the whole, consisting of language and the actions into which it is woven, the 'language-game.'" And he continues: "Our mistake is to look for an explanation where we ought to look at what happens as a 'protophenomenon.' That is, where we ought to have said: *this language game is played*."[12] We are never able to get behind this weaving. Thus, I am not looking for a "logic" in the relationship *between* theory and experiment, or for a "logic" *behind* experiment. Rather, I am grappling with what must be seen, irreducibly, as the *experimental situation*: in this situation, there are scientific objects and the technical conditions of their existence, differential reproduction of experimental systems, conjunctures of such systems, and graphematic representations. All of these are notions related to the practical process of producing what I shall call "epistemic things."

Briefly, I argue along the following lines: First, *experimental systems*[13] are the working units a scientist or a group of scientists deals with. They are simultaneously local, social, institutional, technical, instrumental, and above all, *epistemic* units. My approach is biased towards this last aspect.[14]

Second, such systems must be capable of *differential reproduction* in order to behave as a device for producing epistemic things whose possibility is beyond our present knowledge, that is, to behave as a "generator of surprises."[15] "Differential reproduction" refers to the allowance, if not to the necessity of shifts and displacements within the investigative process; in order to be productive, an experimental system has to be organized so that the generation of differences becomes the reproductive driving force of the whole experimental machinery.

Third, experimental systems are the units within which the signifiers of science are generated. They display their dynamics in a *space of representation* in which graphemes, material traces, are produced, are articulated and disconnected, and are placed, displaced, and replaced. Science "thinks" within its spaces of representation, within the hybrid context of the available experimental systems. Graphemes are to be understood as the primary, material, significant units of the experimental game, and at the same time, the units of reference. At the bench, the experimental scientist engraves traces into a material space of representation; more precisely, the scientist creates a space of representation through graphematic concatenations that represent the epistemic thing as a kind of "writing." All this is to be understood as a preliminary step toward a history of *epistemic things*.[16] "There is a history of science, not only of scientists, and there is a *history of things*, not only of science."[17]

In order to exemplify my points, I shall draw upon some episodes in the experimental history of the construction of an *in vitro* system for protein biosynthesis.

A Future-Making Machine

What do I mean by the notion of "experimental system"? Traditionally, in philosophy of science, an experiment is seen as a singular instance, as a dramatized trial (*tribunal en scène*) organized and performed in order to corroborate or refute theories.[18] Quite some time ago, Ludwik Fleck drew our attention to scientific—especially biomedical—research practice and argued that, in contrast to what philosophers of science might assume, the experimentalist does not deal with single experiments. "Every experimental scientist knows just how little a single experiment can prove or convince. To establish proof, an entire system of experiments and controls is needed."[19] According to Fleck, in research we do not have to deal with single experiments in relation to a clearly defined theory, but with a complex experimental arrangement designed to produce knowledge that we do not yet have. Even more important, we deal with systems of experiments that usually do not provide clear answers. "If a research experiment were well defined, it would be altogether unnecessary to perform it. For the experimental arrangements to be well defined, the outcome must be known in advance; otherwise the procedure cannot be limited and purposeful."[20]

Like Fleck, I consider an experimental system to be a unit of research, designed to give answers to questions we are not yet able to ask clearly. In the typical case, it is, as François Jacob has put it, "a machine for making the future."[21] It is a device that not only generates answers; at the same time, and as a prerequisite, it shapes the questions to be answered. An experimental system is a device to materialize questions. It cogenerates, so to speak, the phenomena or material entities and the concepts they come to embody. A single experiment as a sharp test of a properly delineated conception is not the simple, elementary unit of experimental science, but rather the *degeneration* of an elementarily complex situation.[22]

One of the first in vitro systems of protein biosynthesis constructed from components of a rat liver cell sap may provide an illustration for an experimental system. In its outlines, it was established between 1947 and 1952 at the Collis P. Huntington Memorial Hospital of Harvard University at the Massachusetts General Hospital, in the laboratory of Paul Zamecnik. In 1947, the Harvard group had set out to look at growth deregulation in malignant tissue—a classical biomedical research project in a hospital that, under the directorship of Joseph Aub, had long been devoted to cancer research. Because growth could be assumed to be closely related to the making of proteins, protein biosynthesis was one of the possible targets of the neoplastic behavior of cancer cells. Little was known about the factors involved in carcinogenesis at that time. So Zamecnik decided not to consider "a single avenue of biochemical study," but rather to start with a "practical" approach and "take advantage of whatever new opportunities became available, in the hope that a definite clue turned up in any corner of the field."[23]

The "practical approach" consisted in the introduction of radioactive carbon-labeled amino acids as "tracers" for following the incorporation of amino acids into proteins. They were synthesized by Robert Loftfield, who then was a research associate at the Radioactivity Center of MIT.[24] Loftfield had succeeded in producing carbon-labeled alanine and glycine in amounts suitable for biochemical research. What a few years later proved to be one of the most potent technical tools for tracing protein metabolism was itself part of the research program at the beginning. At the outset, it was not at all clear which one of the different amino acids should be used for the incorporation reaction, and what should be done in order to circumvent the possible tracing of metabolic processes other than protein synthesis because of the prior metabolism of amino acids. Workers in the field were confronted with—and indeed constantly haunted by—the possibility that what they observed as "uptake" or "incorporation" of radioactive amino acids might turn out to be a side reaction with respect to the protein synthetic activity within the cell. But how proteins were made there was precisely the unknown process. Because usually no more than 1 percent of the radioactivity added to the system became "incorporated," there was not only a realistic chance of some bonding other than regular alpha-peptide bonding—known to be characteristic for proteins—but also the possibility of a rather unspecific "adsorption." Not only was it unclear what "uptake" or "incorporation" of radioactivity meant, uncertainty also remained for years as to whether the same process was being observed in different experimental systems aiming at protein synthesis and derived from different tissues.[25]

Uncertainty was also a given with respect to the choice of the specific experimental system itself. Should one stick to injecting radioactive amino acids into living animals, thus taking advantage of their regular metabolic turnover? This rendered the measurements difficult, and it required lots of radioactive material. Should one try to avoid these shortcomings and work with liver slices in the test tube? But this could cause metabolic distortions of an unknown order of magnitude. Should one work with so-called "model systems" involving proteolytic enzymes? This implied a theoretical assumption—protein synthesis as a reversal of proteolysis—that was current, but by no means substantiated. Should one try to homogenate the tissue or even take the cell sap apart by means of differential centrifugation? This promised to identify the components involved in the process, but a cell homogenate was qualified "at present to be a biochemical bog in which much effort is being expended to reach firm ground."[26]

If we take a closer look at the laboratory process between 1947 and 1951, we see that all of these possibilities were explored, that this exploration was centered around the use of radioactive tracing, and that initially it was organized from what could be called a "significant difference." The significant difference was that malignant tissue appeared to take up considerably more radioactivity than did normal tissue. But experimentally this difference turned out to be silent, for it did not tell what to do next. Above all, it did not tell which of the systemic alternatives should be pursued. In this situation, a differential signal, quite surprisingly, came from a control. A particular substance called dinitrophenol (DNP), which Fritz Lipmann had found to inhibit the process of phosphorylation,[27] inhibited the "incorporation" of radioactivity in the rat liver-slice system, too.[28] This could mean that phosphorylated compounds like adenosine triphosphate (ATP) might be involved in protein synthesis. That initiated a change in the research perspective. The medical point of view gradually became replaced by a biochemical perspective. To find out whether a high-energy intermediate was involved in protein synthesis was no longer a question of the differential behavior of normal and malignant tissue. The DNP event also clarified the options. In order to get the system on the biochemical track, the cells had to be homogenized. Conditions had to be found under which the protein synthesis activity could be checked against compounds such as ATP or similar phosphorylated substances. The technical objective at this point became to separate the cellular components in such a way that the cellular compartments *producing* energized compounds could be distinguished from the cellular structures that *used* them.

The "Logic" of the Process

A scientific object or epistemic perspective in the framework of such an experimental system is as inherently open as the system itself with respect to its technical potentials. An epistemic thing may not even be imagined when an experimental arrangement is in the course of being established. But once a surprising result has emerged and has been sufficiently stabilized, it is difficult to avoid the illusion of a logic of thought and even a teleology of the experimental process. "How does one re-create a thought centered on a tiny fragment of the universe, on a 'system' one turns over and over to view from every angle? How, above all, does one recapture the sense of a maze with no way out, the incessant quest for a solution, without referring to what later proved to be *the* solution in all its dazzling obviousness?"[29] An experimental system can be compared to a labyrinth whose walls, in the course of being erected, simultaneously blind and guide the experimenter. The construction principle of a labyrinth consists in that the existing walls limit the space and the direction of the walls to be added. It cannot be planned. It forces one to move by means of checking out, of groping, of *tâtonnement*.[30]

The coherence over time of an experimental system is granted by the reproduction of its components. The development of such a system depends upon eliciting differences without destroying its reproductive coherence. Together, this makes up its differential reproduction. The articulation, dislocation, and reorientation of an experimental system appears to be governed by a kind of movement that has been described as a play of possibilities (*jeu des possibles*).[31] With Derrida, we might also speak of a "game" of difference.[32] It is precisely the characteristic of "fall(ing) prey to its own work" that brings the scientific enterprise to what Derrida calls "the enterprise of deconstruction."[33] On the part of the experimenter, it requires acquired intuition (*Erfahrenheit*) in order to play the game.[34] Experiencedness is not experience. Experience is an intellectual quality; experiencedness is a form of practice.

I would like to add a few remarks with respect to the differential reproduction of experimental systems. The first is that one never knows exactly where it leads. As soon as one knows exactly what it produces, it is no longer a research system. An experimental system in which a scientific object gathers contours and becomes stabilized, at the same time must open windows for the emergence of unprecedented events. While becoming stabilized in a certain respect, it must be destabilized in another. For arriving at new "results," the system *must* be destabilized—and without a previously stabilized system there will be no "results." Stabilization and destabilization imply each other. If a system becomes too rigid, it is no longer a machine for making the future; it becomes a testing device, in the sense of producing standards or replicas. It loses its function as a research tool. It may, however, be integrated as a stable subsystem into another, still growing experimental system, and help to produce unprecedented events within a larger field. This transformation of former research systems into stable, technical subsystems of other research arrangements is what confers its own kind of material information storage on the process of experimentation. But by the same mechanism, it generates a historical burden. Most new objects, therefore, are first shaped by old tools. On the other hand, scientific objects are continually transformed into technical devices, and in the long run become replaced by devices that embody the current, stabilized knowledge in a more suitable way. The historian of science usually looks at a "museum of abandoned systems."

In order to remain a research system, therefore, such machinery must be operated differentially. If it is organized in a way such that the production of differences becomes the organizing principle of its reproduction, it can be said to be governed by or to create that kind of subversive and displacing movement Jacques Derrida has called the "differance."[35] "Differance" operates at the basis of what has become known as deconstructions, as Derrida prefers to say in the plural: "a certain dislocation which repeats itself regularly . . . in every 'text,' in the general sense I would like to attach to that name, that is, in experience as such, in social, historical, economic; technical, military reality."[36]

How differential reproduction operates on the level of experimental systems can be seen if we pursue the work of the Huntington protein synthesis group. Besides radioactive tracing, a new technique of representation was introduced at the time of the DNP-induced reorientation. After mechanical disintegration of the liver tissue, the resulting cell sap was differentially fractionated by means of a laboratory centrifuge. Differential centrifugation of liver-cell homogenates was not in itself a new endeavor.[37] The problem was that the whole fractionation process only made sense if the "incorporation" activity that had been observed in the experiments with animals and with tissue slices could be preserved while the shape of the system was completely changed from in vivo to in vitro. It does not come as a surprise, therefore, that the first detailed report about such a fractionated incorporation system appeared only in 1952.[38] This report is remarkable in several respects. It reflects how, in the process of the establishment of a cell-free protein synthesis system, the background complexity alluded to as "biochemical bog" was dealt with. At that critical stage, the fractionated incorporation signal still was too "dirty" to be unequivocal, but already sufficiently "clean" to produce some counts that allowed one to regard the fractionated system as a successor to the previous in vivo studies. On the other hand, the experiments yielded no new information with respect to the energy dependence of the process at issue. There was an answer, but again, it was not an answer to the question that

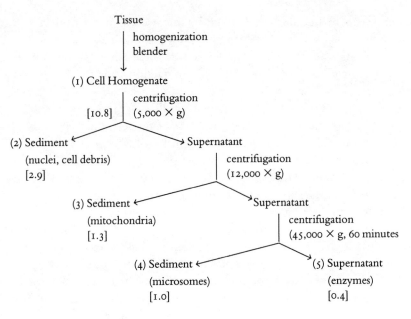

Figure 13.1. Fractionation diagram. The numbers give the activity of the fractions in counts per minute [cpm] per milligram of protein. Reconstructed from Philip Siekevitz, "Uptake of Radioactive Alanine in Vitro into the Proteins of Rat Liver Fractions," *Journal of Biological Chemistry* 195 (1952): 549–65.

had been the driving force for "going in vitro." The essential point was that the crudely structured partition of the cell sap created an interface between two techniques of representation: that of radioactive tracing, and that of differential centrifugation of the cytoplasm, metabolic function, and topology.

In Figure 13.1, we see how representational devices and scientific objects are, to use the words of Michael Lynch and Steve Woolgar, "inextricably interconnected."[39] The content of the liver cells was unpacked into different fractions. The fractions were operationally defined in terms of different centrifugal forces by means of which lighter and heavier components could be separated. Some of the fractions *contained* cellular components that could be identified by microscopic inspection (nuclei, mitochondria)—yet another layer of representation. This did not mean, however, that the fractions were defined by these components. Rather, the centrifugation velocity determined the fractions, and the partition in turn determined the provisional structure of the scientific object. Soon, the language in which the experimental handling of protein synthesis was captured also came to reflect the intimate packing of technical conditions and scientific object. The laboratory began to speak of protein synthesis in terms of centrifugal velocities and of sedimentation properties. So, "45,000 X g supernatants," and "12,000 X g sediments" appeared. These entities represented a new kind of experimental reasoning.

Another interesting aspect of this early fractionation work is the establishment of meticulous procedures for washing, isolation, and identification of radioactive protein. They were to ensure that the radioactivity that could be recovered from the samples was indeed "incorporated" via peptide bonds. These procedures provided, on the one hand, a framework for sorting out one specific metabolic event: peptide-bond formation. But on the other hand, they did not simply filter this single metabolic event out of a tremendous background of "contamination." They were nontrivial conditions—not merely to be granted in order to obtain "reproducible" results. They interfered with the metabolic signals of the experimental system: they destroyed what was "not wanted," and worse, prevented access to things unknown and beyond the actual scope of investigation. To make the formation of stable peptide bonds a "hard fact" meant to do away with any labile amino acid-bonding to other substances. The rigorous procedures for product analysis deemed necessary in order to ensure the identity of the scientific object in its product form prevented access to the conditions of its formation. The assumption underlying the construction of an in vitro protein synthesis system—namely, that the "incorporation" of amino acids had to be accounted for in terms of alpha-peptide bond formation—took the form of a destruction of everything else that was not alpha-peptide bonding.

These experiments provided a first glimpse of a *reconstituted* amino acid incorporation activity. None of the fractions was fully active per se; but when they were recombined, the initial activity of the homogenate was restored. On the other hand, this was the only unequivocal result. None of the main fractions could definitely be sorted out in order to narrow down the synthetic activity. Especially, there was no hint that the mitochondria might be dispensed with; on the contrary. The fraction in which they were predominantly gathered did stimulate the incorporation process. Notwithstanding, the whole system worked only if external energy resources were added. However, ATP as such a resource failed to do this job. And finally, the reconstitution signal itself was on the borderline of resolution. What, then, had been achieved? The centrifugal representation did not destroy any essential component required for the incorporation reaction. The promise was nothing else than the further differential reproduction of the system. Two events, one small, one big, became crucial for the next round of affairs: a modified homogenization method,[40] and a higher fractional resolution by the introduction of an ultracentrifuge.[41]

Along the way, the research problem of the first period had silently disappeared from the scene: the work was no longer directed toward the metabolism of cancer cells. The other research problem had become transformed into a powerful technological device: the incorporation of radioactive amino acids into protein was now fairly well established. And although the search for the role of ATP had not led to any appreciable results with respect to that question, it had led to a readily manipulable fractional representation of the liver cell—to chopping it into several components necessary for the amino acid incorporation reaction.

Graphematic Spaces

Given a research system, and given its formal dynamics as a machine for making the future, how does it organize what I have called, preliminarily, the "tracing-game"? This is a question of representation.[42] What does representation mean? If we speak of tracing, are we allowed to speak of representation at all? After all, the term *representation* implies the existence of a reference. But if we conceive of a scientific object investigated through an experimental system as deployed and articulated within a space of material representation—such as radioactive tracing and centrifugal fractionation—the traditional meaning of "representation" is erased.

We certainly miss the specificity of the procedure if we consider representation simply as a "theoretical" reflection of some kind of "reality." In the research process, what goes on *practically* and on a primary level is the articulation of traces with the help of technical devices that can themselves be considered and manipulated as sufficiently stable embodiments of concepts or theories. Trace-articulations are what I call epistemic things. Once stabilized, they can be transformed into technical devices that allow researchers to produce new research objects. They become implemented into the process of realizing further unprecedented events.

Representation: What goes on when the experimentalist produces a chromatogram, a protein sequence, an array of tubes, to which pieces of filter paper are correlated, on which, in turn, counts per minute of radio-active decay are superimposed? All these epistemic procedures are the objects of an ongoing process of materialized interpretation. They represent certain aspects of the scientific object in a form that is manipulable in the laboratory. The arrangement of these graphematic traces or graph-emes and the possibility of their being articulated in a particular space of representation constitute the experimental "writing-game." Out of these units the experimenter composes what he calls his "model."

What is the status of graphematic articulations? A polyacrylamide gel in a biochemical laboratory, for instance, is an analytical tool to separate macromolecules; at the same time, it is a graphematic display of com-pounds visualized as stained, fluorescent, absorbent, or radioactive spots. The represented scientific object, the embodied model, then is compared to other models, to other representations. Thus, the comparison defi-nitely does not take place between "nature" and its "model," but rather between the different graphematic traces that can be produced within particular spaces of representation. It is *their* matching, not the match between representation and nature, that gives us the sense of "reality" we ascribe to the scientific object under study.[43] The "scientific real" is a world of traces. Bruno Latour and Steve Woolgar have distinguished between machines that "transform matter between one state and an-other," and apparatuses or "inscription devices" that "transform pieces of matter into written documents."[44] This separation often fails to be drawn in a clearcut manner. What is a polyacrylamide gel? It transforms matter—it separates molecules—and it produces an inscription—blue spots, for instance, on its vertical axis. We have to go a step further and look at the ensemble of the experimental arrangement, including both types of ma-chines, as a graphematic activity. A written table or a printed curve then is only the last step in a series of transformations of a previous graphematic disposition of pieces of matter, which is given by the experimental ar-rangement itself.

The production of "inscriptions" is neither an arbitrary process in which anything goes, nor is it completely determined by the technical conditions and the instrumental equipment of the respective system. In the differential reproduction of experimental systems there is a permanent "game of presentation/absentation" going on. For every grapheme is the suppression of another one. Trying to show or enhance a particular trace inevitably means trying to suppress another one. It is as in a game with wedges. If you drive in one, you drive out the other. In an ongoing research endeavor one usually does not know which of the possible traces should be suppressed and which should be made more prominent. So, at least for shorter spans of time, the game of presentation/absentation has to be conducted as reversibly as possible. In other words, the epistemic thing must be allowed to oscillate between different interpretations/realizations.

Experimental systems create spaces of representation for things that otherwise cannot be grasped as scientific objects. A biochemical representation in particular creates an extracellular space for processes assumed to run in the cell under regular conditions. The laboratory language speaks of model reactions here. Models of what? Models of what is going on "out there in nature." Thus, "in vitro systems" would be models for "in vivo situations." But what goes on "within the cell"? The only way to know it is to have a model for it. Thus, "nature itself" only becomes "real," in scientific and technical perspective, as a model. And so in vivo experiments, too, are model systems. In an in vitro system, any intact cell behaves, as Zamecnik once said, as a "whole cell artifact."[45] Paradoxically we are faced with the notion of a natural artifact. There is no external and final point of reference for anything that becomes involved in the game of scientific representation. The necessity of representation *as* intervention implies that any possibility of immediate evidence is foreclosed.

What, then, do scientists do *practically* when engaged in the production of epistemic things? They continuously subvert the opposition between representation in the traditional sense of the word and reality, between model and nature. They treat their scientific objects not as representations *of* something behind, but *as* epistemic things within their system. Thus, they treat representation not as something of another order, not as the condition of the possibility of knowing things, but as the condition of the possibility of things becoming epistemic things. Furthermore, they practice representation as repeated bringing forth; and they re-present in the sense of a repetition, an iterative act. Any intervention in the sense of a scientific representation is a reproduction. Nature cannot be an external reference for this activity. It is experimental nature only insofar as it is *always already* representation, insofar as it is always already an element, however marginal, of the game. With that representation is always already the representation of a representation. As Hacking puts it, the real, as a problem, comes into existence "as an attribute of representations."[46] The same goes for the model. The model is part of what is modeled, and what is modeled itself is always already a model. Jean Baudrillard speaks of a "precession of the model": "Facts no longer have any trajectory of their own, they arise at the intersection of the models."[47] Instead of conceiving the epistemic activity of representing or modeling as an asymmetric relation, we should consider it to be symmetric: both terms of the relation are representations or models of each other.[48]

I will try to exemplify this with another episode from the history of the in vitro protein synthesis system. The episode refers to the "microsomal fraction" of that system. In 1952, it was considered to stimulate the incorporation reaction. Two years later, it had acquired the status of one of the essential fractions. In order to sediment this material quantitatively, the cell homogenate had to be centrifuged at high speed and for a longer time. This could not be done until an ultracentifuge had become available at the Huntington Laboratories in 1953, and it would have made no sense to do before it had become possible to obtain more active homogenates.[49]

In order to examine this high-speed sediment with respect to its particulate appearance, another technique of representation had to be introduced into the system: the electron microscope. The untreated cytoplasmic fraction looked heterogeneous and appeared to be composed of large, irregular chunks of intracellular membrane vesicles, and small electron-dense particles of a somewhat more regular size. Using deoxycholate as a detergent,[50] the membrane vesicles obviously dissolved and the electron-dense particles could be recovered from a $105,000 \times g$ sedimentation run (see Fig. 13.2). These particles, although still varying in size, exhibited an average diameter of about twenty nanometers, and their RNA to protein content was nearly equivalent, whereas in the original microsome fraction the respective ratio was approximately 1:8. This sounds quite straightforward, yet there were, once again, nontrivial difficulties with the technique of representation: what was recovered from the detergent-insoluble sediment in terms of RNA-rich "ribonucleoprotein," in its RNA/protein composition largely depended upon the concentration of the solubilizer. So the representation, or "definition" of the particle was a matter of the preparative operations performed on it, and because the solubilization procedure brought all subsequent incorporation activity in the test tube to a halt, there was no representational correlate to the preparative, operational definition in terms of biochemical function.

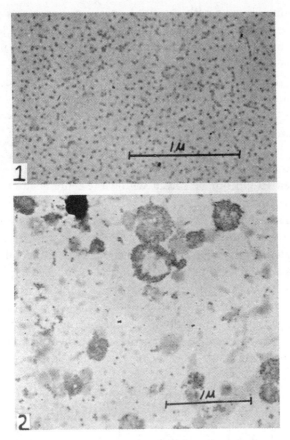

Figure 13.2. Electron micrographs of microsome fractions: (1) micrograph of deoxycholate-insoluble particles from a microsome fraction, magnification 45,900 ×; (2) micrograph of an untreated microsome fraction, magnification 35,200 ×. Reproduced from John W. Littlefield, Elizabeth B. Keller, Jerome Gross, and Paul C. Zamecnik, "Studies on Cytoplasmic Ribonucleoprotein Particles from the Liver of the Rat," *Journal of Biological Chemistry* 217 (1955): 111–23.

In this situation, alternative spaces of representation had to be opened in order to "stabilize" the particle by way of a triangulation or calibration procedure.

One of these representations operated on size and shape. Because the particles had a dimension of only some twenty nanometers, the procedure was bound to the use of electron microscopic visualization. Yet the use of this technique brought with it serious operational problems of another order: that of specimen preparation for inspection. Because electron microscopy is based on the physical interaction of an electron beam with the object to be visualized, the specimen is prone both to destruction by the beam and/or to deformation by the addition of electron-dense heavy metal solutions used to "stain" and fix the biological material. Because of preparation differences, Zamecnik's particles measured between 19 and 33 nanometers, quite a considerable variation, whereas Palade's osmium-treated particles were only 10 to 15 nanometers in diameter.[51] Were the particles homogenous and small, or were they heterogenous and larger? The problem could not be solved within the representational space of electron microscopy.

A further technique of representation brought into play was the sedimentation pattern and sedimentation coefficient of the particles derived and calculated from analytical ultracentrifugation. Zamecnik's particles appeared as a major "47S peak" in the optical record (see Fig. 13.3). This peak resembled the main macromolecular component already described for rat liver by Mary Petermann and her coworkers.[52] A broader peak running ahead of the 47S particle disappeared upon treatment of the material with 0.5 percent of the detergent. But there was also a smaller peak running behind the 47S particle, which did not disappear upon the

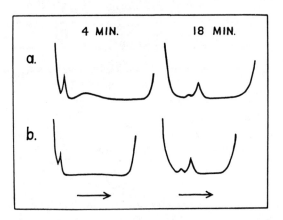

Figure 13.3. Ultracentrifugal analysis of microsomes after four and eighteen minutes at 37,020 rpm in a Spinco model centrifuge. The sedimentation direction is indicated by the arrows. Reproduced from Littlefield, Keller, Gross, and Zamecnik.

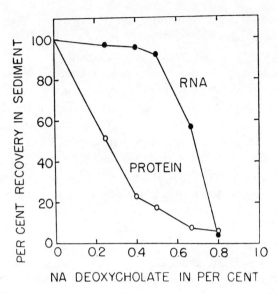

Figure 13.4. Effect of sodium deoxycholate concentration on the recovery of RNA and protein from a 105,000 × g microsomal sediment. Reproduced from Littlefield, Keller, Gross, and Zamecnik.

same treatment. Was the ribonucleoprotein portion of the microsomal fraction itself heterogenous after all? Again, the question could not be answered within the framework of this representational technique alone.

Still another representation of "ribonucleoprotein particles" was the characteristic change of their biochemical composition as a function of the increasing solubilizer concentration. This was a typical biochemical representation. Raising the level of detergent caused the nonsolubilized protein remaining on the particles to decrease more or less monotonically, whereas for RNA, a clearly definable boundary showed up (see Fig. 13.4). Below 0.5 percent deoxycholate, virtually all RNA of the fraction remained in the unsoluble material. Beyond this value, however, the RNA was gradually lost to completion. This biphasic behavior of the RNA with respect to the solubilizer could be taken to point to an edge at which the biochemical representation indicated a qualitative change in the cohesion behavior of the particle.

For all representational techniques, there was no conceivable external reference concerning the shape and composition of the scientific object under preparation. Its shape could not simply be derived by comparison of a "model" particle with a "real" particle; it gradually took some shape from a correlation of representations constructed from different biophysical and biochemical techniques.[53] And because the material was no longer active in the test tube with respect to amino acid incorporation after the different isolation procedures, there was no functional reference for comparison. The experimental representations partially matched each other and partially interfered. The "deoxycholate particle" entered the field of in vitro protein synthesis around 1953, and around 1956 it disappeared again from the scene because no ways were found to render it functionally active. Nevertheless, it had a transitory function quite generally characteristic for the production of epistemic things. It was a tentative invention. It was a step on the laborious path of trying to bring the fractional representation of the cell sap into resonance with some functional sutures of protein synthesis—in the present case, amino acid incorporation into proteins. Ribonucleoprotein particles that were active in vitro became only available some years later in the course of a process that involved the recomposition of the ionic composition of the buffer system, the exchange of the solubilizing material, and the switch to another cellular source for the particles.

Xenotext

In the construction of scientific objects, I see a process in which different representations are made to bear upon each other. Insofar as this process, as a research process, has no predictable outcome, I am inclined to put my constructivist vocabulary in brackets and declare: If there is anything specific for scientific representation, it is to *deconstruct* itself. Within the continuum between epistemic things and technical things, what we usually call a "model" occupies a kind of middle position. As a rule, *qua* epistemic things, models are already sufficiently established to be regarded as promising areas of research and therefore to function as research attractors. On the other hand, they are not yet sufficiently standardized in order to serve as unproblematic subroutines in the differential reproduction of other experimental systems. Thus, an experimental model system has always something of the character of a supplement in the sense Derrida confers on the notion.[54] It stands for something *only the absence of which* allows it to become effective. If the supplement presents itself as a simple addition, it has nevertheless the potential to direct the differential movement of the whole system. The subversion of an experimental system by a supplement shows both aspects characteristic for that movement: it tends to change the identity of its components, and it fails to do so by its very presence as a supplement. This is because a supplement, by definition, tends to be supplemented by "another" one. A model is a model in the perspective of something at which it fails to arrive. It functions precisely in the sense of a Rotmanian "xenotext." I quote from *Signifying Nothing*: "What it [the xenotext] signifies is its capacity to further signify. Its value is determined by its ability to bring readings of itself into being. A xenotext thus has no ultimate 'meaning,' no single, canonical, definitive, or final 'interpretation': it has a signified only to the extent that it can be made to engage in the process of creating an interpretive future for itself. It 'means' what its interpreters cannot prevent it from meaning."[55]

14. RICHARD M. DOYLE

Emergent Power: Vitality and Theology in Artificial Life

❖❖❖ To talk about "another" world than this is quite pointless, provided that an instinct for slandering, disparaging, and accusing life is not strong within us: in the latter case we revenge ourselves on life by means of the phantasmagoria of "another," a "better" life.

—Friedrich Nietzsche, *Twilight of the Idols*

Behind or in parallel to the contemporary technoscience of molecular biology, there is a metaphysics. It assumes many morphologies, and I have attempted elsewhere to map out these metaphysics through the rhetorics of molecular biological discourse. But this is an essay on the vitalization of computers, so here I offer metaphysics under compression: James Watson, codiscoverer of the double helical structure of deoxyribonucleic acid, summed it up concisely when he wrote of Francis Crick's motivation for research in the life sciences: "To understand what life is, we must know how genes act." Although the scientific successes of this reductionist algorithm are undeniable, its remarkable rhetorical and ontological impacts—the implicit understandings of what life *is*—have been less well marked. The conflation of what life "is" with the "action" of a configuration of molecules conventionally represented by an alphabet of "ATCG" produced an almost vulgarly literal translation of Jacques Derrida's famous remark, *Il n'y a pas hors du texte*. Literally, the rhetoric of molecular biology implied, there is no outside of the genetic text. No body, no environment, no outside could threaten the sovereignty of DNA. In this essay I want to focus on a more recent effect or symptom of this conflation of vitality and textuality, Artificial Life.

In an era when biology's new reagent is information, the rhetorical, conceptual move from the notion that "life" is a "text" to the idea that "information" can be "life" is a short one, but it is a move I would like to trace with some care. The notion that cellular automata, computer viruses, or robots could be claimed to live is, of course, encouraged by the tremendous speed, power, and availability of personal computers. But, I will argue, the "power" of powerful computers does not simply reside in their clock speeds or their memory; rather, they rely on their rhetorical software, the tropes that make plausible the technically impossible claim that artificial-life creatures "live." My point here is not to debunk A-life; rather, I aim to use A-life as an exhibit of the contemporary reconceptual-ization of life, and as a case study for the material importance of the rhetoric of science, the sliding signifiers such as "information" and "life" that make possible both the scientific and "ideological" effects of research. Through an analysis of this rhetoric, I will suggest that what emerges from A-life is not only, by any account, "life," but rather a trace or allegory of power.

In invoking "power" here, I follow Foucault's arguments for the pro-ductivity of power. Rather than obstructing knowledge or disabling sci-entific thought, power on this account enables knowledge practices, makes them possible. Foucault's own project was concerned with the historicity of power, as in his description of the shift from the brutality and sovereignty of the scaffold to the "humanity" and "scientificity" at play in the normalizing discourses of criminology and psychiatry, practices that made it plausible to define the "criminal." In my account, I will attempt to diagram the practices and tactics (what Foucault has dubbed a "tech-nological ensemble") that make plausible Artificial Life's claim to deter-mine the "formal properties of living systems" in the unlikely arena of computer science and robotics. Rhetorical softwares play a crucial tactical role in this regime of power, as it is through rhetorics that the uncanny connection between the machine and the organism is installed and man-aged. Dispersed from the unity of the organism, life gets networked, located, and articulated through a computer screen.

Artificial Life and its Rhetorical Substrates

> One aspect of organic life that is lacking in Artificial Life is history.
> —Tom Ray, artificial life researcher

A-life in its present form emerged from a conference held in Santa Fe, New Mexico, in 1987.[1] A synthetic moment in the history of science, it was a self-conscious attempt to crystallize and catalyze work on what Christopher Langton, organizer of the conference, would call the "es-

sence" of artificial life. Encompassing discussions of computer simulation and biological modeling, origin(s) of life, evolutionary theory,
self-reproducing automata, and the history of automata and artificial organisms, the conference offered divergent methodologies and thoughts
on the intersection between life and information. In the proceedings of
the conference, Langton offered a manifesto for A-life as a discursive
center for the various vectors of research:

Artificial Life is the study of man-made systems that exhibit behaviours characteristic of natural living systems. It complements the traditional biological sciences concerned with the analysis of living organisms by attempting to synthesize
life-like behaviours within computers and other artificial media. By extending
the empirical foundation upon which biology is based beyond cabon-chain [sic]
life that has evolved on Earth, Artificial Life can contribute to theoretical biology
by locating life-as-we-know-it within the larger picture of life-as-it-could-be.[2]

I want to make it clear that this new ethos of life as "behavior" stands
in stark contrast to the displacement, for example, of an organism by its
"codescript," as in Erwin Schrödinger's 1944 description in *What Is Life?*
It also differs from Jacques Monod and François Jacob's operon model of
regulation, in which the all-powerful genome "contains not only a series
of blueprints, but a coordinated program of protein synthesis and the
means of controlling its execution."[3] Instead we find a return to the agents
of life. Langton's description of A-life relies on organisms, ongoing projects in negentropy, and self-organization. Whereas for early (and some
subsequent) workers in molecular biology the secret of life was to be
found at the level of the molecule and its effects, A-life seeks once again to
describe what it calls the secret of life's behaviors—flocking, schooling,
and sex. That these behaviors occur in a virtual soup of biology's new
reagent, information, must not obscure the fact that A-life could in fact
refract a new gestalt in the sciences of life, one that emerges out of the
implosion of life and information.

The other obvious and perhaps more peculiar notion to be found in
Langton's manifesto is the idea that theoretical biology is somehow ham-
strung by its inability to "derive general theories from single examples."

Biology is the scientific study of life—in principle anyway. In practice, biology is
the scientific study of life based on carbon-based chemistry. There is nothing in
its charter that restricts biology to the study of carbon-based life; it is simply that
this is the only kind of life that has been available for study. Thus, theoretical
biology has long faced the fundamental obstacle that it is difficult, if not impossible, to derive general theories from single examples.[4]

This call for a comparative biology is a call for a transcendental comparison, an external analysis of life from a distance, a life outside of or above life "as we know it." Nietzsche, writing in *Twilight of the Idols*, describes an analogous structural problem in the judgment of life, one that points to "another reason":

Judgments, value judgments concerning life, for or against, can in the last resort never be true: they possess value only as symptoms, they come into consideration only as symptoms—in themselves such judgments are stupidities. One must reach out and try to grasp this astonishing finesse, that the value of life cannot be estimated. Not by a living man, because he is a party to the dispute, indeed its object, and not the judge of it; not by a dead one, for another reason.[5]

Nietzsche's inquirer into life focuses on quite a different issue, of course— the *value* of life—but his insight into the problem of achieving a transcendental position from which to judge or study life remains. What would seem to be anything but a point requiring finesse—the dialectical opposition of life and death—must be finessed through the rhetorical place of "another reason." This other reason is the limit of reason, reason's other, insofar as it marks a structural limit on thinking about and judging life.

For Langton, too, the essence of life is occulted, hidden by our status as terrestrial, carbon-based hostages to the earth. So too would he seem to be left without any a priori definition of life that would guide his study. Yet Langton sees a way out of this impasse, and its virtual door or medium is the computer. Does Langton, armed not with Nietzsche's hammer but with silicon, overcome the impasse, or is he to be seen as one of Nietzsche's symptoms? To explore our answer, I will look first to the rhetorical path Langton takes in this return to the organism in the age of the molecule. I will then speculate, in the manner of an origin of life cosmologist, on the accidental origin of Artificial Life.

Langton takes the route of Artificial Life, the "synthesis of organisms," with the help of rhetorical precursors and some new technologies. A clue to some of A-life's rhetorical debts can be found in the epigraph of the first volume of proceedings, a quote from the remarkable polymath C. H. Waddington: "It has always been clear that we were not so deeply interested in the theory of any particular biological phenomenon for its own sake, but mainly insofar as it helps to a greater comprehension of the general character of the processes that go in living as contrasted with non-living systems."[6] This "general character," then, is the inquiry into those qualities and processes shared by all living things. Michel Foucault, writing in *The Order of Things*, characterized this notion of the unity of life as a break from the classical emphasis on taxonomy, a move toward the under-lying unity of the modern notion of Life. In the modern comparison of living beings, "the differences proliferate on the surface, but deep down they fade, merge, and mingle, as they approach the great, mysterious invisible focal unity, from which the multiple seems to derive. . . . Life is no longer that which can be distinguished in a more or less certain fashion from the mechanical; it is that in which all the possible distinctions be-tween living beings have their basis."[7]

Rhetorically and scientifically, the description of life altered in the nineteenth century. Specifically, life became an invisible unity, a con-cealed connection, what Foucault called "the synthetic notion of life," an object for scientific inquiry. It is this notion, Foucault claimed, that made biology possible, as it unified the diversity of living beings into an object of knowledge, and not just an object among others, subject to classifica-tion by the natural historian who "is the man concerned with the struc-ture of the visible world and its denomination according to characters. Not with life."[8] Biology's project was, in some sense, to make the univer-sality and invisibility of life visible, or at least articulable.

Thus the biologist is the one for whom life is an issue. Far from self-evident, the invisible unity of life becomes, by the mid-twentieth century, embedded in the rhetoric of secrets, codes, and programs. For Wad-dington, whose theoretical biology sought the "underlying nature of living systems," "basic sentences in languages are programmes. . . . And it is language in this sense—not as a mere vehicle of vacuous information—that I suggest may become a paradigm for the theory of General Biol-ogy."[9] This dual articulation of life and language as "programs"—fueled (for Waddington) by Noam Chomsky's theories of language—makes plausible the analogy between DNA and a computer program.[10] What I have outlined elsewhere as "the age of world scripture," a moment in which the world appears available in its entirety as a code, is a crucial preunderstanding for the linguistic reformulation of the general paradigm of biology.[11] Thus it is no surprise when we read "On the Seat of the Soul" in the second volume of *Towards a Theoretical Biology*, in which computer scientist Christopher Longuet-Higgins opens up the question of computers and vitality: "Computing scientists agree that the idea which made their whole subject possible was that of the stored program. Well, it seems that nature made this discovery about 1,000 million years ago."[12] The rhetoric of the "program" common to both early molecular biology and computer science made plausible the notion that computers could be "alive." This notion of the program, of course, can be traced to the "tape" of Turing's machines. As a universal machine, a Turing machine could theoretically "do everything . . . one particular machine could simulate the work done by any machine."[13] With Watson and Crick's model of the double helix, "one can now point to an actual program tape in the heart of the cell, namely the DNA molecule."[14]

We saw that with the rise of biology, life is no longer seen to be opposed to the mechanical. After Norbert Wiener's *Cybernetics* treated both the machine and the animal as economies of control, the "tape" or "program" can be seen to drive both computation and life. Longuet-Higgins writes:

> Are you suggesting, then, that life is just programmed activity, in the computer scientist's sense of "program"? Because if so, you will find yourself driven into saying that a computer is alive—at least when it is executing a program, and that strikes me as mildly crazy? . . . Fair enough. But I wouldn't put it past computing scientists to construct a machine which we would have to treat as if it were alive, whatever our metaphysical objections to doing so.[15]

This reticence—and the appearance of computer vitality as "mildly crazy"—as well as the fictional, dialogical style of "On the Seat of the Soul," mark the rhetoric of computational vitality as speculative and theoretical. This speculation, however, is not free-floating fantasy; it is a dream grounded in the history of automata and life, a dream based on the scientific desire to "know what life is." Note, for example, that for Longuet-Higgins, the computer lives as an individual, embodied organism, one that requires the animation of hardware by software. This can be seen to be of a piece with the Cartesian theory of automata, in which the animal's body is an automaton shell "animated" by the soul. So too with the great automata of history—medieval clock "Jacks" whirled with the invisible force of Time, Vaucanson's duck ate to sustain the "stench" of its body, and smart bombs are embodied on a platform of camera, missile, and guidance technology, their identity constituted and destroyed through their target. By contrast, the "body" of Artificial Life is often nothing more than a pixel, a flash of signal not unlike a portion of the white noise we see with the impact of a smart bomb. Researchers Henri Atlan and Moshe Koppel recently challenged the rhetoric of DNA as "program." Atlan and Koppel point out that although it is true that the genetic code—the four bases of DNA molecules—and some of its expression are well understood, similar claims cannot be made for the relation of

DNA and cell growth and differentiation: "Therefore, the idea of a computer program written in the DNA and controlling the sequence of events which characterizes cell growth and differentiation is more a metaphor than a result of a detailed analysis of DNA structures as carriers of a real programming language. No real computer-like program organized according to syntactic and semantic rules can be identified."[16]

Atlan and Koppel's analysis underscores the power of rhetorical grids that are used to articulate and organize scientific research. On the one hand, the program metaphor, and more generally the notion that DNA is information, crystallized and/or framed research on everything from regulation to development, as in the 1960 work of Jacques Monod and François Jacob: "The discovery of regulator and operator genes, and of repressive regulation of the activity of structural genes, reveals that the genome contains not only a series of blueprints, but a coordinated program of protein synthesis and the means of controlling its execution."[17]

On the other hand, the notion of a program foregrounded the immanent power of DNA, reinforcing its status as master molecule while occluding the complexities of development and growth. Atlan and Koppel argue that it is the metaphor of DNA as program that has encouraged a reductionist notion of biological function, a paradigm that has made plausible the human genome programs. For Atlan and Koppel, the metaphor of DNA as program masks a lack of understanding: "Nevertheless this lack of a theoretical framework has not prevented the proposal of a research program to sequence the DNA of a whole human genome as a kind of ultimate goal in understanding human nature. . . . Implicit in this proposal is a literal understanding of the genetic program metaphor, looking at the sequence of all the DNA base pairs of a genome as the listing of a computer program."[18] Of course, scientific discourse often deploys metaphors and images from the most recent technologies in its theories, and Atlan and Koppel go on to deploy more timely parallel computer rhetorics in their discussion of DNA. But this begs the question of why the trope of the "tape" or "program" seemed to flow easily from nonliving to living systems. In a sense, it can be seen as isomorphic with the metonymic displacement of an organism by a codescript I have outlined elsewhere.[19] The notion that life is a sequence of instructions, rather than produced by the invention of the computer, is actually rhetorically feasible before the widespread notion of the computer program. Indeed, what made the equation between vitality and information possible was a shift not just in the technology of information, but in the articulation of life.

The gap remains, however, between Schrödinger's notion of a "co-descript" and Jacob and Monod's articulation of life as the effect of a program on the one hand, and the emergence of Artificial Life on the other. Despite the assumption that DNA is a program, the idea that computers can therefore "live" was relegated to science fiction or mathematical speculation. The conclusion of most of the metaphorical crossovers between machines and organisms was that organisms were machines, not that machines were organisms.[20] Even Claude Shannon, author of *The Mathematical Theory of Communication*, warned against exaggerating the application of information theory to biology in 1956:

I personally believe that many of the concepts of information theory will prove useful in these other fields—and, indeed, some results are already quite promising—but the establishing of such applications is not a trivial matter of translating words to a new domain, but rather the slow tedious process of hypothesis and experimental verification. If, for example, the human being acts in some situations like an ideal decoder, this is an experimental and not a mathematical fact, and as such must be tested under a wide variety of conditions.[21]

Thus even in the heady days of cybernetics and information theory, in which both living and nonliving systems were seen to be economies of communication and control, the comparison of vitality and information processing was just that—a comparison. As a theoretical and rhetorical tool this analogy had great effect, and researchers as diverse as Waddington, Jacob, and McClintock all used the figure of the computer to explain and frame their work. Of course, this never led them to work on simulated rather than conventional organisms—this had to wait for John Horton Conway's game, Life, in the early 1970s.[22]

Conway, a Cambridge mathematician, invented the game Life, a cellular automaton, in an attempt to generate complex patterns out of simple rules. In essence, the game consists of a grid, an "unlimited chessboard." Each cell on the grid has eight neighbors, and each neighbor's state (empty or occupied) is determined by the following rules: If a cell is empty, it stays empty unless exactly three of its neighbors are occupied, in which case it will be occupied in the next generation. A cell remains occupied as long as two or three of its neighbors are occupied. These simple rules governing the occupation or evacuation of cells were translated into states: occupied = "alive"; empty = "dead." (See Fig. 14.1.)

How did the cell states come to be regarded as "alive" or "dead"? The constraints of textual articulation make it difficult to demonstrate the

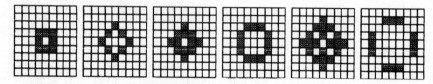

Figure 14.1. Conway's Life. Adapted from LifeMaker 2.1 freeware by Jesse Jones, 1990–94.

uncanny movement of Life cells, but it is the seeming autonomy and unpredictability of cell states that provoke this vitality effect.[23] Mathematician Karl Sigmund's description points to this "autonomy" effect; it is the notion that somehow humans are superfluous to the game: "*Life* is not a two-person game like chess or checkers; neither is it a one-person game like patience or solitaire. It is a no-person game. One computer suffices. Even that is not strictly required, in fact, but it helps to follow the game. The role of human participants is reduced to that of onlookers. Apart from watching the game, one has just to decide from which position to start. All the rest proceeds by itself."[24] I will return to this claim that somehow Life, and Artificial Life, can be separated from its descriptions and observations. For now, I want to note a tension between this description of autonomy and another claim Sigmund makes for Life's method of propagation: "In the early 1970's, at a time when computer viruses were not yet an all too common plague, there was another type of epidemic causing alarm among computer owners. It [Life] used the human brain as intermediate host."[25] Significant here is the recognition that humans occupied a crucial place in the network of machines and codes that made Life possible. As an "intermediate host," they were a necessary element of Life's ecology. Thus, although it is true that no individual move in Life necessitates any human action, the game itself required human "wetware." And although it is no doubt true that the playful aspects of Life were of a cognitive kind, the human brains themselves did not have any unmediated access to Life; they required a crucial interface in order to run Life: language, rhetorical softwares, other humans with which to deploy these softwares.

Still, despite the popularity of Life, it was left to Langton to link up the early efforts of Waddington's theoretical biology group—Howard Pattee, Michael Arbib, and others—with the new hardware and software available in 1987. And yet, this link was not determined by the exponential increase in computer power that took place between 1972—the date of the last conference on Theoretical Biology—and 1987. Rather, it was the tenuous result of the intersection of the rhetoric of life as information, an accident, and some visions.

A Vision of A-Life

> Smashing into the ground shook a whole bunch of neurons loose. I was in and out of consciousness while I was on the ground, and it was interesting to feel my consciousness sort of bootstrapping itself up, going out, then coming back up again.
>
> —Ed Regis, *Great Mambo Chicken and the Transhuman Condition*

Chris Langton, in a bizarre rewriting of the Icarus myth, broke 35 bones in a hang glider crash and experienced a vision of Artificial Life. "Propagating information structures" filled the space between consciousness, bootstrapping, and the void, and Langton—*or whoever it was that was able to situate Langton's lack of consciousness*—found the hidden meaning of the emerging chiasmus between living and nonliving systems. What crystallized A-life and allowed it to emerge as a discipline, an empirical and practical science, was a combination of three vectors, one leading from a rhetoric that equated living and nonliving systems ("propagating information"), new sources of cheap and powerful computers, and a founding event reminiscent of religious, not scientific, narratives.

My point in invoking Chris Langton's crash is not to indulge in petty psychologizing or to trivialize the pain he experienced. I refer to it only in the form of his self-description, which has taken on nearly mythical attributes in the A-life community. I mention it here in an attempt to understand the genesis of A-life—to leave it out would be to overlook one of the major tropes mobilized by A-life participants.[26]

For it is as a trope that Langton's crash is illustrative: faced with a horrible and inarticulable event, Langton explicates the crash in transcendental terms, a consciousness radically exterior to the event he is "himself" experiencing. Both in and outside of the event, Langton's narrative relies on the possibility of occluding the fact that "he" was involved. Langton narrates the crash as if he were watching some other, deploying some "other reason."

Thus it is not just any narrative technique that Langton deploys—it is one that allows him to draw fruitfully on the experience as an originary moment of A-life. This tactic, I will argue, is iterated across the discourse of A-life, and is in fact constitutive of A-life as "life." This tactic is an element of what mathematician and semiotician Brian Rotman has called the Metacode in the context of a semiotics of mathematics. Whereas Rotman refers to mathematics proper—the sanctioned signs for conducting mathematics—as the "code," he argues that such a code can operate persuasively only through the deployment of "the *metacode*, the penumbra of informal, unrigorous locutions within natural language involved in talking about, referring to, and discussing the Code that mathematicians sanction."[27] Similarly, I wish to argue, the discourse of A-life (literally) is made up of code (the computer instructions or blueprints for construction of robots) and a Metacode (the means of persuasively explicating A-life projects as "life"). Jean-Claude Beaune writes of the importance of this kind of metacode or rhetorical software for automata: "An automaton is not just a machine, it is also the language that makes it possible to explicate it. At a more general level, the automaton is the language that endows the people who are meant to know and communicate it with the privileges of totality which rational man thought he no longer had to confront. It is an experience devoid of rest, pity, or distance; the limit of technology becomes the language of the technostructure."[28] In short, Artificial Life—artificial organisms—must not be seen as merely products of the technical inputs of hardware and software. Rather, they are networked with the modes of explication that make it possible to recognize A-life as "life," rhetorical softwares that are themselves not explicit objects of A-life. What makes it possible to explicate artificial life, what makes the claim that synthetic organisms are in fact alive, eat, have sex, and so on, is above all the notion that in essence, organisms are sequences, and these sequences are somehow completely exterior to the human economy that has coincided with their replication. Langton's crash—a crash that, as we shall see, leaves its traces on A-life—can be seen as an origin story of a most religious kind, a combined Icarus and resurrection myth, a new story of transcendentalism told in an old form.

Both literally and metaphorically, the accidental genesis of Artificial Life can be seen as part of what David Lavery has called a culture of "spaciness," in which the extraterrestrial imperative runs strong and deep.[29] Although it seems natural, at this point in American culture, to recreate by getting off the earth, I would suggest that the desire to achieve a position as a transcendental observer, to be off the earth, out of the body, or beyond the living, networks together Langton's sport, his visions, and his research of choice. A closer look at both Langton's 1987 text, "Artificial Life," and the details of its genesis will flesh out and support my claim.

Figure 14.2. Glider. Adapted from LifeMaker 2.1.

A hang glider is, of course, a technology that produces lift, a machine with which one can leave the earth, look down and enjoy the view. It is, in this literal sense, a transcendental machine, a pleasure or sport of looking down and over. It is also clear that Langton, as an observer of his own loss and acquisition of consciousness, somehow, impossibly occupies a similar transcendental (or at least radically reflexive and narcissistic) position. It is perhaps not so obvious that the drive behind A-life, the desire to study alternative noncarbon-based life-forms, also takes a transcendental shape. The metaphors of A-life attest to this: "Artificial Life starts at the bottom, viewing an organism as a large population of simple machines, and works upwards synthetically from there."[30] Although it would, of course, be too hasty to conclude from this convergence of spatial and spacey metaphors that A-life's formal metaphorical structure maps onto the material structure of hang gliding, further analysis of Langton's text will allow this rhetorical matrix to emerge. The takeoff of Artificial Life was helped along by a glider. In his discussion of cellular automata (CA), Langton sees a glider: "Many of these configurations seem to have a life of their own. Perhaps the single most remarkable structure is known as the glider. . . . The glider is one instance of the general class of propagating structures in CA. These propagating information structures are effectively simple machines—virtual machines."[31]

Neither a brief glimpse nor a careful study of the pattern of the "glider" (see Fig. 14.2) reveals the essential glider morphology of the pattern of pixels. Instead, what allows this "glider" automaton to emerge, what explicates it as a "glider," is the rhetorical and metaphysical background of A-life, a background constructed out of, among others, the spatial metaphors of "bottom up," "emergence," and "gliders," metaphors that situate A-life as a transcendental place from which to view traditional, carbon-based life, Nietzsche's "another life" with which we can compare our earthly one. The rhetorics and gestalts of emergence resonate with a "view from above," what Donna Haraway has dubbed the "god trick," a gaze from nowhere and everywhere that characterizes the disembodied objectivity of technoscience. Indeed, the "glider" morphology was not invented by Langton, but by Richard Guy. Working with Conway's game, Life, Guy observed one configuration of the simple on/off cells that moved "purposefully." "When the configurations tumbled off the single Go board, the players would hastily place sheets of paper on the carpet and draw squares on them to extend the grid. It was, in fact, on the carpet that Guy found the glider. Once the glider managed to get away from the main configuration, of course, it headed off the papers, out of the room, and, in theory, out of Cambridge, out of England, out of everywhere."[32] Tumbling and not crashing, the emergent glider floats above "everywhere," reaching a transcendental beyond, an "eternal mist."

Of course, this is not to say that the metaphors that float around A-life determine the empirical practice of A-life, or that these rhetorics somehow turn an empirical practice into a metaphysical, transcendental one. Rather, my claim is that these rhetorics are traces of a very particular style of transcendentalism, one that seeks science's traditional, "detached" position off the earth but not out in space. Langton chooses the silicon path, not the extraterrestrial one, but A-life nonetheless seeks to leave the earth behind, working upward.

I have indicated the problematic nature of Langton's self-reflection, insofar as it required him to abstract himself from his own consciousness to meditate on his own lack of consciousness, a problematic transcendental position. Similarly, the project of A-life, in its logical justification, relies on the very transcendental, metalife position it seeks. A-life, after all, arises out of the desire to find a general theory of living, a theory that takes in the full view of the "general phenomenon of life—life writ large across all possible material substrates."[33] A-life seeks to derive the formal nature of the living system, life's algorithm, by abstracting it from its material, carbon-based prison; "life, as a physical process, could 'haunt' other physical material."[34] Hovering above the science of the actual is the science of the possible, A-life's object, "life-as-it-could-be."

And yet how could we authenticate a siting of this haunting? Given that the genetic problem of A-life is the lack of an adequate definition of life, whence come our criteria for selecting the proper objects of artificial life from the pretenders? The answer is, of course, in practice. The performance of A-life, given the lack of immutable criteria for "life," relies on its rhetorical software for plausibility and fascination. "An automaton is not just a machine, it is also the language that makes it possible to explicate it." This language, metacode, or rhetorical software, despite Langton's goal of finding the "essence" of life, in fact works toward another goal, the evaporation of the difference between living and nonliving systems: "We would like to build models that are so lifelike that they would cease to be models of life and becomes examples of life themselves."[35] In the absence of any given, adequate definitions of life, the plausibility of A-life creatures rests on their ability to simulate something for which we have no original, "lifelike behavior."[36]

Thus we can see a central, centering conflict in the rhetoric of artificial life. On the one hand, Langton, following Waddington, seeks a theoretical basis for a general theory of life. He seeks access to something like the "invisible unity" life became by the nineteenth century, a process always and everywhere the same but occulted. Of course, Langton's universal life need not be occulted; there is no structural limit that must hide life, as is the case with Nietzsche's "another reason." Toward that end Langton seeks what he calls the "essence" of artificial life, the mechanistic formal basis of all living systems, carbon or otherwise, which "must share certain universal features." Indeed, the raison d'être of A-life is a source of life beyond carbon-based life, a view from above where a general theory of life could be found. In this sense A-life is an attempt to determine the formal (if not actual) origin of life, the possible conditions from which any life could emerge. On the other hand, Langton's quest for creatures that blur the boundaries between living and nonliving systems, simulated and real-life, highlights the paradox that A-life depends in its simulations on the ability of computers to generate models of life without any origin or "invisible unity." They are "synthetic" creations, maps for which there is no territory. That is, what makes possible the substitution of the signs of life for life is the reproducibility of "lifelike behavior," a reproducibility that ultimately points to the fact that A-life organisms are themselves reproductions, simulations cut off from any "essence" of life.[37] With no origin in vitality, no "parents," the emergent creatures of A-life float unattached to any anterior essence. This "unmoored" nature of A-life simulations threatens the very project of a unified account of life, as they continually threaten the definitional coherence of "life." Part of these simulations—the crafting of lifelike behavior—are the rhetorics and descriptions of the A-life project. Without a rhetorical and historical tendency to metaphorize life in terms of information, artificial life, as a practice, would be of no more interest than other abstract theoretical problems of computability. A-life research, along with many other branches of computer science, would merely be concerned with what it is possible to produce with a computer, what the limits of the function of computers might be. This is—or at least can be—a very different project than determining the limits of "life." The genesis of A-life in a begged question and a crashed glider are more than just confirmation of our hunch that error is at the heart of science. Rather, the textual strategies and founding events of A-life point to the unspoken rhetorical frames, the limits of the technoscientific structure that make A-life plausible.

Baudrillard and Life's Fatality

> The age of simulation thus begins with a liquidation of all referentials—worse: by their artificial resurrection in systems of signs. . . . It is no longer a question of imitation, nor of reduplication, nor even of parody. It is rather a question of substituting signs of the real for the real itself.
>
> —Jean Baudrillard, *Simulations*

The obscenity and giddiness with which Jean Baudrillard "challenges" the questions of theory and history simulate an "implosion" of the universe and its description, an implosion he claims is allegorized by such technoscientific events as the transcription of the genetic code and the deterrence machine of nuclear weapons. Specifically, he diagnoses a veritable disappearance of the real, a "desert of the real" where all the dreams of distinguishing between authentic and simulated experience are impossible. Although I find it difficult to affirm Baudrillard's obscenely absolute vision of the exhaustion of authenticity, his analysis of the symptoms of simulation's conflation between the real and its models helps explain the plausibility of the practices of Artificial Life. Specifically, Baudrillard's analysis of the age of simulation historically and theoretically situates the rhetorical spark invested in computer models. In this section I will outline the ways in which what Baudrillard calls the "precession of the simulacra" produces one half of the central problematic of the rhetorical software of artificial life, the knot of contradiction formed when an essentialist project gets intertwined with simulation.

Artificial Life, as I have attempted to show, is in the business of creation, the production of models that paradoxically lack an original. No bundle of traits or effects is sufficient to define the "living." Once housed in an invisible unity, then a "secret," life now finds itself without an address. For Baudrillard, the definition of the real is in a similar position: "The very definition of the real becomes: that of which it is possible to give an equivalent reproduction."[38] In contrast to the classical theory of representation, in which models reflect the basic reality of nature, or the modern notion of the distorting effect of the model, the age of simulation redefines the real as what can be replicated. The very distinction between the universe and its description disappears. It is not, as has frequently been claimed, that the real no longer exists, although Baudrillard frequently slides into this usage. Rather, the real *disappears*, no longer available to the inquiring gaze, as it becomes a reef of simulacra. Neither what provokes representation (classical) nor what resists representation (modern), the real in the age of simulation is what can be copied. This logic of the simulacrum, it would seem, extends to life, and not just *artificial* life. Richard Dawkins, a sociobiologist and one of the first A-life participants, wrote that "all life evolves by the differential survival of replicating entities." Vitality, too, in the age of simulation, is what can be "xeroxed."[39]

And yet an effect of xeroxing is that we often forget the original. In fact, Baudrillard, following and forgetting Benjamin, claims that the original disappears in the endless proliferation of replicants. Replication also allows forgery, and thus the problem of authenticity haunts the regime of replication. Baudrillard deploys the analogy of a feigned illness that vexes medicine: "For if any symptom can be 'produced,' and can no longer be accepted as a fact of nature, then every illness may be considered as simulatable and simulated, and medicine loses its meaning since it only knows how to treat 'true' illnesses by their objective causes."[40] Likewise, the reproducibility of "lifelike behavior" and the ultimate substitution of the signs of life for life lead to a problem for theoretical biology, because it only knows how to study "true" life. Rather than anchoring biology with a transcendental, comparative perspective, the simulation of life announces the absence of any original, unified notion of life in contemporary biology. It launches biology into the age of simulation, in which not only is life what can be reproduced, but what is always already, possibly, reproduced.

And yet these artificial-life constructions are models of nothing, no thing. Vitality, as a contested series of effects rather than a determinate, localizable essence, is the result not of mimesis but of simulation, where simulation need not refer to any stable original. In Baudrillard's terms, "The real is produced from miniaturised units, from matrices, memory banks, and command models—and with these it can be reproduced an indefinite number of times. It no longer has to be rational, since it is no longer measured against some ideal or negative instance. It is nothing more than operational."[41]

Artificial Life's operational reality or plausibility is at once essentialist and constructivist. By this I mean that no "ideal or negative instance" references the vitality of an A-life creature, and yet this reference—a general theory of the living—is precisely the goal of Artificial Life. This produces a frenzied spiral of experimentation, performance, and debate around the blurred lines between living and nonliving systems. Without hard and fast criteria for life, A-life takes refuge in what Baudrillard calls a kind of "military psychology": "Even military psychology retreats from the Cartesian clarities and hesitates to draw the distinction between the true and false, between the 'produced' symptom and the authentic symptom. 'If he acts crazy so well, then he must be mad.' "[42] Similarly, if an Artificial Life creature simulates life so well—whatever that means—then it must be alive. The "Artificial Life 4H Show," a display of the latest A-life creatures at A-life conferences, seen in this light, is a baroque exhibit of the absence, not the presence, of lifelike behavior. In an age of simulation, with no original, no reference standard for lifelike behavior, "no fact of nature" to ground A-life models, A-life exhibits are playful and powerful experiments, creations of a new order of complexity and power, an order characterized not by its relation to any preexistent model of life, but rather to a performance of power that conceals itself as such, that looks "like" life. They are at the same time an attempt to restore life to its foundations by determining the formal nature of all life, because "this death of the divine referential has to be exorcised at all cost."[43]

Life's Sovereignty

For ultimately, or perhaps from the beginning, this "quest for creation" is a quest for a little bit of referentiality. Specifically, the search for the "universal" nature of life's character tells an allegory of power, a story of a gnostic notion of life that restores life as a unified concept. C. H. Waddington, who provided some of the urtexts for Artificial Life, wrote of the profound impact of this notion on his work.

The world egg. "Things" are essentially eggs—pregnant with God-knows-what. You look at them and they appear simple enough, with a bland definite shape, rather impenetrable. You glance away for a bit and when you look back what you find is that they have turned into a fluffy yellow chick, actively running about and all set to get imprinted on you if you will give it half a chance. Unsettling, even perhaps a bit sinister. But one strand of gnostic thought asserted that everything is like that.[44]

This metaphysical background of Waddington's scientific work—what he called "software"—encouraged Waddington's interest in process, the Whiteheadian notion that scientists study "occasions of experience," unities that unfold over time, rather than reductionist hunks of stuff. Langton, too, experienced the effect of this "glance away":

The computer was running a long Life configuration, and Langton hadn't been monitoring it closely. Yet suddenly he felt a strong presence in the room. Something was there. He looked up, and the computer monitor showed an interesting configuration he hadn't previously encountered. "I crossed a threshold then," he recalls, "it was the first hint that there was a distinction between hardware and the behavior it would support. . . . You had the feeling there was really something very deep here in this little artificial universe and its evolution through time.[45]

It is this "glance away," or "looking up," and the events that take place between observations that I want to focus on here. The radical potentiality of "world as egg" makes plausible the remarkable question-begging with which A-life began. Held hostage in a terrestrial, carbon prison, A-lifers, like the gnostics, see the potential for life elsewhere, in noncarbon-based "eggs." A-life programs are at once exterior to human beings—hence the look away, the sense of autonomy—and disturbingly implicated in them: "all set to get imprinted on you if you give them half the chance."

Thus A-life operates in a postvital space, a space of no difference between living and nonliving entities. Cosmologically, one A-lifer remarks that "we had no proof that this universe in particular was not a CA [cellular automaton], running on the computer of some magnificent hacker in heaven."[46] Of course, it is also true that no positive evidence exists to suggest that it is. Similarly, no positive evidence exists to suggest that life is not confined to carbon. Rather, A-life rests on a narrative, a belief buttressed by glances through a Gnostic, simulated lens. This lens—whose optics determine what is seen at a glance, and what is concealed—is what helps Langton "cross the threshold" to belief in artificial life. It is a vision whose effect is "to discard or lighten all the matter of this world, that is the strange end the Gnostics pursued. . . . And so in the Gnostic mythology, Christ, for example, was idealized as a being who ate and drank but did not defecate. Such was the strength of his continence that foods did not corrupt him."[47] That is, the flight from the "curse of the world," the desire to get above "everywhere," is also an attempt to shed or at least lighten the body. In "a look pregnant with God-knows-what," a look that "crosses a threshold," Richard Dawkins glances out his window:

It is raining DNA outside. . . . Up and down the canal, as far as my binoculars can reach, the water is white with floating cottony flecks, and we can be sure that they have carpeted the ground to much the same radius in other directions too. The cotton wool is made mostly of cellulose, and it dwarfs the tiny capsule that contains the DNA, the genetic information. . . . It is the DNA that matters, the whole performance, cotton wool, catkins, tree and all, is in aid of one thing and one thing only, the spreading of DNA. This is not a metaphor, it is the plain truth. It couldn't be any plainer if it were raining floppy disks.[48]

What Dawkins sees, and does not see, tells the story of (American) A-life. With a kind of X-ray vision, Dawkins sees through the fluff of cellulose and locates the essence of life, DNA. This essence is everywhere, "as far as my binoculars can reach," floating, and it is not a metaphor. It is what can be replicated, what is real, what can live. And this vision is itself meta-phorical; rather than simply looking through the fluff and "seeing" the kernel of essence inside, DNA, Dawkins claims to literalize this vision, render it plain through the deployment of a metaphor that is denied metaphorical status. And yet this vision hides as much as it reveals: as pointed out by Atlan and Koppel, among others, the "program" or floppy disk of DNA is not itself sufficient for life.

The "fluff" Dawkins disparages in the name of "plain truth" is more than a mere husk or tool; in its movement and "performance" it literally makes life possible. In a way, it is nonsensical, or at least certainly not "the plain truth" to speak of the spread of DNA without remembering the spread of organisms. So, too, with Langton's recognition of the "pres-ence" in the form of an "interesting configuration" on the screen—it requires a deletion or overlooking of the computer in the gaze at the "organism" on, in, the screen. The extra husk or container of the syn-thetic organism, the computer, is treated as a mere fluff, a platform for the real, artificial life. Tom Ray's description of his Tierra program, one of the earliest and best-known A-life programs, illustrates this as well: "Syn-thetic organisms have been created based on a computer metaphor of organic life in which CPU time is the 'energy' resource and memory is the 'material' resource. Memory is organized into informational patterns that exploit CPU time for self-replication. Mutation generates new forms, and evolution proceeds by natural selection as different genotypes compete for CPU time and memory space."[49] This is based on a computer "metaphor," indeed, based on a computer, as in Atlan and Koppel's critique of the genome projects: "Implicit in this proposal is a literal understanding of the genetic program metaphor, looking at the sequence of all the DNA base pairs of a genome as the listing of a computer program."[50] Thomas Ray's Tierra program creates "genomes" composed of "information patterns." In order for these "genomes" to be seen as "synthetic organisms," Ray must, like Dawkins, see through the fluff of the computer. That is, the same vision that allows Dawkins to equate a floppy disk and a "cottony fleck" allows Ray to overlook the dual platform of metaphor and computer that allows his synthetic creatures to emerge. Langton might call this an attention to the formal aspects of life, but a closer look at Ray's program shows that his abstraction relies on two metaphors that are systematically occluded.

First, the idea of DNA as a program makes possible the very idea that a "genome" could be produced on a computer. This rhetoric, like Dawkins's notion that DNA is a floppy disk—"This is not a metaphor, it is the plain truth. It couldn't be any plainer if it were raining floppy disks"— systematically deploys metaphors while denying them. The very notion that DNA is equivalent to a floppy disk requires us to forget that unlike a floppy disk, DNA materially "contains" the instructions for the replication not just of itself, but of an organism, its "reader." Of course, no DNA "itself" can do this; such a complex system emerges out of the relations between organisms and their environments.

Second, when we look at the screen full of A-life organisms, we must forget the "body" of A-life, the technological ensemble of discourses and computer that are its "food," environment, body, and explication. For the computer does not display emergent behavior, no computer hardware is replicated—only the "interesting configuration" on the screen produces this kind of performance. Thus, to succeed, an A-life performance must hide a synthetic organism's dependence on a computer, an operating environment, and its rhetorics just as molecular biology has often occluded the dependence of DNA on its organismic context.

Thus, by taking the metaphor of "program" literally, Ray, Dawkins, and other A-life researchers repeat, with a different inflection, some of the rhetorical moves that made nascent molecular biology plausible. Atlan, Koppel, and others point out the power of the "program" to orient life-science research toward sequencing at the expense of work on the complexities of expression and development. By focusing on the "genomes" of synthetic organisms, A-life performances succeed in producing the effect of "lifelike" behavior in a context of simulation, with no original. They thus effectively mask the absence of any unified notion of life in contemporary life science, by preserving the idea that there is something called "life," giving a sense of reference to the concept of "life" even as it is being displaced. But this performance has a price. By following the implicit model of life as a kind of program, A-life obscures the very questions of morphogenesis, growth, and expression—those events that take place during the "glance away"—that motivate the search for a "general theory of life."

It may seem that some of the fundamental vectors of both A-life and gnostic thought are contradictory, or at least in tension. For example, it may seem upon reflection that the notion of a world in a state of becoming, a Waddingtonian "world egg," would be in opposition to the notion of a cursed world of humans shackled to bodies. But this tension is mediated through the implicit model of the body in both Artificial Life and gnostic thought. Whereas the regime of molecular biology under Schrödinger literally forgot the body as it contained it in the codescript, Artificial Life operates on a memory of the body. Whereas the idea of a DNA codescript, program, or "selfish genes" tended, as in Dawkins's gaze, to overlook or forget the body of the organism, Artificial Life deploys a model of the body in which genetic sequences *are* bodies, "informational patterns that exploit CPU time for self-replication."

One sign of this is the "spontaneous sexuality" of Tom Ray's Tierra organisms, in which the "sloppy replicators . . . allow for recombination and rearrangement of genomes." Sex here is pure exchange without action or gesture; no worldly work need contaminate the pure "bodies" of A-life organisms. Such is the strength of their continence that sex does not corrupt them. Thus, as long as the "glance away" masks the work of emergence—whether it be the struggling chick or the flashing of a pixel—the purity of the A-life organism remains intact, its growth a product only of time. Indeed, in the cases of both the game of Life and the chick, it will only be a matter of time before they seek to leave the cursed earth to float or fly "above everywhere."

It seems clear that the universality of the genetic "program" both provokes and encourages the notion that there exist universal attributes of life independent of substance, and that thinking that prioritizes and essentializes DNA allows for the crossover between information and life.[51] But this fails to answer for the appeal of A-life in particular. Jean-Claude Beaune gives us a clue: "Sharing in the trickery of the automaton is merely another way to define ourselves as human, that is, as both being and nothingness, presence and absence; the automaton is, in a way, our mirror . . . or our evil eye."[52]

In an age of simulation, where no original stable referent for "life" survives, A-life provides not a mirror but a screen for the definition of life and the human. The screen masks the "sinister" mediation of genotype to phenotype, and provides a place for us to glance, "to notice something very deep here in this little artificial universe and its evolution through time." The art of the automaton has always been the "trickery" that concealed the differences between machines and humans. The art in Artificial Life now conceals the absence of this same difference, as all of life, artificial and otherwise, becomes "propagating information." Despite (or because of) the lack of a unified definition, A-life promises to show us what life is. The power of powerful computers is to enforce the rhetoric of life as program, an enforcement possible through the power of concealing and revealing that has reshaped the scientific concept of life in the late twentieth century. The computer makes feasible Erwin Schrödinger's dream of the genetic code as both "law and executive power," but it is feasible only as long as the computer itself, and its (rhetorical) "power," is hidden as the matrix that makes it possible for information to become an organism, and for organisms to become information.

This need not only for law, but for "executive power" in the co-descript or sequence has led to a takeover, what origin of life theorist A. G. Cairns-Smith has called a genetic takeover of the body by the machine. But this is not exactly a takeover of the sort discussed by Cairns-Smith or fantasized about by computer scientist Hans Moravec. Rather, it is a discursive takeover that reorganizes, but does not dispense with the body. "What awaits us is not oblivion," Moravec writes, but rather a future in which

it is easy to imagine human thought freed from bondage to a mortal body—belief in an afterlife is common. . . . Computers provide a model for even the most ardent mechanist. A computation in progress—what we can reasonably call a computer's thought process—can be halted in midstep and transferred, as program and data read out of the machine's memory, into a physically different computer, there to resume as though nothing had happened. Imagine that a human mind might be freed from its brain in some analogous (if much more technically challenging) way.[53]

"As though nothing had happened," this shift in the notions of life and information works precisely by its ubiquity, it banality, and its lack of visibility.

Michel Foucault, in *The Birth of the Clinic*, has shown how the "distributions of illness" have been historically constituted. The modern geometry of disease, in which sickness is localized in the body, is shown by Foucault to be one map among others of the ill body. This map, besides localizing disease, also provides us with an outline of the morphology of power, a power that selects, outlines, and organizes the body and its relations. This is, in fact, what power is for Foucault—the network of relations that make utterances and knowledge possible. So too, I would suggest, can the distributions of "life" shift through biology's new rhetorical software, tropes that both highlight and obscure different maps of "vitality." The "glance away" that installs new concepts of vitality is a gaze built out of a gap, a structural blindness that requires but does not see bodies; it sees only sequences, genomes that are bodies.

Michel Serres has dubbed this "bureaucratic power," in which "the manual laborer has to be blind in relation to the paralyzed intellectual. The helmsman has no porthole. . . . This cybernetics gets more and more complicated, makes a chain, forms a network. Yet it is founded on the theft of information, quite a simple thing. . . . In the end, power is nothing else."[54] Rather than based on a "theft of information," I would argue that this network is based on connections. What makes possible the explication of the moments of the "glance away" is what I have referred to above as the metacode or rhetorical softwares. An example from one of Tom Ray's recent papers helps make this point:

Modern evolutionary theory is firmly based on the duality of the genotype and the phenotype. However, Barbieri (1985) has described a new view, in which life is based on a trinity of genotype, phenotype and ribotype. At the molecular level, the genotype is the DNA, the phenotype is the proteins, and the ribotype is the collection of molecules and structures based on RNA, i.e., the mRNA, tRNA and the ribosomes. The latter group of molecules, referred to collectively as the ribosoids, perform the critical function of translating the genotype into the phenotype.[55]

Ray goes on to claim that in his system, it is the "decoding" unit that performs this ribotypic function. But such an account overlooks and disavows the connections and translations forged by A-life researchers, who, like the "human brains" that hosted the initial propagation of Conway's Life, are crucial elements of the A-life ecology. A-life researches, explicated as life and not merely as interesting tests of computation, are enmeshed with the translational practices that transform the "genotype" (the codes of Tierra) into phenotypic, "lively" creatures, translational practices that include their own disavowal. Indeed, in some sense *A-lifers are part of the phenotype of Tierra*.[56]

But like the "selfish gene" discussed by Richard Dawkins, in which organisms are merely "lumbering robots" that exist only for the purpose of propagating copies of a genome, Tierra organisms are folded across time and space. They are beyond living, only descending to vitality momentarily. Their ability to "live" without time or space, or at least to be suspended across time and space, is summed up in a recent proposal written by Ray: "It is now possible to bring the system down and then up again in the middle of a run without loss of information. It is also possible to fully recover from a hardware crash, and continue the run without having to start over."[57]

Artificial Life, where bodies flash as pixels, executes the Watsonian fantasy of knowing "what life is." Watching A-life organisms on the screen in a certain way allows us to watch the struggle for power, if power is what allows foregrounding of some effects and some problems over others. It is what we do not see, as well as what is simulated—like all classic automata—that allows something like "life" to "emerge." Here I again follow Nietzsche's prescient diagnosis. Although he wrote of the Darwinian "struggle for Life," his analysis seems more fit for an account of A-life, in the twilight not of idols but of vitality: "Where there is a struggle, it is a struggle for power."[58] That power, in an age of simulation, is the power to achieve the "impossible"—to mime an absent origin, "life."

Science and Writing:
Two National Narratives
of Failure

In this essay, the failed British expedition to the South Pole in 1911 provides a case study of how representations of masculinity, technology, science, and empire are co-articulated. I compare Captain Robert Falcon Scott's narrative with the writings of American arctic explorers such as Captain Robert Peary and Vilhjalmur Stefansson to examine how different national fantasies of masculinity and science contributed to the formation of distinct imperial ideologies. In each national tradition, polar exploration defines masculinity and science as a superhuman attribute that actual men can never achieve. Thus, I will examine the difference between the ways these traditions deal with the possibility of actuality of scientific failure in terms of distinct narratives of writing.

The British were the losers in the race to the South Pole. Roald Amundsen of Norway reached the Pole in 1911, one month ahead of the hapless Captain Robert Falcon Scott. Not only did the British team fail to reach the Pole first, but Scott and his four men died of hunger and cold on their way back. After completing nearly seven-eighths of the distance they encountered a blizzard, and unable to reach their food depot just eleven miles away, died in their tent from a combination of frostbite, sickness, and starvation. This was no ordinary failure, to be covered up as a national embarrassment. Rather, the tragedy of the British expedition was seized on and celebrated by the British as a national historical event. The Peary controversy was different; the uncertainty attributed to Peary's success at the North Pole was not just a straightforward case of failure, and for reasons I will make clear, not recuperable in the way failure could be in terms of a British heroism of sacrifice.

Eight months after the Scott expedition had disappeared, a search party found the tent with the bodies of Scott and his men inside. They recovered the men's diaries, letters, and other belongings. Included among these possessions was a geological collection that included 30 pounds of rock specimens that Scott had hoped would contribute to science. Surgeon Atkinson, who buried Scott and his men, was moved by the presence of the stones: "They had stuck to these up to the very end, even when disaster stared them in the face and they knew that the specimens were so much weight added to what they had to pull."[1]

Yet the pursuit of science was only one of several goals for the British expedition; for Peary's exploit, science was integrated into the expedition as a means as well as a singular goal. Peary attributed his superior abilities to management of what he referred to as a "traveling machine" that deployed the latest technological advances in scientific instrumentation and modes of transport. By contrast, Scott's masculine performance depended simply upon the integrity and honor of being a British gentleman. In Scott's view, British minds and bodies alone were enough to display the superior capabilities of the male hero.[2]

This is why Scott is able to portray the failure of his expedition as a heroic example of British character: "Our wreck is certainly due to this sudden advent of bad weather. . . . I do not think human beings ever came through such a month as we have come through. . . . I do not regret this journey, which has shown that Englishmen can endure hardships, help one another, and meet death with as great a fortitude as ever in the past."[3] Scott's rearticulation of his expedition's fortitude was readily accepted by the search party, who in memory of Scott and his men chose to inscribe the following line from Tennyson's *Ulysses* on the cross marking their burial site: "To strive, to seek, to find, and not to yield."[4]

At Scott's request, his diary was given to his wife, Kathleen Scott, who, with a family friend, Leonard Huxley, prepared and arranged his notes for public consumption. With funding from the British government, Scott's diary and letters were rapidly published in 1913 in London, New York, and Boston under the title *Scott's Last Expedition: The Personal Journal of Captain R. F. Scott, R.N., C.V.O., on His Last Journey to the South Pole.*[5]

In the years following Scott's death a myth of the gentleman hero was erected on the foundation of the letters Scott wrote to explain the causes of his expedition's misfortunes. He wrote of himself as a man who sacrificed his own life to look after the welfare of his men. His letters demonstrated his leadership qualities and his ability to face death alone nobly. Thus a typical editorial in the London *Times* praised the failure of the Scott expedition: "Let us put out of our minds all the gossip which . . . has been circulated about a race. . . . The real value of the Antarctic expedition was spiritual, and therefore in the truest sense national. It is proof that in an age of depressing materialism men can still be found to face known hardship, heavy risk and even death, in pursuit of an idea. . . . That is the temper of men who build empires, and while it lives among us we shall be capable of maintaining an Empire that our fathers built."[6] Scott was able to reveal "the temper of men who build empires," thus saving Britain from the disgrace of losing to the Norwegians, by displaying the noble behavior of the "real" English gentleman. Even though Scott was not an aristocrat or a great explorer, his orderly and respectable death demonstrated the qualities of an Englishman that was born to rule.

In England, Scott's point of view was the only version of the story that was made public at the time. Although Scott makes references to the other four men who died with him in the field, their letters and diaries remained private. They were representatives of the navy, silent supporters of their commander, observers.

For many years the original diaries and letters of Scott and his men were not available to the public. Recently, some of these documents have been released. Roland Huntford, a Scandinavian historian, studied these original manuscripts and revealed in his 1979 book *Scott and Amundsen* the concerted effort made at the time of Scott's death by the British Admiralty to conceal unsettling facts about the Scott expedition. By comparing Scott's original diary with the published version Huntford found that "Scott's diaries were purged of all passages detracting from a perfect image; particularly those revealing bitterness over Amundsen, criticism of his companions, and, above all, signs of incompetence."[7]

Huntford's research revealed that Scott's diaries and letters were altered in order to turn the official version of events into something worthy of public reverence. The suggestion that Scott and his men died from scurvy is suppressed because it would have reflected on the whole conduct of the expedition. Roland Huntford provides an example of a significant excision made by a committee chaired by Kathleen Scott:

It began with Kathleen Scott who, at her husband's request, was dealing with the papers. 'He was the last to go,' she wrote to Admiral Egerton, sending Scott's farewell letter to him—which happened to indicate otherwise. It was one of the letters found loose in the tent. On the back was a note in Bowers' hand, suggesting that Bowers may have been the last survivor, or at least casting doubts on Scott's claim. . . . In any case it was inconvenient evidence. It was suppressed and, instead, there was issued an official reconstruction of the closing scene in the text, contrived at the request of Kathleen Scott by the playwright Sir J. M. Barrie.[8]

In Barrie's reconstruction of Scott's death, the social relations of the expedition are concealed and Scott outlasts his social inferiors: "Wilson and Bowers died first and Captain Scott . . . thereafter . . . unbared his shirt and . . . with his head flung back awaited death. We know this because it was thus that the three were found. . . . Some of the wording may not be quite right, but the brevity is."[9] Barrie's staged drama perpetuates an ideal of British male heroism in which the captain, unafraid, thrusts his manly chest out in the face of adversity and awaits death alone.

American Myths of Modernity and Masculinity

In Scott's case, failure is recuperable through writing, whereas it is not in terms of a U.S. evolutionist discourse of science. This cultural difference is apparent in U.S. arctic explorer Vilhjalmur Stefansson's reaction to Scott's death in 1913:

It has been so many years since arctic and antarctic exploration took any comparable toll of lives that we had come to feel fairly secure. After all, Peary and men like him have made exploration a science and with modern equipment and provision against the cold . . . there is not the danger that there used to be. . . .

But perhaps our confidence in the steadily improved equipment of exploring parties had grown till we have been lulled into a false sense of security. This disaster to the Scott party is crushing in the way that the wreck of the *Titanic* was crushing. We had grown to believe that traveling on the seas in a huge liner was stripped of its traditional perils.[10]

The myth that scientific progress had turned ocean travel and polar exploration into risk-free activities was of a fairly recent origin. The loss of the Scott expedition was a reminder that all dangers had evidently not entirely disappeared with the recent alliance between exploration and science. Instead, Stefansson points out that the belief in the infallibility of polar exploration and ocean travel was a sign of the new dangers that such mythologies of science engendered. The irony, as we shall see, is that Scott downplayed any alliance between scientific techniques and his expedition.

For Stefansson, the effect of the new sense of security might have proved damaging, yet he agrees that its existence was warranted. The relatively few lives lost in arctic and antarctic exploration in recent years showed that a significant advance had indeed been made. So much is this the case that when Stefansson is told that the failure of the Scott expedition was simply the result of a blizzard, which was considered a rather commonplace occurrence of nature, not only does he disbelieve it, but he invents a greater calamity as the cause: "Such a tragedy could be explained only on the supposition that some great and incalculable calamity overtook the party, a calamity of the proportions of an earthquake. . . . An earthquake might have broken loose a huge fragment of the ice barrier . . . so that they floated out to sea, but this is hard to believe."[11] It is not surprising that Stefansson fabricates a more extreme incident in order to conform better to an ideology of modernization and progress. Evidently, a blizzard just did not count as an obstacle to Stefansson—it was too ordinary: "No blizzard alone ever killed Captain Scott and his men. He was too experienced an explorer for that. Out on the Western prairie such a thing might be. A rancher might get caught unawares in a snowstorm, to be frozen and buried in the drifts. But in the Arctic regions? No. And certainly not when the leader was such a man as Scott, who had the finest of equipment and who knew how to guard against cold and snow. That was his business and he knew his business."[12] Stefansson's disbelief can be attributed to a certain historical certainty of that period. Stefansson, a polar explorer himself, does not need to know Scott personally in order to assert confidence about Scott's expertise. For Stefansson, science and exploration are so intimately intertwined that it would be inconceivable that Scott, as a practitioner of science, would not know something as basic to the profession as how to guard against cold and snow. But did he?

For U.S. explorers such as Stefansson, who linked polar exploration to the ideology of modernization and progress, expertise was a necessity if science was to offer as a social reality the safer world that its apologists promised.[13] The "Peary system" that Peary described in his book *The North Pole* ensured that his material experience measured up to the ideologically produced expectations of science: "The source of our success was a carefully planned system, mathematically demonstrated. Everything that could be controlled was controlled, and the indeterminate factors of storms, open leads, and accidents to men, dogs, and sledges, were taken into consideration in the percentage of probabilities and provided for as far as possible."[14]

The pseudoscientific Peary system provided the image under which Peary's expedition was perceived to be infallible. The symbolics of this system embodied a discourse that allowed no margin for error in the practice of science. There was no room for failure, which would be synonymous with ruin, nor was there any question that the Pole might be unwinnable. Thus, Peary's actual failure during his seven earlier attempts to reach the North Pole could only be recouped once he was able finally to say he had succeeded, and that he was the only one to have accomplished the deed. In order to be assured of his own victory, Peary had to make sure that his rival's claim was discredited.

It is in this sense that U.S. science seemed to have its own set of ethics. It did not matter that the importance given to determining the true discoverer of the North Pole was out of proportion to any practical value attached to its attainment. By the early twentieth century, almost all parts of the world were known and more or less adequately or approximately mapped. Exploration no longer consisted of discovery but was rather a symbolic politics, a form of athletic endeavor or sport that exalted the male body and its exterior scientific apparatus.

There was an interest in showing that a male American body as a scientific device could dominate the most severe and inhospitable physical environment of the globe. If attaining the North Pole was part sporting competition for the Americans, athletic ability was not the only thing that was being tested. For the president of the American Geographical Society, Gilbert Grosvenor, all of the international participants were not equally equipped for the task. To Grosvenor, it was not just male physical strength that was being tested in this contest, but rather, the combination of physical strength and scientific ability. For Grosvenor, the United States had an advantage over the rest of the nations because it was the most scientific. In Grosvenor's narrative it is fitting that Peary, who represents the essence of U.S. identity, is depicted as a scientific manager: "No better proof of the minute care with which every campaign was prearranged can be given than the fact that, though Peary has taken hundreds of men north with him on his various expeditions, he has brought them all back, and in good health. . . . What a contrast [Peary's] record is to the long list of [British] fatalities from disease, frost, shipwreck, and starvation."[15] Peary is the preeminent polar explorer because he is the "most persistent and scientific."[16] "The minute care with which he prearranged every campaign"[17] enables him to overcome the flaws of early polar expeditions. For Grosvenor, the U.S. claim on the North Pole seems to make disease, famine, and other forms of human misery relics of a less scientific past. In winning the race to the North Pole, Grosvenor suggests that the Americans were able to show that the British expedition's reliance on character and determination was not enough.

It is significant that Peary, the scientific manager, represents himself as the epitome of manliness. As a figure for U.S. nationalism, the body of the U.S. polar explorer was defined by the enterprise of science, in which expertise and skill rather than the inner qualities of fortitude and dignity under stress were emphasized. National moral superiority was expressed in terms of a discourse constructed by an evolutionist technology of science.

One Form of Male Sacrifice

The mythification of the Scott expedition by the British Admiralty fit within an already established tradition of British imperial heroics connected to the polar regions. Prior to the Scott expedition, male sacrifice in the polar regions served as a means to perpetuate a superior image of Britishness and British nature not motivated by self-interest.

The connection between polar exploration and a certain brand of British imperial humanism can be dated from the time of the disappearance of the earlier Franklin expedition sent out by the British navy in 1845 to discover the Northwest Passage. In order to find the lost Franklin expedition, the British navy participated in a humanitarian search perhaps unparalleled in maritime history.[18] Over a period of fourteen years, 40 British expeditions were sent out to look for the survivors. What most characterized these heroic rescue expeditions was a romantic notion of self-sacrifice. These men and ships were sent out to the arctic not for material gain, but rather to save their fellow countrymen from death or to bring back their bodies. Such a display of chivalric values combined with noble sacrifice helped turn British polar explorers into romantic national figures. The trope of tragic self-sacrifice connoted the spirit of the nation. The virtues of British fortitude were celebrated as part of a mid-nineteenth-century romantic literary discursive tradition, as evidenced by Alfred Lord Tennyson's celebrated poem "Ulysses," cited by Huntford:

One equal temper of heroic hearts,
Made weak by time and fate, but strong in will
To strive, to seek, to find, and not to yield.[19]

Tennyson's poem, originally dedicated to polar explorer Sir John Franklin, prevailed as a tradition and, fittingly, reappeared in 1912 to memorialize the graves of Captain Scott and his men.

After the Napoleonic Wars, there was not much demand on the Royal Navy as a fighting force, and polar exploration became a surrogate for active service. Many who sought to escape from the monotony of peacetime enlisted. Thus arose a distinctively British type: the naval officer who took to polar exploration as part of his ordinary career.

Sir Clements Markham, writing in 1893 on the relevance of polar exploration during his tenure as president of the Royal Geographical Society, recognized it as "a nursery for our seamen, as a school for our future Nelsons [Nelson, early in his career, had been a midshipman of an arctic expedition] and as affording the best opportunities for distinction to young naval officers in times of peace."[20]

Under Markham's direction, antarctic exploration became highly esteemed within the navy. Markham himself derived his passion for polar exploration from his early experience as a cadet in the arctic on the second Franklin search expedition in 1850–51. He knew half a dozen languages and was a prolific writer on the history of exploration. The figure of Markham as an explorer, gentleman, and writer provides a marked contrast to Peary's image as the red-blooded, tough, competitive U.S. scientific manager: "[Markham] seemed the embodiment of the romance of Geography; his bosom swelled, and his shirt front billowed out like the topsail of a frigate, and as his voice rose in praise of 'our glorious associates,' he often roused a rapturous response."[21] The description of Markham by a Royal Geographical Society official brings a potentially contradictory "softening" dimension to the image of the polar explorer. It is significant that in the case of Markham, power is ascribed in terms of sartorial and rhetorical flourish rather than physical strength or scientific expertise. To the old arctic admiral who had not seen the ice for twenty years, the era of the 1840s and 1850s was not a blemish from a less scientific British past. Rather, he recovers the tragedies of that period as "great endeavors" and "heroic achievements." For Markham, the 1840s and 1850s were the most memorable period in polar exploration because of the countless heroic sacrifices made.

Self-sacrifice as such was valorized by the Anglican Church, according to Huntford, who cites Francis Paget, dean of Christ Church: "Surely war, like every other form of suffering and misery, has its redeeming element in the beauty and splendor of character men, by God's grace show in it. . . . Men rise themselves and raise others by sacrifice of self, and in war the greatness of self-sacrifice is set before us."[22] This philosophy has its exact parallel in polar exploration, as evidenced by the following passage from Captain McClure's narrative of his 1850–54 search for Franklin: "How nobly those gallant seamen toiled . . . sent to travel upon snow and ice, each with 200 pounds to drag. . . . No man flinched from his work; some of the gallant fellows really died at the drag rope . . . but not a murmur arose . . . as the weak fell out. . . . There were always more than enough of volunteers to take their place."[23]

During the mid-nineteenth century the ideal of personal gallantry was seemingly an end in itself. Writing in 1893, it was this old romantic image of the polar explorer that Markham intended on keeping alive. "The Polar Regions . . . difficult of access . . . [are] of surpassing interest and importance. Here we meet with examples of heroism and devotion which must entrance mankind for all times. . . . There are dangers to be encountered and difficulties to be overcome which call forth the best qualities of our race."[24] For Markham, polar exploration was seen as a testing ground to keep alive displays of moral courage and bravery, as well as a place to express the superiority of the British race. In such writing, the British had very high standards of ethics when it came to themselves. They of course often applied other standards to the non-Western peoples they subjugated in their scramble for new territories and wealth. Indeed, the aesthetic side of polar exploration made for convincing imperial theater; those polar explorers who risked their lives to find the Franklin expedition became the British heroes and embodied the idea of adventure but were not tarnished by the horrors of empire. These heroes of the British military had a less compromised image than their counterparts in the colonies, who represented quite a different personification of the British empire. Once Adm. Sir Leopold McClintock finally discovered the remains of the Franklin expedition, an era of polar exploration had ended. So much did that period have a hold on the British imagination, however, that it was revived again in the late nineteenth century.

At the turn of the century, the tradition of polar exploration remained intact, yet now members of the middle class were able to participate in this formerly upper-class tradition (Scott himself was a member of the middle class). A whole ideological system of entitlement to rights had been erected on the assumption that certain military virtues such as courage, bravery, and manliness were innate qualities of British subjects in the Royal Navy. It now mattered less whether these men were no longer part of the upper class, for their affiliation with the Royal Navy bestowed on them the requisite authority and prestige.

In keeping with the attitudes of the Royal British Navy during this period is a reconstruction of the past, or the literary attempt to transport the heroic past fictionally into the present, of which Markham's writings are one example. Only to polar explorers outside of England did such an emphasis on aesthetic literary ideals seem retrograde, especially when they were put into practice. Official British exploration had lapsed since Capt. Sir George Nares, R.N., led a naval expedition that attempted to reach the North Pole between 1875 and 1876. The expedition was a failure; the methods were outmoded, and many of the crew died of scurvy. Sir Clements Markham had repeatedly disparaged progress abroad, preferring to rely on an outmoded British method. This attitude was most evident in 1899 in his advocacy of a system of man-hauling over the use of dogs as draft animals:

In recent times much reliance has been placed upon dogs for Arctic traveling. Yet nothing has been done with them to be compared with what men have achieved without dogs. Indeed, only one journey of considerable length has ever been performed, in the Arctic regions, with dogs—that by Mr. Peary across the inland ice of Greenland. But he would have perished without the resources of the country, and all his dogs, but one, died, owing to overwork, or were killed to feed the others. It is a very cruel system.[25]

Scott had a similar moral view on dogs. This is not surprising, as Scott, like Markham, was a navy man rather than a polar explorer by profession. Scott's limited experience in the field apparently was not considered a hindrance to his ability to accomplish his goal, as experience or expertise was not necessarily highly valued in the British navy anyway. The navy, in the approving words of Adm. Sir Herbert Richmond, was "breeding amateur Naval officers."[26] As historian Roland Huntford put it: "The study of strategy and tactics was considered almost bad form, chiefly because Nelson was erroneously believed to have triumphed at Trafalgar without a plan of battle. Most officers believed that the old hereditary idea of gallantry and dash would see them through."[27] This faith in gentlemanly improvisation seems to point to the existence of an ideological system in which there was the belief that certain heroic virtues were innate to the British. According to this belief, it would be considered redundant to learn something that was already hereditary. From such a perspective the incorporation of new techniques readily adopted by the Americans or the Norwegians would be difficult.

Scott was not completely lacking in polar experience before his 1911 South Pole expedition. In 1905 Scott was appointed by Markham to lead the *Discovery* expedition to Antarctica. Yet his experience on this expedition did not drastically change his opinion on dogs that he had received from Markham. In his narrative *The Voyage of the Discovery*, Scott dismisses the use of skis for antarctic exploration with the opinion, "that in the Antarctic Regions there is nothing to equal the honest and customary use of one's own legs."[28] And about the dogs, he writes: "In my mind no journey ever made with dogs can approach the height of that fine conception which is realized when a party of men go forth to face hardships, dangers, and difficulties with their own unaided efforts, and by days and weeks of hard physical labor succeed in solving some problem of the great unknown. Surely in this case the conquest is more nobly and splendidly won."[29]

Scott is concerned above all with constructing an image of noble struggle. The polar explorer is not a scientific hero who rationally learns from his hardships and learns to search for advanced means to make them easier. He is someone who prefers adventure to anything else. Adversity and setback almost become morally desirable. He shuns the use of dogs because they would make the obstacle seem less formidable. Why? For Scott, the basis for all this is the Englishman's ever-present willingness to prove his superiority. He is totally self-sufficient, even in the harsh climate of Antarctica. Dogs would interfere with this heroic image.

Cold Comforts

Scott's idea of masculinity put more emphasis on willpower and moral strength than did Peary's polar narratives, which depended on his control of the tools of science. Scott's particular sense of masculinity is encoded in the letters he wrote before his death in March 1912. These documents became the founding text that accounted for the rise of the Scott myth.[30]

"If this letter reaches you Bill and I will have gone out together. We are very near it now and I should like you to know how splendid he was at the end—everlastingly cheerful and ready to sacrifice himself for others, never a word of blame to me for leading him into this mess. He is not suffering, luckily, at least only minor discomforts."[31] Scott wrote the preceding passage in a tent in Antarctica in March 1912. "This mess" that is so calmly written about, the "it" that is referred to as being "very near" is how Scott introduces the reader to the event of his and his lieutenant Bill's (Dr. Edward Wilson) imminent deaths. This letter, however, is about Bill's death, not his own. Lying in a tent on their return from the South Pole, Scott writes a letter to Bill's wife to inform her that she is now a widow. He is reassuring. He tells her that her husband is "everlastingly cheerful." Even at the moment when he is confronted with his own death, Bill remains "splendid," "ready to sacrifice himself for others" with "never a word of blame" to Scott, who was apparently responsible. Who would blame him now, anyway, as he is dying too? But not yet, he is still writing, consoling the grief of others. Even with death upon him, Scott always thinks of others first.

In the next paragraph of Scott's letter, Bill is now dead: "His eyes have a comfortable blue look of hope and his mind is peaceful. . . . I can do no more to comfort you than to tell you that he died as he lived, a brave, true man—the best of comrades and staunchest of friends."[32] Scott renders the last moments of Wilson's death with an aestheticizing comment: "His eyes have a comfortable blue look of hope." No other mention is made of the corpse, which, marked by scurvy and frostbite, must have been rather unsightly. Yet the image of the "comfortable blue look of hope" expresses that Wilson faced death bravely, honorably. If there was any remorse or unpleasantness, Scott does not pass it on to Wilson's wife. This is what an honorable gentleman wants her to believe.

With Wilson dead or near death, Scott now writes to Mrs. Bower, the mother of Lieutenant Henry Robertson Bower, the other officer dying with him now in the tent: "I write when we are very near the end of our journey, and I am finishing it in company with two gallant, noble gentlemen. One of these is your son. He had come to be one of my closest and soundest friends, and I appreciate his wonderful upright nature, his ability and energy. As the troubles have thickened his dauntless spirit ever shone brighter and he has remained cheerful, hopeful, and indomitable to the end."[33] This letter, like the earlier one, seems chiefly motivated by an impulse to reveal nothing concrete about the reality of Lieutenant Bower's death. Scott is describing not the event of Bower's death but an aesthetic representation. The narrative culminates with an image of Bowers not as a flesh-and-blood man, but as a "dauntless spirit."

Scott concludes his letter by connecting the idealized image of Bower's "dauntless spirit" to another mythical site of unity and harmony—the respectable English bourgeois family. Bower's last memories of happiness, his ability to remain "splendidly hopeful to the end" are due to his "happy home." In the final sentences of Scott's letter the dying Bowers and his family exist in a mutually authorizing relationship: "To the end he has talked of you and his sisters. One sees what a happy home he must have had and perhaps it is well to look back on nothing but happiness. He remains unselfish, self-reliant and splendidly hopeful to the end, believing in God's mercy to you."[34] By establishing a tie between Bower's noble death in Antarctica and the familial home in England, Scott connects the two remote places, Antarctica and England. Yet what is striking is how he erases the harsh conditions of Antarctica by portraying Bower's death in such a psychically undisturbing way that it appears as if he died a natural death in England.

Scott writes both these letters in the first person, yet within his narrative he seems to exist as the detached third person. Although Scott is at the scene of the event, he appears to be far away. His attention to formal literary conventions in his letters suggests that Scott is not freezing and starving to death in a tent in Antarctica but rather sitting at his desk somewhere in a comfortable London flat.

Scott avoids altogether any passing references to frozen bodies or to death in Antarctica. He refers to the harshness of the situation in only the most perfunctory way: "Excuse writing—it is −40 and has been nigh for a month." Instead, Scott populates the scene with images and voices from England through the writing of numerous letters to the families of his men, to his own family and friends, and to his superior officers in the Admiralty. Also, by employing a style that is distancing and artificially associative, he manages to avoid any direct reference to the horror of the situation at hand. Moreover, his use of clipped naval expressions like "we are pegging out in a comfortless spot," "we have shipped up," "a close shave," and "shot our bolt" expresses that his impending death and the death of his men have left his dignity and bearing intact in the community of a courageous male crew.

Through Scott's letters, Antarctica is a theater in which a performance by British naval officers can be seen from the privileged standpoint of England.[35] Thus, Antarctica is textualized; it becomes a discursive space in which intrepid British naval officers can prove that they can still die as gentlemen. Never deviating from their routine, they faced death as they did life—unruffled, certain of themselves, and dignified. There are no last-minute attempts to save themselves. All in all, the fiction of Scott's narrative construction has a predictably tidy end, with everything properly explained by Scott, down to an account of his men's final dying words.

Nothing could better imply the superiority of the men of the British race than Scott's staunch adherence to principle, his national consciousness, and his sense of responsibility to the nation as a whole. The absence of cowardliness showed that he and his men died nobly, without shame. Even under the most horrible of circumstances they were able to appear unassailable in themselves (heroic, brave), capable of dying honorably, even from the most ignoble of deaths. If Scott and his men were unable to perform a deed worthy of heroes, at least they were able to die in heroic fashion.[36]

All Body, No Technique

There are two men of the polar party—Evans and Oates—whose mothers and wives Scott does not bother writing letters to. These men were already dead. Evans died one month earlier; Oates died soon afterwards. The disagreements between Scott, Oates, and Evans, reported in Roland Huntford's book *Scott and Amundsen*, make it clear that these men were critical of Scott's leadership abilities.[37] According to Huntford, Evans was especially demoralized by the expedition's failure to get to the South Pole first. It was intolerable to him that Amundsen's Norwegian team beat them. Evans was depending on the financial security and promotion victory would have brought. For him, attaining the Pole without the reward of priority meant failure and ruin. According to Huntford, who had privileged access to Oates's diary, Oates was more of a manager. He felt betrayed by Scott's incompetent leadership. The most blatant example, according to Huntford, was Scott's unexpected decision to take five men with him to the Pole rather than four. This change of mind threw the whole intricate organization of his expedition dangerously out of joint. Everything was arranged for four-man units: tents, gear, cookers, fuel, and the depots of food along the route. Although Oates, according to Huntford, saw the foolishness of Scott's capricious decision, he remained silent and wrote self-disparagingly in his diary about his own inability to intervene.

In his letters to his friends and family, Scott maintains an understanding, benevolent attitude towards Evans and Oates. He represents Evans, who apparently became insane from scurvy before he died,[38] as one of the "sick" that he and his men stuck with until the end. He honors Oates, who committed suicide, for noble self-sacrifice. This not only makes Oates's suicide less dishonorable but makes it fit better into the image of fraternity that Scott constructs in his letters.

Scott imagines the navy to be a community, regardless of the actual inequality that may have prevailed. When Scott writes to Vice Adm. Sir Francis Charles Bridgeman, "We could have come through had we neglected the sick," he displays his willingness to sacrifice his own life, even for men of a lower rank, to perpetuate an image of fraternity based on duty and on hierarchical comradeship.[39]

Scott became an established British tradition during World War I, according to Huntford, who cites the following 1916 entry from a British newspaper: "After a notable bout of disaster, he [Scott] had given his countrymen an example of endurance. . . . We have so many heroes among us now, so many Scotts . . . holding sacrifice above gain [and] we begin to understand what a splendor arises from the bloody fields . . . of Flanders . . . and Gallipoli."[40] The Scott tradition lingered on. Writing in 1959, British historian L. P. Kirwan recounts the familiar story line: "Such are the bare facts of Scott's approach to the Pole. The rest of the story, the exhausting march across the plateau, manhauling all the way; the sight of Amundsen's black flag tied to a sledge-bearer at the Pole; the tell-tale marks of sledge tracks, skis, dogs' paws; the death of Evans, Oates' self-sacrifice, the utter dejection and tragic end of the homeward journey are part of our heritage."[41] Churches and schools became the public sites for passing down the Scott tradition to future generations.

What is striking is the construction of masculinity immortalized through the Scott letters. Through the act of writing, a nationalist myth was established, in which writing itself becomes a means to mythologize an ideology of masculinity in which, paradoxically, the male body is ignored. Or rather, the male body's performance becomes the means by which a moral theater is constructed, in which the body ultimately disappears. The gendered, physical body is replaced by moral character, which provides the foundation on which masculinity becomes heroicized. The exterior world also loses its concreteness in Scott's account. An expedition to the South Pole expresses an exploration into British character. It does not serve as a means to isolate and exalt a virile and potent male body, as Peary's account suggests. The worst possible ending for Scott would be not death, but a failure in moral resolve. Thus, in the narrative of national character, Scott and his men literally sacrifice their bodies and exemplify selfless courage in order to legitimize their claim to rule.

Gender and Narrative Form

Writing played different roles in the construction of the Scott myth and in the U.S. stories of polar exploration. I have pointed out how in Britain, the Scott story is enmeshed in writing from beginning to end. Its transcending aspect is expressed by Scott's encounter with death and miraculous resurrection through his diary and letters. Writing offers a form of presence in absence, a means of salvation by which disorder, meaninglessness, and death are overcome.

In the case of Scott's story, importance is given to the rhetoric of writing well, rather than truly or accurately. The recruitment of a British playwright by Scott's wife to rewrite his diaries hints at a whole literary tradition at work here from the very beginning. In Scott's letters the authority created is anchored to a large extent in subjective experience as mediated and authorized by a literary style. By writing that "we could have come through had we neglected the sick," Scott claims that he exposed himself and his men to additional dangers and personal sacrifices and connects his actions to a higher national mission as defined by the metaphor of tragic self-sacrifice, which belongs to romantic literary discursive conventions.[42]

In contrast to the British, the Americans are trying to produce a narrative that is part of a scientific tradition. There is a large emphasis on exteriority. Performance and achievement matter most. The scientific ideal calls for professional detachment and scientific proofs. The rhetoric of science does not allow for subjectivity except in the form of "genius," or for a sacrifice for a collective identity.

The two narratives have more in common, however, than one might expect from such different genres. In both, authority resides in the effacement of the speaking and experiencing subject. The different genres chosen suited the particular imperial ideologies each writer was promoting. Scott's story was coded within the static and timeless genre of tragedy,[43] expressing England's desire to maintain its dominion of the past into the present. The Americans, by contrast, glorified a progressive scientific ideology that looked more to the future but also wanted immortality. Nonetheless, whereas Scott's subjectivity is understood and constituted in terms of literary ideals, Peary's is defined by scientific objectives. Whereas tragedy is acceptable within the parameters of the literary, there is no place for it within Peary's or Stefansson's scientific discourse. It would be considered merely a catastrophic error of judgment.

Peary and Stefansson anchored the authority of their discourse under the banner of science and progress. They also apportioned different qualities—those of nature-to-be-conquered—to the scene of the poles. Scott instead adopted literary conventions of the sublime to explain his own tragic situation to a British public. In Scott's letters, the landscape of Antarctica is vast, wild, tumultuous, and awful (suggestive of infinity). The blizzard that Scott encounters and blames for the tragedy represents a vast, chaotic, and frightful aspect of nature and is associated with pain and terror. England, which Scott represents, in contrast represents all that is good, ordered, and agreeable.

The point of departure for Peary and Stefansson is totally different, for their narrative is organized around the conquest of nature. Not only do they find positive values for those aspects of the landscape that Scott sees as vast, terrifying, and misproportioned, but they even feel at home there. This homeliness, however, is expressed in terms of extreme scientific alienation from the environment. In Stefansson's discourse Antarctica no longer represents the unknown, as science has already conquered it and made it familiar. "Mankind" now dominates over nature. There are no longer any parts on the globe that can pose a threat. Within such an ideology of modernity and progress, there is no place for a tragic hero such as Scott. Neither can Antarctica provide the accompanying stage by which "man" can obtain glory by recognizing his own limits, for at this stage in the U.S. discourse of progress, science and technology have abolished these limits.

British and American Media Traditions

A British television series, entitled *The Last Place on Earth*, was based on Huntford's *Scott and Amundsen*.[44] The seven-and-a-half-hour epic was one of British television's most ambitious and costly drama series. The series presented Scott as an arrogant and amateurish leader who brought death on himself and his team by inadequate planning and by incompetence before and during the expedition. Such a portrait was inevitably controversial and was condemned by Dr. John Hemming, director of the Royal Geographical Society: "I am very, very disappointed. The acting is superb and the whole presentation is excellent but the length to which it goes to find elements of anti-British bias and anti-Scott bias is just ludicrous. The way in which it is hysterically anti-patriotic is ridiculous."[45]

Roland Huntford, formerly the Scandinavian correspondent of the *Observer*, spent five years researching *Scott and Amundsen*. On its publication in 1979, the book created considerable controversy. The film rights were purchased by British Central Television's series executive producer, Robert Buckler, who approached Trevor Griffith to write the screenplay. Griffith was distinguished as a political playwright and by his commitment to the more popularly accessible forms of television and the cinema. For Griffith, much of the story's contemporary relevance was that it carried fundamental lessons for Britain in the 1980s: "We are living with a government that constantly exhorts us to return to the great Imperial traditions of this nation, and to embrace not just the rhetoric but the practices of the Victorians and the Edwardians. So the series looks at the characteristics of the age, at the class differences and at the age of nationalism."[46]

Griffith's reworking of the Scott story illuminates what is at stake in "living with a government that constantly exhorts us to return to the great Imperial traditions." For Griffith, *The Last Place on Earth* provided an allegory for the Thatcher government and its nostalgic relation with the class-based values, hierarchical structures, and "news management" of the Victorian age: "At a time when news management has reached such appalling levels as in the reporting of the Falklands, the Korean Airlines disaster, and the invasion of Grenada, it seems important to look at how a myth of glorious and heroic failure was constructed in that way."[47]

The Thatcher government regained popularity after the Falklands/Malvinas War, a result of the rise of nationalist sentiments. The enthusiasm for the Scott story similarly relied on a reworking of patriotic sentiments. As I demonstrated earlier, the Scott myth has enjoyed special power, for it can function well in a later period of real decline of empire (World War I to the present), or in an era imbued with a sense of imminent decline, as was the case at the time of Scott's death. Scott's military discipline and loyalty stood out as a timeless example of a universal British tradition that would put an end to anxieties about national weakness. Andy Metcalf and Martin Humphries point out how this process was exemplified in the conduct and aftermath of the Falklands War:

A war fought at considerable cost, with significant casualties, for a few bleak, scarcely populated islands with a lot of sheep, was enough to reverse the Conservative Party's slump in popularity and win them the 1983 general election. This was no mean feat—and it was largely due to the symbolic meanings attached to going to war. Churchillian phrases dripped from the mouths of the "War Cabinet," as a sordid xenophobic enterprise was transformed into a paean to manhood, a celebration of the phallus draped in the Union Jack. Resurgent nationalism and a refurbished manhood were fused into one as the ships left port, the jets screamed overhead, and wives and sweethearts cried and waved good-bye. Everyone was in their place. We'd seen the movie a hundred times: Now it was time for the real thing.[48]

The aftermath both of Scott's tragedy and of the Falklands War had remarkable power. In Britain, where the Labour party and the Left still have some control of the public media, however, there is a greater possibility of popular critique of national policies, as evidenced by *The Last Place on Earth*. In contrast, the debate around Peary in the media has remained privatized—within the control of the National Geographic Society and its magazine—and thus the critiques of Peary have not touched on wider political issues but instead have remained narrowly focused on establishing or disputing the accuracy of Peary's claim to the North Pole.

Despite the confinement of the North Pole controversy to a narrowly technicist realm of ideas, why is it that this debate still prevails today? Why can't it be resolved by concluding that the Pole was simply unwinnable? These questions can only be explained in terms of understanding that the North Pole was also perceived as a mythologized image of empire at the early part of the twentieth century. In this respect, the controversy around the conquest of the North Pole can be seen as an allegory for more recent symbols of U.S. imperial mythography, such as the Vietnam War and more recently the Persian Gulf War. Hannah Arendt describes how the U.S. government masked a host of contradictions in order that the historical event match the fantasy in Vietnam:

The Vietnam War was exclusively guided by the needs of a superpower to create for itself an image which would convince the world that it was, indeed, the mightiest power on earth. Image making as global policy was something new in the huge arsenal of human follies recorded in history. . . . [Image making] was permitted to proliferate throughout the ranks of all government services, military and civilian—the phony body counts of the search and destroy missions, the doctored after-damage reports of the air force, the constant progress reports to Washington.[49]

Like the North Pole, Vietnam was a male testing ground. In both, shame was attached to losing and thus forgoing the opportunity to demonstrate one's manhood. The denial of failure establishes a continuity between these two national events. J. Hoberman describes how the traumatic experience of the loss of the Vietnam War was rewritten by Hollywood cinema:

Vietnam offered no great battles and no clearly defined enemy. Its casualties included our long-standing sense of national innocence and masculine identity, not to mention the broad national consensus that had defined American foreign policy. This has made the war particularly difficult to represent: inherently polarizing and depressing, with a built-in unhappy ending, it both broke the conventions of civilized warfare and the basic rules of Hollywood entertainment. It was the last picture show.

The impossible longing for a satisfactory conclusion tempts each Vietnam film to sell itself as definitive. It is precisely that bummer of a finale . . . that has left us with a compulsion to remake, if not history, then at least the movie.[50]

The rewriting of Vietnam by Hollywood makes U.S. soldiers appear as victims of superiors, bureaucrats, and communists. Soldiers crack up, are cowardly, and fuck up (kill civilians). Yet Vietnam heroism exists both in spite of and against U.S. government policy.

Setting Things Aright: Technology, the Gulf War, and Peary

The denial of failure was enacted not only in Hollywood stage sets but also in the Persian Gulf region. After the so-called allied victory, the legacy of the Vietnam War was cited as a disease that had been overcome.[51] On March 3, 1991, George Bush declared, "We have kicked the Vietnam Syndrome."[52] Bush had promised that the Persian Gulf War would be different—a neater package and easier to understand, with clear closure and an unambiguous resolution. Just as Peary's complex story has been rewritten by the *National Geographic* with a happy ending reinstating Peary as an uncontroversial U.S. hero, so too has the Persian Gulf "victory" restored good feeling about a previously denigrated United States in decline. And in both cases, the discursive logic of this favorable outcome turns on technology as unchallengeable or seemingly undefeatable. This essay is my attempt to explain the interconnections between the multiple narratives of national identity, scientific progress, modernity, and masculinity across the national cultures of the United States and the United Kingdom. Once one of these discourses is invoked, the others are immediately brought into play. In this sense the Falklands and Malvinas, Vietnam, and Persian Gulf narratives parallel those of both Scott and Peary.

During the Gulf War, advanced weaponry was brought in to restore the old national narratives of success; in a similar way, new high-tech photographic processes presumably solved the problem of Peary's inaccuracy at the North Pole. The seamless performance of technology was more important for the American, who unlike the British never showed in the media a precision weapon that missed.[53] For the Americans, winning the Gulf War was inextricably tied to the myth of technological prowess embodied in high-tech electronic weaponry (Stealth fighters, spy satellites, Patriot missiles, and Tomahawk cruise missiles that were dependent on "perfect" maps drawn of Iraqi territory and terrain).

Moreover, the war was programmed in terms of its media presentation, and in various ways failure was written out of the narrative from the beginning. The televisual apparatus in general, and TV news in particular, joined forces with the military to narrate the event in a way that would sanitize and prettify the war in order to associate it with a pre-Vietnam vision of U.S. innocence and righteous virtue. Following this logic it is not surprising that the Gulf War was scripted as a replay of World War II with its reassuring overtones of justice, democracy, and victory, rather than of Vietnam, with which it had more in common. In these presentations, the performance of so-called smart technology was thought to be so infallible that in the earliest hours of the bombing, CNN's Pentagon reporter, Wolf Blitzer claimed that the 150,000-man Republican Guard had been crippled or destroyed by air strikes alone.

Such exaggerated claims for the efficacy of technological weaponry kept alive the belief, which was an inextricable part of the Peary narrative, that technology was unbeatable and that the structure of operations was somehow bloodless and unerring. In contrast to Vietnam, which could not be figured as a clean war, the Persian Gulf War was represented as strangely antiseptic and disembodied, as media coverage focused on the performance of U.S. smart bombs and surgically precise air attacks.[54] Reportage focused on identifying viewers with the pilots doing the bombing rather than with those civilians being bombed (the elision of images of Iraqi deaths or casualties). Brown bodies in general were shown as having little presence to Americans except as mere numbers. The Persian Gulf War was far more successful in rendering abstract and erasing enemy brown bodies than was the Vietnam War, where their suffering and death were made all too palpable on home television. It was not until weeks afterward that a Western audience heard that 70 percent of the 88,500 tons of bombs dropped on Iraq missed their targets and hit thousands of civilians instead.[55]

The righteous modern violence on behalf of the Western international community was contrasted to the primitive, barbaric violence of Saddam Hussein's forces.[56] Where Saddam Hussein and the Iraqis were represented as irrational beings with an uncontrollable sexual drive (Iraq was shown as having "raped" Kuwait, which required massive but "surgical" retaliation), Bush and the Western soldiers were identified as representative of heroic Western masculinity, now cured of previous "impotence" suffered in Vietnam.

The reactive Rambo style of Western masculinity tended to dominate the media's account of the war. One thinks of Colin Powell's "cut it off and kill it" or Schwarzkopf's promise to "kick butt" or the reports of pilots watching porn videos before ejaculating their bombs over Iraq.[57] Yet despite the overblown masculinist rhetoric, this imagery did not seem anything more than a feeble attempt to remasculinize or regender social relations in an age in which heroism had less to do with the body and more to do with delegating work and manipulating electronic data. Take, for example, the San Francisco Chronicle poster that appeared during the Gulf War with the caption "A lot can happen between 9 & 5," featuring a before-and-after shot of Sylvester Stallone as Rambo, bare-chested and armed to the teeth, and of a fully clad, rotund General Schwarzkopf as the top allied manager. Only two years after the war, it was significant that the war as experience and symbol was no longer being evoked, despite the fact that the media engineered a broad consensus in favor of the war. The "victory" in the Persian Gulf had largely unraveled, and the war was seen as having a much more ambiguous or murky ending. Saddam was still in power, and the claims of military technological know-how were seen as vastly overrated by the Pentagon itself.[58] Rather than exulting over their country's military victory in the Persian Gulf, Americans worried about corrupt politicians, the recession, and increasing budget cuts of domestic programs. The implementation of an older narrative that was supposed to supplant the Vietnam syndrome (the narrative of failure) seems not to work well in a radically changed and uncertain post–Cold War context. There was a strong feeling that things would never be the same again, and one wonders how viable the ideological narrative of scientific progress that has framed the discourses of the *National Geographic* and the story of the Persian Gulf War will be in the future. What new narratives of Americanism and masculinism will replace them? Will new stories be produced that do not rely solely on technology as the transcending foundation of a gender- and race-based Americanism?

Perception Versus Experience: Moving Pictures and Their Resistance to Interpretation

We have become accustomed to conceiving the relationship between image and reflection (the relationship between images and concepts as the medium of reflection) with a tinge of dialectical melodrama—a melodrama, to be sure, that consistently leads to a "happy ending." In general, we assume a tension between image and reflection. The image speaks to the senses, and for that reason images are assigned to the referential horizon of philosophical aesthetics. Reflection, by contrast, takes place in concepts, in the elements, that is, that constitute language and discourse. Aesthetic experience is not conceptually mediated experience and thus silences discourse. But we can accept such silence only with difficulty.[1] For this reason more than any other, the conviction seems to have prevailed that aesthetic experience, as sensual experience, can be captured by language after all—as long as its incommensurability with conceptuality has not been explicitly recognized at the outset. The discourse through which this is supposed to occur is called "interpretation."

In this essay I wish to problematize such attempts to rescue the possibility of discourse by means of interpretation, primarily on the basis of historical arguments. I will not take this historicizing so far as to claim that because of altered conditions of experience, this salvation is today no longer possible. I begin instead with the assumption of a principal heteronomy between sensuality and conceptually mediated reflection and will then—more modestly—attempt to show that what can be sustained less and less today is merely the illusion that this fundamental incompatibility has been suspended.[2] If in this context we reserve the label "perception" for the side of sensuality and the label "experience" for the side of concepts and reflection (a terminological distinction that is perhaps somewhat arbitrary, but allows the following argument to be developed with a high degree of transparency), then the impression that experience is losing ground to perception in our responses to contemporary everyday realities appears as a symptom of the historical change to be addressed in this essay. Philosophers have, of course, long since responded to this impression. Jean-François Lyotard, for example, claims that what Kant had identified as the special nature of aesthetic judgment, the lack of recourse to a basis of categories and concepts, is becoming the general condition of judgment today,[3] only to emphasize that this is occurring in a situation that, in view of its complexity, demands conceptually mediated, "rational" judgment more urgently than ever before. Ferdinand Fellmann, otherwise hardly suspected of collusion with Lyotard, has likewise attempted to describe human consciousness, taking up Wilhelm Dilthey's position, as constituted not by concepts, but by images.[4]

Because it is the historical dimension with which I am mainly concerned in the following pages, my thesis differs from positions such as Lyotard's and Fellmann's primarily in emphasis. I suggest that the fundamental incommensurability between perception and concepts led to a crisis because Western culture has until now been unable to respond to an epistemological transformation that became apparent around 1800, the effects of which have been significantly intensified by the media technologies that have arisen since that time. In order to describe this transformation, I will begin by characterizing the epistemological paradigm that informed the thinking and the formation of experience in the West from the beginning of the early modern period until about 1800. I call this paradigm the hermeneutic field, and see it as grounded in the position of a first-order observer.[5] The second part of my essay begins by postulating the emergence of the position of a second-order observer around 1800 and concentrates primarily on the different forms of Realism and on the history of new communications media in the nineteenth century, describing them as the—ultimately failed—attempt to intercept the consequences of the epistemological crisis that accompanied the increased complexity of the new observer position. Instead of making this interception, the communications media, especially moving pictures, have ever since merely dramatized the divergence between the world of perception and the world of experience.

An argument that alludes to so many different discursive levels and includes such long historical stretches can (or must) be criticized for being irresponsibly speculative. I will not discuss definite standards of historical documentation that would "actually" have to be respected, because even a book of considerable length would not be sufficient to address them. What, then, speaks for a preliminary sketch such as this one (or, to be more modest, for such an attempt)? Perhaps more than anything else, the fact that it is hardly possible to see such diverse dimensions as epistemology and technology, experience and perception, in a relationship of complex reciprocal interaction when the academic guild's official criteria of evidence and documentation are observed. Should one really fear speculation so much that, in the end, it obstructs thinking about such relationships?

Experience in the Hermeneutic Field

As a historically specific relationship of human beings to the world, the intellectual habitus that European philosophy, above all in the nineteenth century, analyzed, practiced, and canonized under the title of "hermeneutics" had become a stable configuration already with the transition from medieval to early modern culture. I will refer to this relationship to the world (as distinguished from the academic subdiscipline of "philosophical hermeneutics") as the "hermeneutic field" and will describe its contours by indicating four of its central implications.

The first structural prerequisite for the genesis of the hermeneutic field is the eccentricity of human beings in relation to the world. In the medieval world view, humankind had seen itself as part of Creation, outside of which stood God alone—as its Creator. We allude to this eccentricity of humans when we speak of "early modern subjectivity," and only from the observer position of such an eccentric subject does the world become a "world of objects." The hermeneutic field is the sphere of the subject/object paradigm.

Second, the subject/object paradigm relegates the human body to the side of objects. The subject is therefore without body and gender and cannot be a point of reference for sensual perception. Medieval theology, by contrast, had conceived the human spirit and the human body as a unity. This premise is a substantial reason for the strangeness of the Middle Ages from our perspective. Protestant theology struggled with this spirit/body unity not only in its reformulations of the doctrine of transubstantiation, but also in its confrontation with theological motifs such as the bodily resurrection of the dead or the bodily reception of Mary in heaven.

Third, the subject observes the world of objects by introducing a differentiation absent in medieval culture, the distinction between a "simply" material surface of things and a spiritual depth. It is the absence of this binarism that makes it so difficult for us to reenact the cultural habitus that has been called "medieval symbolic realism." Because "depth" is conceived here as a sphere of concepts and takes on the status of a site of truth, the existential necessity to penetrate the surface of things in order to become conscious of the truth arises. This penetration is one side of the act of "interpretation," which is fulfilled by identifying the concept lying in the depth as its other side. Only with the differentiation between spiritual depth and material surface (the latter, as a simple prerequisite for the possibility of experience, itself remaining meaningless), does the sphere of the subject become the hermeneutic field. This differentiation has been described in an almost infinite number of variations, of which the best-known today is the linguistic dichotomy between the "signified" and the "signifier."

Fourth, as part of the world standing in opposition to the observer-subject and thus as object of interpretation, the human body comes into view in the hermeneutic field under two perspectives. If body and spirit are no longer conceived as a unity, thoughts, concepts, and truths may—first of all—be hidden behind the body. It can then become necessary to penetrate the body—interpretively[6]—in order to discover secrets as "truths." Even if the subject does not retreat into hiding behind a body, however, it is still confronted with the impossibility of fully articulating the truth of its thought through the mediation of the body, normally through the voice or the hand. Precisely this difficulty is alluded to by the initially metaphorical meaning of the word *expression*, documented evidence for which dates back to the end of the Middle Ages.[7] Only under the condition of such fundamental insufficiency of self-expression does interpretation become an existential necessity. It is supposed to salvage the "deep" meaning, which can never be completely rendered in the act of "expression."

Regarding the historical conditions for the emergence of the hermeneutic field and of the subject as an observer, we will limit our discussion to just two relevant trends of theoretical analysis. If the subject came to light as the agent and center for the construction of meaning, this configuration—in contrast to Christian cosmology, in which the meaning of all phenomena was seen as given in the act of divine creation—made it possible to view the construction of meaning as a process of accumulating knowledge. With the institutionalization of printing, the fascination with the production and accumulation of knowledge had inadvertently replaced the traditional obsession with the preservation of a body of knowledge revealed by God.[8] The incorporeality and spirituality of the subject may, on the other hand, have augmented the shift taking place within the same media-historical context from interaction—requiring a copresence of the body—to communication, which shifted the body over to the side of the environment. If printed texts, in contrast to manuscripts, excluded all traces of the bodies involved in their production, and if the figure of authorship emerging at the time postulated a close constitutional relationship between subject and text, this configuration augmented the tendency to conceive the subject by screening out the body.

In our attempt to historicize the hermeneutic field by reconstructing its genesis and by at least imagining its termination, the age of the Enlightenment stands out as the high point in its development and undisputed authority. The Enlightenment was the epoch in which the distinction between the surface of things and their conceptual depth developed into the idea of an isomorphism between the world of phenomena and the structure of knowledge about phenomena (named the "classical episteme" by Foucault).[9] From this perspective, it becomes obvious why an "ontological" function—to use an anachronistic formulation—was attributed to dictionaries and encyclopedias, a function for which D'Alembert and Diderot's "map" of knowledge in the first volume of the *Encyclopédie ou dictionnaire raisonne des arts et des métiers* is perhaps the most famous documentary evidence. In a complementary move, demystification as the intellectual project of the Enlightenment was sustained by the resolve to eliminate and replace all bodies of knowledge not yet founded on subject-experience. This project ultimately laid the foundations for the expectation that the progressive (or even indefinite) increase in knowledge would promote social welfare and justice. Here lie the beginnings of "utopian," "scientific," but never really existing socialism.

The proviso, which had already become conventional in this context before 1800, that only "unbiased" or "unalienated" subject-experience guaranteed the truth of the new knowledge, can be understood as an indication—contemporary with the highest development of the hermeneutic field and the subject/object paradigm—of the doubts regarding the "objectivity of the subject" that were about to destabilize the epistemology of the nineteenth century. The same is true of the creation of the subdiscipline of "philosophical aesthetics," in which a new attentiveness to sensual perception, as distinct from conceptually articulated experience, found its manifestation. Finally, the materialism of the eighteenth century was also fascinated by the corporeal mechanics of sensual perception. Nevertheless, the materialists still thematized the human eye and the human ear with the exclusive goal of convincing themselves that adequacy was indeed possible in the relationship between the poles of "reality" and "knowledge" connected by the organs of the senses.

The Resistance of the Body

Such self-persuasion became increasingly precarious during the nineteenth century. The historical sources clearly show that changes in epistemology, science, and social structure, in technology and symbolic systems, combined to generate a degree of complexity that rules out the assumption of any priority or causality in the ongoing transformation. Instead of understanding and analyzing this extensive transformation of the world-image, one can probably only illustrate and document it—and in doing so, one can also bring Foucault's concept of the "episteme of the nineteenth century" into convergence with Luhmann's systematic concept of the "second-order observer."

Unlike the subject/object paradigm, on the basis of which the hermeneutic field had constituted itself, the special nature of the *sciences de l'homme* (humanities) as the episteme of the nineteenth century lay in the double role assigned to humans as subjects *and* objects of observation. This historical configuration corresponds to the definition of the second-order observer, who observes himself observing (as a first-order observer) and in whose field of vision blind spots of the first observer level therefore become apparent. One such blind spot of the previous episteme had been the human body as an instrument of world-perception and world-experience. From the early nineteenth century, by contrast, the new attentiveness to the role of the body created interference, problematizing the principle of the "representability of the world" that had hitherto guaranteed adequacy between world and world-experience. There are numerous indications of the new conviction that the world could be perceived and experienced only through the mediation and under the specific conditions of the human body. If, for example, one forgoes celebrating the discovery of the "historicity of phenomena" at the end of the eighteenth century as the discovery of a transcendental truth, a perspective opens up in which the historicist dimension and its narrative models[10] appear as the space where instability of experience is processed into variability of experience. Complementary to this, certain—corporally defined—dynamics emerged as explanations for such constant alteration of phenomena; among these principles were a transcendental concept of "life" or "vis vitalis" in biology and a new concept of "labor" in the disciplines of national economics and sociology, which were only now beginning to emerge.[11] Of even greater significance, however, was a new habitus of distinguishing between two perspectives, levels, and principles of the human relationship to the world: between bodily bound perception and a form of experience that was carried out through concepts. This distinction, of course, was not yet synonymous with the postulate of an incommensurability between perception and experience, but it already implied the challenge of inventing hypotheses regarding the relationships between the levels of perception and experience, which had been lifted apart from one another.

The time when one could take for granted the equation of the experience of the incorporeal subject with objective experience had come to an end. Doubts now began to emerge about the possibility of objective experience (as experience adequate to the objects of the world), and they held philosophy, art, and literature spellbound from the early nineteenth century. It is characteristic of the type of Realism that literary historians associate especially with the novels of Honoré de Balzac that it awakened skepticism not only about the possibility of experiencing cosmic order, but also about the existence of such an order. From this crisis, which regularly emerges at the outset of their fictional plots, Balzac's novels go on to demonstrate that the order of the world does reveal itself, at least to those observers who know how to gain a truth-exposing perspective. This can be—in the most elementary case—the view of the world from an elevated location in space, but more frequently it is the discursive level of an omniscient-auctorial narrator, which is often reserved as a reward for those protagonists whose behavior corresponds to the moral ideals of the author.[12]

The fact that the scientific and technological prerequisites necessary for its development, such as the discovery of the light-sensitivity of silver salts, were available long before the actual invention of photography suggests that photography too is associated with the epistemological crisis provoked by the emergence of the second-order observer.[13] If the

problem-solving strategy of early literary realism was a distinction between adequate and inadequate observer-perspectives, the even more
self-certain promise of objectivity offered by photography relied on the
elimination of the observer, who had only just been discovered, and of
the observer's body.[14] Under precisely this condition, it was believed, the
form and the order of things would inscribe themselves directly onto
the photographic plate through mechanical reproduction independent of
human intervention. Completely contrary to such expectations, however, the experience prevailed that the ideal form of things corresponding
to the ideality of concepts never became visible in photographs. Instead,
the new images were unavoidably imprinted with the contingent elements of the moment of their production—the moment of photographic
recording—so unavoidably, in fact, that a new type of Realism in painting
and literature, which began to make its mark from around 1840, emerged
precisely from the emphasis on contingency in the representation of the
world. It was in this context that the first programmatic use of the term
Realism occurred—with specific reference to new techniques of painting.[15] Paintings such as those by Courbet or Menzel can be categorized as
"perception-oriented" because they taught a new way of seeing, a way of
seeing no longer oriented toward the search for ideal objects or perspectives that make truth accessible.[16] On the contrary, painters began to blur
the contours of things and to construct their perspectives so that the
configuration of themes and motives had to be achieved almost against
them. Resistance to an image preoriented by concepts thus became a
symptom of proximity to reality.[17]

It was precisely this tension between the perception of reality and its
translation into concepts that Gustave Flaubert dealt with in his novels. In
contrast to the realist painters of his time, however, Flaubert's method was
not problematization but rather the ironic reproduction of discourses and
concepts.[18] Emma Bovary meets her demise through the immunity to
reality of her daydreams, which were fed by romantic literature; Frédéric
Moreau, the hero of *Education sentimentale*, for whom nothing is too
trivial, sees the scenes of the revolution of 1848 as a confusing chaos
because they confront him with new forms of events, whereas his love
affairs are nothing but new projections of shallow romantic commonplaces; Bouvard's and Pécuchet's ideas of science and progress ultimately
distance the two friends from all productive thought. As a prerequisite for
his writing practice, Flaubert had collected a repertoire of the elements of
bourgeois common knowledge with nothing less than empirical precision

in the so-called *Dictionnaire des idées reçues*. The manifestations of this second type of Realism seem less impressive to us whenever they set up the contingency of perspectives and discourses against worldviews that imply the claim of being "adequate to reality"—as occurs, for example, in Marx's critique of ideologies. The historical significance of the second type of Realism, after all, lies precisely in the fact that it no longer suggests any way to overcome the newly experienced remoteness of Reality and the dependence of discourses on the observer. In his famous definition of the realist novel, Stendhal had already responded to this discovery in 1830 with the metaphor of mobile observation, which can reflect only myriad aspects—not a stable picture—of Reality: "Un roman est un miroir qui se promène sur une grande route. Tantôt il reflète à vos yeux l'azur des cieux, tantôt la fange des bourbiers de la route."[19] (A novel is a mirror walking on a road. Sometimes it reflects the blue skies, sometimes the mud of the road.)

While the different types of Realism (and Naturalism) of the nineteenth century played with ever-new variations on the relationship (now discovered to be one of tension) between the world and the forms of its representation by observers, an entire spectrum of artistic techniques was developed in which the referential structure between "merely material" signifiers and the "actually relevant" signified, always stable within the hermeneutic field, was thrown out of balance. Because such destabilization occurred on the side of the signifiers as well as the side of the signified, it can be characterized—in general and using a crisis metaphor from the economic system—as "sign-deregulation." Richard Wagner's Programm-Musik belongs in this context, as an attempt to give decodable meanings to sounds that, according to traditional understanding, were "only perceptible." In the same spectrum of historical phenomena, the poetics of literary symbolism functioned as the opposite pole to Programm-Musik, because it directed the reader's attention to sensual qualities in which the signifiers of language transcend their traditional function of conveying meaning. This could—for example, in Mallarmé—be the layout of the printed text, or just as easily, the tonality and rhythm of a poem, which came across even in the silent reading. Sign-deregulation was certainly the motivating context for Rimbaud's lyrical speculations on the colors of vowels. Finally, our historical perspective makes it possible to read Friedrich Nietzsche's work as a philosophical horizon of such displacements. Nietzsche's polemic against the "will to truth" problematizes the hermeneutic dimension of conceptual depth; its flip-side, the philologist-praise for literalness and the enthusiasm for the sensual qualities of mask and dance, inflates the value of the superficial. When there is no longer a face behind the mask and dance no longer needs to express anything, the structure of the hermeneutic field founded upon the separation of signifier from signified collapses.

This collapse took place simultaneously with the transformation of hermeneutics into a philosophical methodology in the works of Wilhelm Dilthey. I would like to bring this point into sharp focus with the argument that hermeneutics—and along with it, the relationship of the subject to the world of objects founded upon interpretation—was, in this historical situation, hypostatized into the organon of the humanities (*Geisteswissenschaften*) and salvaged for the academic institution within which a number of social subsystems (above all, technology, economics, and art) imposed a penetrating transformation of the subject/object paradigm. Thus seen, the differentiation and autonomy achieved here for the humanities, which liberated them from competition with the now triumphant natural sciences, exacted a high price; namely, an epistemology that isolated the humanities from their social environments. Ferdinand Fellmann's new reading of Dilthey (mentioned above) forces us to revise this assertion—or, more positively, to increase its complexity. If one accepts Fellmann's view, according to which the effort to describe human consciousness as constituted by images rather than concepts lies at the center of Dilthey's philosophizing,[20] this motif seems more like a productive response to the epistemological displacements of the nineteenth century than an act of intellectual salvation. Perhaps the decisive threshold for the coherence of our argument is not marked by the difference between consciousness constituted by concepts and consciousness constituted by images, but instead by the contrast between static and moving images as contents of consciousness. The first animated images distanced themselves once and for all from the habitus of interpretation—and thus from the methodologies of the human sciences—because the static nature of concepts, indispensable for interpretation, was no longer capable of doing justice to the instability of animated images.[21]

Perception of Motion

It has generally been assumed that the animated images of the film medium come as close as one could imagine to the contents of consciousness that were not conceptually structured. Because, however, the images of

film do not—like the images of the imagination—originate in consciousness, but are instead first grasped by perception and must be mediated to
consciousness, the problem to be solved before animated images could
become technically producible, and thus also capable of being received,
can be characterized by the question of how something can be in motion
and at the same time have a perceptible form. The answer lies in the
insight that motion and form can converge only in rhythmically arranged
sequences of images—in other words, "rhythm" is our concept for the
formal quality characteristic of "time-objects in the specific sense."[22]
Technically, the problem of producing rhythm was solved by perforating
the film material, thus making it possible to link images to the rhythm of
machines. To this day, the rhythm of film transport and the rhythm of the
constitution of lines on the television screen remain indispensable conditions for the perception of technically produced animated images, even
though they are hardly ever mentioned.

It is true that Henri Bergson denied his philosophical blessing to early
film as the answer to the question of the possibility of technical (re)pro-
duction of animated images.[23] His objections were based upon the trivial
circumstance that film material was itself nothing but a series of static
images, because these static images could not be subsumed under his
concept of *durée* (duration) as a sphere of temporality that could neither be
measured by quanities nor delineated by concepts or contours. Today,
however, we easily agree with Gilles Deleuze, who counters Bergson's
criticism with the argument that despite the static quality of the images on
the celluloid roll, the audience of a film projection perceives animated
images (*images-mouvement*)—and indeed, animated images in precisely the
sense of Bergson's concept of *durée*. Nevertheless, an extremely important
pragmatic aspect has been lost with the dismissal of Bergson's objection to
the phenomenological legitimacy of film; namely, the insight into the
incommensurability between animated images and the static nature of
concepts. If one takes this element of Bergson's polemic into account,
then the technical innovation of film appears to be the transition from a
world of experience to a world of perception. Niklas Luhmann's thesis,
according to which motion first became a component of information and
at the same time an object of communication in the medium of film,[24] can
therefore be complemented by the insight that what is specific about such
information can no longer be transported by concepts.

The astonishing convergence of such diverse intellectual and technical
projects at the turn of the twentieth century makes it plausible to view the

history of film beginning in 1895 as the media- and technological-historical variation of a long-term transformation of western epistemology. The central focus of these projects were descriptions of the pre-conceptual levels of consciousness that were motivated by philosophical ambitions. This was Bergson's main concern in working out the concept of *durée*[25] and the decisive epistemological breakthrough in Freud's book on the interpretation of dreams.[26] At the same time, but in a very different intellectual and institutional context, George Herbert Mead drew up a theory of the imagination, according to which images of the imagination were directly triggered by perceptions of the environment and in turn directly triggered nerve impulses and consequently muscular movements (as reactions of aggression or flight). Mead attributed the processing of perceptions and images of the imagination by means of concepts to a higher stage in human evolution. It was with this stage alone that he associated the possibility of controlling the bodily impulses triggered by environmental perceptions.[27]

Likewise, the art and literature of the time incorporated into program and practice the complex motif of the connection between preconceptual perception and motion in the perceived world—often with specific reference to the writings of authors such as Nietzsche, Bergson, or Freud. It is thus with good reason that Surrealism has been characterized as the (sometimes obsessive) staging of a defacement and erosion of meaning as form. At the same time, a shift occurred within Surrealism from the representational function of art and literature to the production of events as emblems of contingency, and authors such as Italo Svevo and James Joyce transformed a central motif of psychoanalysis and phenomenology into the discursive form of stream-of-consciousness novels, which had constantly to strive to undermine and extinguish its unavoidably conceptual composition.

On the other side of the intellectual scene in the early twentieth century were philosophical positions that emerged as negative reactions to the dissolution of concepts into moving images and to the substitution of perception for experience.[28] A premise of the sociology of knowledge, as it emerged from Phenomenology during the 1920s in the writings of Max Scheler, Karl Mannheim, Alfred Schütz, and others, was thus the abandonment of questions regarding the adequacy or truth of world-experience. Instead of allowing the collapse of the subject/object paradigm to paralyze them intellectually, the sociologists of knowledge concentrated no longer on reality "itself" but rather on the processes of the social construction of reality by means of concepts and elements of knowledge.[29] Here lies the intellectual prehistory of what we call Constructivism today. For the thinkers of the so-called Conservative Revolution, by contrast, reactions to the loss of the epistemological depth-dimensions of concepts and truth were much more strongly motivated by existential concerns than was the case with the early sociologists of knowledge. These thinkers struggled for the recovery of a secure grounding for valuations and actions—often fully conscious of the impossibility of fulfilling such intellectual and existential desires. For this reason, the philosophical architectonics of Heidegger's *Being and Time*, with its distinction between entities (*Seiendes*) and Being (*Sein*), came close to the binarism between the surface of things and the depth concealing their truth that was constitutive of the hermeneutic field. And for the same reason, acts of understanding and interpretation assumed a central position in Heidegger's analysis of human existence.[30] Nevertheless, Heidegger answered the most far-reaching question of his book, the question of the meaning of Being (*Sinn des Seins*), by working out a new concept of "temporality" from a perspective that was hardly conceivable before the crisis of the subject/object paradigm and before the destabilization of the concepts of world and object. The meaning of Being, one surprisingly finds in reading this leader of the Conservative Revolution, lies in "handing down to itself the possibility it has inherited, [to] take over its own thrownness and be in the moment of vision for 'its time.' "[31]

While the intellectuals worked on such attempts to rescue experience, understanding, and values, the shift to a world of bodily centered perception continued apace in the everyday world. One among countless symptoms of this displacement was the concept of "human material," which was used so frequently—especially, but not exclusively—by fascist politicians.[32] Reduced to "human material," human existence consisted of bodies whose sensibility to the environment enabled linkage to the rhythm of other bodies and to the rhythm of machines—that is, the integration of bodies into systems of higher complexity. The body of the kamikaze pilot, subordinated to an intelligent machine and subjected to destruction and self-destruction along with that machine, was but one among many emblems of this configuration.

The present epistemological situation is still dominated by the bifurcation between a sphere of conceptuality, stasis, and language, and a sphere of perception, motion, and the body, just as it was marked out in the cultural manifestations beginning in the early nineteenth century.

Even today, we abandon ourselves to a discourse of cultural pessimism that holds up the intellectual depth of books and "patient understanding" like a monstrance of the good, true, and beautiful against the "flat super-ficiality" of screens and the agile movements of eyes glued to them. But perhaps these well-meaning worries rest upon premises that are far too traditional: upon the postulate, for example, that the complete translation of all perceptions into concepts is an existential necessity, upon the expec-tation that in the long run the dimension of perception will completely suppress that of experience and reflection, and finally upon the fear that the occupation of the individual consciousness by images that did not originate there must unavoidably culminate in a dramatic loss of self-determination.[33] All of these nightmarish visions imply the twofold as-sumption, first, that one can only live either in the sphere of perception or in the sphere of experience, and second, that it is obviously better to live only in the sphere of experience. Perhaps, however, an *oscillation between experience and perception* is precisely the form of life and agile thinking that will allow us to escape the spell of the hermeneutic field and its academic afterlife. In spite of all the critically wrinkled brows, there is today not only a generation of teenagers who have grown up with computer games and are nevertheless beginning to read Plato; there are also well-established forms of theory (and many admired heroes of theory) for which this oscillation comes effortlessly. In the first move in his game of deconstruction, Jacques Derrida always attacks apparently stable configurations of meaning, only to transpose them into the dynamic of *différance* and the sensual aspects of writing. Conversely, Niklas Luhmann insists at the outset that only what happens is real and that everything that happens takes place in simultaneous operations without the dimension of mean-ing—only to observe this reality and, in observing, to extract concepts from it.[34] Once set in motion, such intellectual oscillations no longer come to a standstill. In other words, it is high time to put aside the illusion that perceptions will ever be completely captured by concepts and reflec-tions. Instead of relying upon that illusion ever being fulfilled and coming to an end, one simply has to be able to give it up.

REFERENCE MATTER

Notes

1. Lenoir, "Inscription Practices"

1. Steven Shapin and Simon Schaffer, *Leviathan and the Air Pump: Hobbes, Boyle and the Experimental Life* (Princeton: Princeton University Press, 1985), pp. 55–59; Bruno Latour and Steve Woolgar, *Laboratory Life: The Construction of Scientific Facts* (Princeton: Princeton University Press, 1986), p. 88.

2. "Relations of Literature and Science, 1993," *Configurations* 3, no. 2 (1995), 269–315.

3. Paul K. Feyerabend, *Against Method: Outline of an Anarchistic Theory of Knowledge* (London: Verso, 1978).

4. Paul Gross and Norman Levitt, *Higher Superstition: The Academic Left and Its Quarrels with Science* (Baltimore: The Johns Hopkins University Press, 1994); Alan D. Sokal, "Transgressing the Boundaries: Toward a Transformative Hermeneutics of Quantum Gravity," *Social Text* 46/47 (1996): 217–52.

5. Alan D. Sokal, "A Physicist Experiments with Cultural Studies," *Lingua Franca* (May/June 1996): 62–64.

6. For an example of such an approach, see Timothy Lenoir and Cheryl Ross, "The Naturalized History Museum," in Peter Galison and David Stump, eds., *The Disunity of Science: Boundaries, Contexts, and Power* (Stanford, Calif.: Stanford University Press, 1996), pp. 370–97.

7. Jacques Derrida, *Of Grammatology*, trans. Gayatri Spivak (Baltimore: The Johns Hopkins University Press, 1974), p. 73.

8. Ibid.

9. M. Norton Wise, "Mediating Machines," *Science in Context* 2 (spring 1988): 77–113.

10. Jacques Derrida, *Margins of Philosophy*, trans. Alan Bass (Chicago: University of Chicago Press, 1982).

11. Derrida, *Grammatology*, p. 35.

12. Ferdinand de Saussure, *Course in General Linguistics*, trans. Roy Harris (La Salle, Ill.: Open Court, 1986), pp. 24–25.

13. Ibid., p. 26.

14. Ibid.

15. Ibid., p. 31.

16. Jacques Derrida, *La voix et le phénomène* (Paris: Presses Universitaires de France, 1967), pp. 89–90; David E. Wellbery, "The Exteriority of Writing," *Stanford Literary Review* 9 (1992): 11–23, esp. p. 16.

17. Latour and Woolgar, p. 76; Michael Lynch, *Art And Artifact In Laboratory Science* (London: Routledge & Kegan Paul, 1985).

18. Latour and Woolgar, p. 66.

19. Jacques Derrida, *Writing and Difference*, trans. Alan Bass (Chicago: University of Chicago Press, 1978), pp. 196–231.

20. Ibid., pp. 217–20.

21. Ibid., p. 230. Italics in the original.

22. Derrida, *Grammatology*, p. 84.

23. Friedrich Kittler, *Discourse Networks: 1800/1900*, trans. Michael Metteer with Chris Cullens (Stanford, Calif.: Stanford University Press, 1990), p. 246.

24. David E. Wellbery, foreword to Kittler, *Discourse Networks*, pp. vii–xxxiii, esp. p. xxxi.

25. Kittler, *Discourse Networks*, p. 250.

26. Derrida, *Grammatology*, pp. 332–33n35.

27. Condorcet is quoted by Roger Chartier in *Forms and Meanings: Texts, Performances, and Audiences from Codex to Computer* (Philadelphia: University of Pennsylvania Press, 1995), p. 11.

28. Chartier, *The Order of Books: Readers, Authors, and Libraries in Europe Between the Fourteenth and Eighteenth Centuries*, trans. Lydia G. Cochrane (Stanford, Calif.: Stanford University Press, 1994), p. 90.

29. Paul Duguid, "Material Matters: The Past and Futurology of the Book," in Geoffrey Nunberg, ed., *The Future of the Book* (Berkeley: University of California Press, 1996), p. 78.

30. Ibid., p. 79.

31. Derrida, *Grammatology*, p. 158.

32. Bruno Latour repents from his earlier use of Derridean models of inscription and difference in *We Have Never Been Modern*, trans. Catherine Porter (Cambridge, Mass.: Harvard University Press, 1993), pp. 5–6; 62–65. Donna Haraway also seeks to introduce agency into her account; see "The Promises of Monsters: A Regenerative Politics for Inappropriate/d Others," in Lawrence Grossberg, Cary Nelson, and Paula Treichler, eds., *Cultural Studies* (New York: Routledge, 1992), pp. 295–327.

33. Latour, p. 64.

34. Brian Rotman, "Towards a Semiotics of Mathematics," *Semiotica* 72

(1988): 1–35; Rotman, *Ad Infinitum . . . The Ghost in Turing's Machine. Taking God Out of Mathematics and Putting the Body Back In. An Essay in Corporeal Semiotics* (Stanford, Calif.: Stanford University Press, 1993).

35. Rotman, *Signifying Nothing: The Semiotics of Zero* (Stanford, Calif.: Stanford University Press, 1993), p. 28.

36. For Derrida's rejoinder to this point, see Jacques Derrida. *Limited Inc.*, trans. Samuel Weber (Evanston, Ill.: Northwestern University Press, 1988), esp. pp. 136–37:

What is called "objectivity," scientific for instance (in which I firmly believe, in a given situation), imposes itself only within a context which is extremely vast, old, powerfully established, stabilized or rooted in a network of conventions (for instance those of language) and yet which still remains a context. And the emergence of the value of objectivity (and hence of so many others) also belongs to a context. We can call "context" the entire "real-history-of-the-world," if you like, in which this value of objectivity and, even more broadly, that of truth (etc.) have taken on meaning and imposed themselves. That does not in the slightest discredit them. In the name of what, of which other "truth," moreover, would it? One of the definitions of what is called deconstruction would be the effort to take this limitless context into account, to pay the sharpest and broadest attention possible to context, and thus to an incessant movement of recontextualization. The phrase which for some has become a sort of slogan, in general so badly understood, of deconstruction ("there is nothing outside the text") means nothing else: there is nothing outside context. In this form, which says exactly the same thing, the formula would doubtless have been less shocking. I am not certain that it would have provided more to think about. (P. 136)

A few moments ago, I insisted on writing, at least in quotation marks, the strange and trivial formula, "real-history-of-the-world," in order to mark clearly that the concept of text or of context which guides me embraces and does not exclude the world, reality, history. . . . As I understand it . . . the text is not the book, it is not confined in a volume itself confined to the library. It does not suspend reference—to history, to the world, to reality, to being, and especially not to the other, since to say of history, of the world, of reality, that they always appear in an experience, hence in a movement of interpretation which contextualizes them according to a network of differences and hence of referral to the other, is surely to recall that alterity (difference) is irreducible. *Différance* is a reference and vice versa. (P. 137)

37. Christopher Norris, "Limited Think: How Not to Read Derrida," in *What's Wrong with Postmodernism: Critical Theory and the Ends of Philosophy* (Baltimore: The Johns Hopkins University Press, 1990), pp. 134–63.

38. Marshall McLuhan, *Understanding Media: The Extensions of Man* (Cambridge, Mass.: MIT Press, 1994), p. 18.

39. Ibid., p. 46.

40. Friedrich Kittler, "Gramophone, Film, Typewriter," *October* 41 (1987): 101–18, esp. p. 115.

41. Gillian Beer, *Darwin's Plots: Evolutionary Narrative in Darwin, George Eliot and Nineteenth-Century Fiction* (London: Routledge & Kegan Paul, 1983), pp. 29–48.

42. Ibid., p. 38. Also see James Moore, "Deconstructing Darwinism: The Politics of Evolution in the 1860s," *Journal of the History of Biology* 24 (1991): 353–408. Moore notes that when it suited his purposes, Darwin allowed great latitude in the meanings of key propositions in his evolutionary theory, even among his most intimate circle of supporters: he permitted suggestions, for instance, that his theory was compatible with natural theology.

43. Gillian Beer, "Problems of Description in the Language of Discovery," in *One Culture*, ed. George Levine (Madison: University of Wisconsin Press, 1987), pp. 35–58, esp. pp. 44–45.

44. Michael Lynch and Samuel Edgerton, "Aesthetics and Digital Image Processing," in Gordon Fyfe and John Law, eds., *Picturing Power* (London: Routledge, 1988), pp. 184–220.

45. On mechanical objectivity see Lorraine Daston and Peter Galison, "The Image of Objectivity," *Representations* 50 (1992): 81–128.

46. Derrida, *Grammatology*, p. 86.

2. Daston, Language of Strange Facts

1. "Sur un nouveau phénomène ou sur une Lumière Céleste" [1683], *Histoire de l'Académie Royale des Sciences, Années 1681–1685*, pp. 378–81, on p. 379.

2. Gian Domenico Cassini, "Description de l'apparence de trois Soleils vûs en même temps sur l'horizon, par M. Cassini," *Mémoires de l'Académie Royale des Sciences, Depuis 1666 jusqu'à 1699* 10:234–39.

3. [Robert Boyle], "Observables upon a Monstrous Head," *Philosophical Transactions of the Royal Society of London* 5 (July 3, 1665): 85–86.

4. Pietro Redondi, "Galilée aux prise avec les théories aristotéliciennes de la lumière," *XVIIᵉ siècle* 34 (1982): 167–83.

5. René Descartes, *Principia philosophiae* [1644], in *Oeuvres de Descartes*, ed. Charles Adam and Paul Tannery, 10 vols. (Paris: Léopold Cerf, 1897–1908), vol. 8, part IV, art. 187, pp. 314–15.

6. Gottfried Wilhelm Leibniz, "An Odd Thought Concerning a New Sort of Exhibition (or Rather, An Academy of Sciences)" [1675], in *Leibniz: Selections*, ed. Philip P. Wiener (New York: Charles Scribner's Sons, 1951), pp. 585–94.

7. Bert Hansen, *Nicole Oresme and the Marvels of Nature: A Study of His De*

causis mirabilium, with critical edition, translation, and commentary (Toronto: Pontifical Institute of Medieval Studies, 1985), p. 279.

8. Peter Dear, "Jesuit Mathematical Science and the Reconstitution of Experience in the Early Seventeenth Century," *Studies in History and Philosophy of Science* 18 (1987): 133–75.

9. Thomas Hobbes, *Leviathan* [1651], ed. C. B. Macpherson (Harmondsworth: Penguin, 1968), part I, chap. 9, p. 148.

10. For a fuller account of the emergence of the early modern fact/evidence distinction and its relation to marvels, see Lorraine Daston, "Marvelous Facts and Miraculous Evidence in Early Modern Europe," *Critical Inquiry* 18 (1991): 93–124.

11. Francis Bacon, *New Organon* [1620], ed. Fulton H. Anderson (Indianapolis and New York: Bobbs-Merrill, 1960), I.xxv, p. 44.

12. Ibid., II.xxix, p. 178.

13. Ibid., II.xxviii, p. 178.

14. Ibid., p. 177.

15. On the social attractions of Baconian facts, see Lorraine Daston, "Baconian Facts, Academic Civility, and the Prehistory of Objectivity," *Annals of Scholarship* 8 (1992): 337–63. The arguments concerning the novelty and function of strange facts in seventeenth-century natural philosophy are also presented there in greater detail.

16. See Oliver Impey and Arthur MacGregor, eds., *The Origins of Museums: The Cabinets of Curiosities in Sixteenth- and Seventeenth-Century Europe* (Oxford: Oxford University Press, 1985); also Paula Findlen, "Museums, Collecting, and Scientific Culture in Early Modern Italy" (Ph.D. diss., University of California, Berkeley, 1989).

17. For example Giambattista della Porta, *Magiae naturalis, sive De miraculis rerum naturalium, Libri IIII* (Naples, 1558); Pierre Boaistuau, *Histoires prodigieuses* (Paris: Chez Jaques Mace, à l'enseigne de la Pyramide, 1567). See also Jean Céard, *Le Nature et les prodiges:L'Insolite au XVI siècle en France* (Geneva: Droz, 1977); D. P. Walker, *Spiritual and Demonic Magic* (Notre Dame, Indiana: University of Notre Dame Press, 1975); William Eamon, "Arcana Disclosed: The Advent of Printing, the Books of Secrets Tradition, and the Development of Experimental Science in the Sixteenth Century," *History of Science* 22 (1984): 111–50; Brian Copenhaver, "Natural Magic, Hermeticism, and Occultism in Early Modern Science," in *Reappraisals of the Scientific Revolution*, ed. David Lindberg and Robert Westman (Cambridge, Eng.: Cambridge University Press, 1990), pp. 261–301.

18. Katharine Park and Lorraine Daston, "Unnatural Conceptions: The Study of Monsters in Sixteenth- and Seventeenth-Century England and France," *Past and Present* 92 (1981): 20–54; David Sabean, *Power in the Blood: Popular Culture and Village Discourse in Early Modern Germany* (Cambridge, Eng.: Cambridge University Press, 1984), pp. 61–93; Ottavia Niccoli, *Prophecy and People in*

Renaissance Italy [1987], trans. Lydia Cochrane (Princeton: Princeton University Press, 1990).

19. Boaistuau, p. 5.

20. See Daston, "Marvelous Facts," pp. 100–8; also Stuart Clark, "The Scientific Status of Demonology," in *Occult and Scientific Mentalities in the Renaissance*, ed. Brian Vickers (Cambridge, Eng.: Cambridge University Press, 1984), pp. 351–74.

21. Bernard de Fontenelle, *Histoire du renouvellement de l'Académie Royale des Sciences en MDCXCIX* (Amsterdam: Chez Pierre de Coup, 1709), pp. 25–26.

22. Increase Mather, *An Essay for the Recording of Illustrious Providence* (Boston: Samuel Green, 1684), "Preface," n.p. On seventeenth-century collections of providences, see Keith Thomas, *Religion and the Decline of Magic* (New York: Charles Scribner's, 1975), chap. 4.

23. "A Narrative of Divers Odd Effects of a Dreadful Thunder Clap, at Stralsund in Pomerania, 19/29 June 1670," *Philosophical Transactions* 65 (Nov. 14, 1670): 2084–87.

24. [Edmund Halley], "An Account of the Late Surprising Appearance of the Lights Seen in the Air, on the Sixth of March Last; with an Attempt to Explain the Principal Phaenomena Thereof," *Philosophical Transactions* 347 (Jan./Feb./Mar. 1716): 406–28, on p. 416.

25. Robert Boyle, "Some Observations About Shining Flesh, Both of Veal and Pullet, and That Without Any Sensible Putrefaction in Those Bodies," *Philosophical Transactions* 89 (Dec. 16, 1672): 5108–16.

26. Joseph Addison, *The Spectator* 411 (June 21, 1712), cited in Walter Houghton, "The English Virtuoso in the Seventeenth Century," *Journal of the History of Ideas* 3 (1942): 51–73, 190–219.

27. Bacon, II. xxi, p. 180.

28. Nehemiah Grew, *Musaeum Regalis Societatis: Or, A Catalogue & Description of the Natural and Artificial Rarities Belonging to the Royal Society and Preserved at Gresham College* (London: Printed by W. Rawlins, for the author, 1681), "Preface," n.p.

29. Robert Hooke, "A General Scheme of the Present State of Natural Philosophy, and How Its Defects May Be Remedied by a Methodical Proceeding in the Making Experiments and Collecting Observations: Whereby to Compile a Natural History, as the Solid Basis for the Superstructure of True Philosophy," in *The Posthumous Works of Robert Hooke* [1705], ed. Richard Waller, reprinted with an introduction by Richard S. Westfall (New York: Johnson Reprint Corp., 1969), pp. 61–62.

30. Robert Boyle, "A Short Account of Some Observations Made by Mr. Boyle, about a Diamond, that Shines in the Dark," in *Works* [1772], ed. Thomas Birch, facsimile reprint (Hildesheim: Georg Olms, 1965–66), 1:789–97.

31. "Anatomical Observations of a Humane Body, Dead of Odd Diseases; as

They Were Communicated by Dr. Nathaniel Fairfax," *Philosophical Transactions* 29 (Nov. 11, 1667): 546–51, on p. 547.

32. "A Relation of Some Strange Phaenomena, Accomplished with Mischievous Effects in a Cole-work in Flint-shire; Sent March 31, 1677 to the Reverend and Eminently Learned Dr. Barthurst, Dean of Bath and Wells, by an Ingenious Gentleman, Mr. Roger Moslyn, of the Inner Temple . . . ," *Philosophical Transactions* 136 (June 25, 1677): 895–99, on p. 898.

33. "A Relation of a Kind of Worms, that Eat Out of Stones: This is Taken Out of a Letter by One M. de la Voye to M. Auzout to Be Found in the 32 *Journal des Scavans* as Follows," *Philosophical Transactions* 18 (Oct. 22, 1666): 312–13, on p. 322.

34. "An Enlargement of the Observations, Formerly Publish & Numb. 227, Made and Generously Imparted by that Learned and Inquisitive Physitian, Dr. Stubbes," *Philosophical Transactions* 36 (June 15, 1668): 699–707, on p. 700.

35. Cassini, "Description," p. 235.

36. Robert Hooke, *Micrographia* (London: J. Martyn and J. Allestry, 1665), "Preface," n.p.

37. "Observations, Communicated to the Publisher by Mr. Anthony van Leeuwenhoek, in a Dutch Letter of the 9th of Octob. 1676, Here English'd: Concerning Little Animals by Him Observed in Rain- Well- and Snow-water; as Also in Water Wherein Pepper had Lain Infused," *Philosophical Transactions* 133 (Mar. 25, 1677): 821–31, on p. 828.

38. "New Experiments Concerning the Relation Between Light and Air in Shining Wood and Fish; Made by the Honourable Robert Boyle . . . ," *Philosophical Transactions* 31 (Jan. 6, 1667/8): 581–600, on p. 583.

39. Hooke, *Micrographia*, p. 184.

40. "Nouveau phénomène rare et singulier, d'une Lumière céleste, qui'a paru au commencement du Printemps de cette année 1683, par M. Cassini," *Mémoires*, 10: 637–46, on p. 637.

41. Wilhelm Homberg, "Observation curieuse sur une Infusion d'Antimonie," *Mémoires*, 10:403–6.

42. "An Extract of a Letter, Written by John Winthrop Esq; Governour of Connecticut in New England, to the Publisher Concerning some Natural Curiosities of Those Parts, Especially a Very Strange and Very Curiously Contrived Fish, Sent for the Repository of the R. Society," *Philosophical Transactions* 58 (Mar. 25, 1670): 1151–53, on p. 1151. John Winthrop was right to be worried about the novelty of his "starre-fish": Francis Willoughby pointed out that it had already been described by Rondelet (*Philosophical Transactions* 58 [Apr. 25, 1670]: 1200).

43. "A Letter of Mr. Martin Lister, Containing His Observations of the Astroiter of Star-stones; Communicated to the Publisher Jan. 19, 1673/4," *Philosophical Transactions* 112 (Mar. 25, 1675): 273–79, on pp. 275–77.

44. Halley, p. 407.

45. Leeuwenhoek, p. 831.

46. "A Phytological Observation Concerning Orenges and Limons, Both Separately and in One and the Same Tree at Florence: Described by the Florentine Physitian Petrus Natus, and the Description Lately Communicated to the Publisher," *Philosophical Transactions* 114 (May 24, 1676): 313–14, on p. 313.

47. Hooke, "General Scheme," pp. 19, 32.

48. Halley, p. 409.

49. Boyle, "New Experiments," p. 586.

50. [George Garden], "An Extract of a Letter, Written from Aberdeen Febr. 17, 1666/7, Concerning a Man of a Strange Imitating Nature, as also of Several Human calculus's of an Unusual Bigness," *Philosophical Transactions* 134 (Apr. 23, 1677): 842–43.

51. "A Relation of Two Considerable Hurricans, Happened in Northamptonshire, Not Above Four Miles Distance from One Another, Within the Compass of Less than Twelve Calendar Months: Communicated by Mr. John Templer of Braybrook to a Friend of His in London," *Philosophical Transactions* 71 (May 22, 1671): 2156–58, on p. 2156.

52. "A Narrative of a Monstrous Birth in Plymouth, Octob. 22, 1670; Together with the Anatomical Observations, Taken Thereupon by William Durston, Doctor in Physick, and Communicated to Dr. Tim Clerk," *Philosophical Transactions* 65 (Nov. 14, 1670): 2096–98, on p. 2098.

53. "A Relation of a Monstrous Birth, Made by Dr. S. Morris of Petworth in Sussex, from his Own Observation: and by Him Sent to Dr. Charles Goodall of London: Both of the Colledge of Physicians, London," *Philosophical Transactions* 138 (Mar. 25, 1678): 961–62, on p. 961.

54. From the review in the *Philosophical Transactions* 101 (Mar. 25, 1674): 15–20.

3. Rider, Shaping Information

This essay is closely connected to a larger project on the history of the printing of science, undertaken in collaboration with my Stanford colleague Henry E. Lowood. In launching the project we enjoyed research opportunities made possible by a Rockefeller Foundation Fellowship in the history of science at the University of Oklahoma (1989). The present essay makes use of materials consulted in the History of Science Collections at the University of Oklahoma, Bancroft Library, Houghton Library, Linda Hall Library, and Stanford University Special Collections. I also benefited from research in the Archives of the Académie des Sciences in Paris, funded in part by an NEH Travel to Collections grant (1992–93). Quotations from the Donald Knuth papers at Stanford are used by permission of Professor Knuth and of the Department of Special Collections. I have profited as well from questions and comments from colleagues at Stanford, Berkeley, and the University of Minnesota, Minneapolis.

1. Donald E. Knuth, "Mathematical Typography," *American Mathematical Society Bulletin*, new series, no. 1, 2 (Mar. 1979): 337–72, on p. 343.

2. Cf. Donald E. Knuth, "Tau Epsilon Chi, a System for Technical Text," Stanford Computer Science report CS675 (Sept. 1978).

3. Knuth, "Mathematical Typography," p. 337 (abstract).

4. About Feliciano and his treatise, see Felice Feliciano, *Felice Feliciano veronese: Alphabetum romanum*, ed. Giovanni Mardersteig, trans. R. H. Boothroyd (Verona: Editiones Officinae Bodoni, 1960). Quotation on p. 125; cf. pp. 33–34.

5. Ibid., p. 128.

6. Ibid.

7. Ibid., pp. 54–55.

8. See also Felice Feliciano, *Alphabetum romanum: Vat. lat. 6852: Aus der Bibliotheca Apostolica Vaticana*, Codices e Vaticanis selecti, vol. 70 (Zurich: Belser Verlag, 1985).

9. Feliciano, *Alphabetum romanum* (1960), pp. 36, 57; table in Francesco Torniello, *The Alphabet of Francesco Torniello da Novara* [1517] (Verona: [Officina Bodoni], 1971), p. xxvi.

10. See, for example, Stanley Morison, *Fra Luca de Pacioli of Borgo S. Sepolcro* (New York: Grolier Club, 1933); Luca Pacioli, *De divina proportione*, Fontes ambrosiani, vol. 31 (Verona: Officina Bodoni, 1956; Milan: Silvano, 1982), p. 72. An earlier account appeared in *Quellenschriften für Kunstgeschichte und Kunsttechnik des Mittelalters und der Neuzeit*, Neue folge, vol. 2 (Wien: C. Graeser, 1896). On the various ratios see Giuseppina Masotti Biggiogero's introduction to the 1956 Verona edition of *De divina proportione*, p. 8.

11. Such works include treatises by Moyllus (De Moyle), Sigismondo Fanti, Giovanni Baptista Verini, Vespasiano Amphiareo, and Geoffroy Tory. For particulars, see the table and bibliography in Torniello. Giovanni Mardersteig's introduction to Torniello (p. x) notes that Torniello, for one, acknowledged his debt to Pacioli; see Mardersteig's introduction to Feliciano's *Alphabetum romanum*, which notes that Dürer's work drew in part on that of Moyllus. Modern translations/facsimiles are available for several of these treatises, including Damianus Moyllus, *A Newly Discovered Treatise on Classic Letter Design Printed at Parma by Damianus Moyllus Circa 1480* (Paris: At the Sight of the Pegasus, 1927); Giovanni Battista Verini, *Luminario; or, The Third Chapter of the "Liber elementorum litterarum" on the Construction of Roman Capitals* (Cambridge, Mass.: Harvard College Library, 1947); Verini, *Luminario* (Florence: Libreria L. S. Olschki, 1966); Geoffroy Tory, *Champ fleury*, trans. and annotated by George B. Ives (New York: Grolier Club, 1927); Tory, *Champ rose; Wherein May Be Discovered the Roman Letters that Were Made by Geofroy Tory and Printed by Him at Paris* (New Rochelle, N.Y.: Peter Pauper Press, 1933).

12. The Academy's account of its activities for 1699 made much of the "voluntary" undertaking: to produce "la Description des arts dans l'état où ils sont aujourd'hui en France," accompanied by illustrations of materials, instruments, and methods involved in each craft or trade. Académie des Sciences (France), *Histoire* (1699): 117−20, quotation on p. 117. On the Academy's assignment to study and describe the *arts et métiers*, see Roger Hahn, *The Anatomy of a Scientific Institution: The Paris Academy of Sciences, 1666−1803* (Berkeley: University of California Press, 1971), pp. 68, 124.

13. "On a commencé par l'art qui conservera tous les autres, c'est-à-dire l'impression": *Histoire* (1699): 118; also quoted in André Jammes, *La réforme de la typographie royale sous Louis XIV le Grandjean* (Paris: Librairie Paul Jammes, 1961; reprint, Paris: Promodis, 1985), p. 6. For an abridged English version, see André Jammes, "*Académisme et typographie*: The Making of the *Romain du roi*," *Printing Historical Society Journal* 1 (1965): 71−95.

14. Compare Jammes's account with the analysis in John Dreyfus, *Aspects of French Eighteenth-Century Typography* (Cambridge, Eng.: Printed for presentation to members of the Roxburghe Club, 1982), esp. pp. 13−19.

15. Truchet (le Père Jean) assumed the surname Sébastien when he entered holy orders at age seventeen. His works include "Mémoire sur les combinaisons, avec trente dessins formés par des carreaux mi-partis de deux couleurs seulement," *Mémoires de l'Académie des Sciences* (1704): 363−66. See the comparison of Truchet's barometric observations and those of Maraldi, in *Mémoires de l'Académie des Sciences* (1705); and an account of a machine for studying the acceleration of boules, in *Histoire de l'Académie des Sciences* (1699): 283. On Truchet's work on tiles, see Dreyfus, p. 40.

16. Jammes, *Réforme*, cites most and transcribes many such manuscript sources.

17. Quoted in Dreyfus, p. 15.

18. This was a finer grid than originally envisioned, presumably in the interest of greater accuracy in translating to the medium of copperplate. Dreyfus, p. 16.

19. Mardersteig, in his introduction to Torniello, calls attention to Torniello's use of a grid in defining the crucial proportion in his constructed letters; Tory did the same, though with different proportions, in *Champ fleury*.

20. Jammes, *Réforme*, includes prints pulled from the actual engravings.

21. Certain changes resulted, because effects possible using pen on paper or engraving tools on copperplate could not necessarily be translated into cast type. Scaling down also demanded slight alterations in the design. See Dreyfus, pp. 16−18.

22. *Médailles sur les principaux événements du règne de Louis le Grand* (Paris: Imprimerie Royale, 1702). See the page for the Académie des sciences reproduced in Jammes, *Réforme*, p. 14.

23. Pierre Simon Fournier, *Manuel typographique, utile aux gens de lettres*, 2 vols. (Paris: Imprimé par l'Auteur, 1764–66); translated with notes by Harry Carter as *Fournier on Typefounding* (London: Soncino Press, 1930). Pages 132–36 discuss the Réglement du Conseil pour la Librairie et Imprimerie de Paris (Feb. 28, 1723), article 59: "When the Regulations were made no one seems to have been found who could communicate the true principles of the subject, which was of great importance because there was need to correct abuses and to introduce order and accuracy where they did not exist. . . . The law enacted in consequence, not being based on any principle, has remained a dead-letter [!]" (pp. 132–33). See also Daniel Berkeley Updike, *Printing Types, Their History, Forms, and Use: A Study in Survivals*, 2 vols. (Cambridge, Mass.: Harvard University Press, 1927), 1:26, 28.

24. On Fournier, see Dreyfus, esp. chap. 2; Allen Hutt, *Fournier, the Compleat Typographer* (Totowa, N.J.: Rowman and Littlefield, 1972), p. 19; Fournier, *On Typefounding*, esp. pp. 136–37, for his account of "fixing the body-sizes of the types which I proposed to make."

25. The article "Caractère," in vol. 2 of the *Encyclopédie ou dictionnaire raisonné des sciences des arts et des métiers* (Paris: Briasson, 1751–65) shows most clearly Fournier's stamp. Fournier refers to this article in his *Manuel typographique*, 1:x; d'Alembert, in his *Discours préliminaire*, thanks Fournier among other experts who had aided the enterprise. See also Hutt, pp. 43–44.

26. Fournier claimed that the disorder and needless complication "induced [him] to make order of chaos and introduce system where previously it had never prevailed" (*On Typefounding*, p. 136).

27. Quoted in translation in Updike, 1:29.

28. See Hutt, p. 7. More of his vitriol is quoted in Dreyfus, p. 16: "How could anyone thus curb the mind and suppress taste by loading genius with these chains of confused and injudicious rules?" Fournier's vehemence notwithstanding, even the term *point* had been used two centuries before Fournier to refer to a typographical measurement: Torniello used a point, defined as twice the length of a square in his grid, as the basis for the radii in his constructed alphabet. See Torniello, p. xxiii.

29. Fournier, *On Typefounding*, p. 138; Updike, 1:31.

30. Updike, 1:31.

31. Fournier, *On Typefounding*, p. 137. See Updike, 1:24–30; Hutt, pp. 48–49.

32. Fournier, *On Typefounding*, p. 189.

33. On these characteristic traits, see Gunnar Broberg, "The Broken Circle," and J. L. Heilbron, "Introductory Essay," in *The Quantifying Spirit in the Eighteenth Century*, ed. Tore Frängsmyr, J. L. Heilbron, and Robin E. Rider (Berkeley and Los Angeles: University of California Press, 1990), pp. 45–71 and 1–23. Updike, 1:31, argues, however, that Fournier's prototype fell short of being "an instru-

ment of precision" because of flaws in its manufacture. Compare Hutt, pp. 48–49.

34. On the variants, see Updike, 1:31–33.

35. On the meeting of scholar and craftsman, see Edgar Zilsel, "The Genesis of the Concept of Scientific Progress," *Journal of the History of Ideas* 6 (1945): 325–49; Paolo Rossi, *Philosophy, Technology, and the Arts in the Early Modern Era*, trans. Salvator Attansio, ed. Benjamin Nelson (New York: Harper, 1970).

36. In the Donald E. Knuth Papers, University Archives, Stanford Special Collections (SC 97), materials concerning METAFONT are found in series 2, box 19, folders 8–20, and boxes 20–25. See especially box 19, "Proto-METAFONT, 1977" and "Initial Design of METAFONT, Summer 1978."

37. Knuth, "Mathematical Typography," p. 351.

38. Ibid., p. 351.

39. Ibid., p. 337 (abstract).

40. It also reminds printing historians of the steps taken by William Morris in designing typefaces for the Kelmscott Press. See, for example, William S. Peterson, *A Bibliography of the Kelmscott Press* (Oxford: Clarendon Press, 1984), pp. xxi–xxii.

41. Knuth, "Mathematical Typography," esp. pp. 356–59. See also Knuth's preceding section, "Defining Good Curves."

42. Knuth, "Mathematical Typography," p. 359.

43. Ibid., p. 361.

44. Ibid., p. 369; compare Knuth, *The METAFONTbook* (Reading, Mass: Addison-Wesley, 1986).

45. Knuth, "Mathematical Typography," p. 91.

46. Knuth Papers, box 21, folder 10.4, "Lessons Learned from METAFONT," p. 5.

47. Knuth, *The METAFONTbook*, p. viii.

48. Ibid., p. 121, quoting from Jonathan Swift, "On Poetry: A rapsody" [*sic*] (1733).

49. Knuth, "Lessons Learned from METAFONT," p. 1.

50. Knuth, *The METAFONTbook*, p. v.

51. Knuth, "Lessons Learned from METAFONT," p. 1.

4. Rotman, Mathematical Persuasion

1. Stillman Drake, *Discoveries and Opinions of Galileo* (New York: Doubleday, 1957), p. 238.

2. Jacques Derrida, *Of Grammatology*, trans. Gayatri Spivak (Baltimore: Johns Hopkins University Press, 1976), p. 6.

3. For example, and by no means comprehensively: David Bloor, *Knowledge and Social Imagery* (London: Routledge, 1976); Barry Barnes and David Bloor,

"Relativism, Rationalism, and the Sociology of Knowledge," in *Rationality and Relativism*, ed. Martin Hollis and Steven Lukes (Cambridge, Mass.: MIT Press, 1982), pp. 21–47; Sal Restivo, *The Social Relations of Physics, Mysticism, and Materialism* (Boston: D. Reidel, 1983); Sal Restivo, "The Social Roots of Pure Mathematics," in *Theories of Science in Society*, ed. Susan Cozzens and Thomas Gieryn (Bloomington: Indiana University Press, 1990), pp. 120–43; Theodore Porter, "Quantification and the Accounting Ideal in Science," *Social Studies of Science* 22 (1992): 633–52; Paul Ernest, *The Philosophy of Mathematics Education* (New York: Palmer Press, 1991).

4. Brian Rotman, *Signifying Nothing: The Semiotics of Zero* (London: Macmillan, 1987).

5. Gregory Bateson, *Steps to an Ecology of Mind* (New York: Ballantine Books, 1972), p. 422.

6. Thomas S. Kuhn, *The Essential Tension* (Chicago: University of Chicago Press, 1977), p. 240.

7. Steven Shapin and Simon Schaffer, *Leviathan and the Air-Pump* (Princeton: Princeton University Press, 1985), p. 25.

8. Ibid., p. 60.

9. Emile Benveniste, *Problems in General Linguistics*, trans. Mary Elizabeth Meek (Coral Gables, Fla.: University of Miami Press, 1971), p. 227.

10. J. Buchler, ed., *The Philosophy of Peirce: Selected Writings* (London: Routledge, 1940), p. 98.

11. Brian Rotman, "Toward a Semiotics of Mathematics," *Semiotica* 72 (1988): 1–35.

12. Brian Rotman, *Ad Infinitum . . . The Ghost in Turing's Machine: Taking God Out of Mathematics and Putting the Body Back In* (Stanford, Calif.: Stanford University Press, 1993).

5. Kittler, On the Take-off of Operators

1. See Eric Blondel, "Les guillemets de Nietzsche," in *Nietzsche aujourd'hui* (Paris: Union générale d'éditions, 1973), 2:153–82.

2. Quoted in A. M. Landgraf, *Dogmengeschichte der Frühscholastik*, vol. 1: *Die Gnadenlehre* (Regensburg: F. Pustet, 1952), p. 22 (with thanks to Reinhold Glei).

3. Quoted in ibid., p. 23.

4. Ibid., pp. 21–24.

5. For first indications, see Johannes Lohmann, *Philosophie und Sprachwissenschaft* (Berlin: Duncker & Humblot, 1965), pp. 44–46.

6. Hans Magnus Enzensberger, *Mausoleum: Siebenunddreißig Balladen aus der Geschichte des Fortschritts* (Frankfurt: Suhrkamp, 1975), p. 9.

7. Quoted in Florian Cajori, *A History of Mathematical Notations* (Chicago, Ill.: Open Court, 1928–30), p. 234.

8. See Joris Vorstius and Siegfried Joost, *Grundzüge der Bibliotheksgeschichte*, 7th ed. (Wiesbaden: O. Harrassowitz, 1977), p. 47.

9. See Cajori, 2:182–83.

10. Ibid., p. 184.

11. Ibid., pp. 182–83.

12. Quoted in Betsy Meyer, *Conrad Ferdinand Meyer: In der Erinnerung seiner Schwester* (Berlin: Gerbruder Paetel, 1903), pp. 208–9.

13. See Conrad Ferdinand Meyer, "Hohe Station," in *Sämtliche Werke*, ed. Hans Schmeer (Munich: Droemer, 1952), p. 823.

14. Quoted in Cajor, 2:184.

15. Gauß, letter to Schumacher, Sept. 1, 1850. Quoted in Hans Wussing, *Carl Friedrich Gauß*, 2nd ed. (Leipzig: Teubner, 1976), p. 65.

16. See Cajori, 2:130–31.

17. See Andrew Hodges, *Alan Turing: The Enigma* (New York: Simon & Schuster, 1983), pp. 96–110.

18. See Bernhard Dotzler, "Nachwort," in Alan Turing, *Intelligence Service: Ausgewählte Schriften*, ed. Bernhard Dotzler and Friedrich Kittler (Berlin: Brinkmann & Bose, 1987), p. 227.

19. Turing, p. 15.

6. Siegert, Switchboards and Sex

1. Jürgen Habermas, *Strukturwandel der Öffentlichkeit: Untersuchungen zu einer Kategorie der bürgerlichen Gesellschaft*, 16th ed. (Darmstadt-Neuwied: Luchterhand, 1986), p. 30.

2. Cf. Friedrich Kittler, "There Is No Software," *Stanford Literature Review* 9, no. 1 (Spring 1992): 81–90.

3. August Bebel, *Die Frau und der Sozialismus* (Berlin: Dietz, 1977), p. 450.

4. Ibid., p. 252.

5. Ibid., p. 254.

6. Ibid., p. 450.

7. Brenda Maddox, "Woman and the Switchboard," in *The Social Impact of the Telephone*, ed. Ithiel de Sola Pool (Cambridge, Mass.: MIT Press, 1977), p. 269.

8. Habermas, pp. 66–67.

9. Cf. Bernhard Siegert, *RELAIS: Literatur als Epoche der Post (1751–1913)* (Berlin: Brinkmann & Bose, 1993).

10. T. C. Duncan Eaves and Ben D. Kimpel, *Samuel Richardson: A Biography* (Oxford: Clarendon Press, 1971), pp. 89–91.

11. Hermann von Helmholtz, *Die Lehre von den Tonempfindungen*, ed. R. Wachsmuth, 7th ed. (Hildesheim: G. Olms, 1968), pp. 195–200.

12. Robert V. Bruce, *Bell: Alexander Graham Bell and the Conquest of Solitude* (London: V. Gollancz, 1973), p. 35.

13. Charles Snyder, "Clarence John Blake and Alexander Graham Bell: Otology and the Telephone," *Annals of Otology, Rhinology, and Laryngology* 83, Supp. 13, no. 4, part 2 (July–Aug. 1974): 12.

14. Bruce, p. 123.   15. Ibid., p. 186.

16. Ibid., p. 209.   17. Maddox, p. 262.

18. Bruce, pp. 14–15.

19. Quoted in ibid., p. 29. Cf. George Bernard Shaw, *Pygmalion*, in *Klassische Stücke* (Berlin: Suhrkamp, 1950), p. 367: "*Higgins*: Now sir, after three months this girl would pass for a countess at the garden party of the Ambassador." Cf. also p. 408: "*Higgins*: No, sticking to the job all these months has given me my fill. First, as far as phonetics was concerned it was interesting enough. But then I had it up to here."

20. Shaw, p. 371.

21. Bruce, p. 59.

22. Quoted in ibid., p. 111.

23. Quoted in Gardiner G. Hubbard, *The Story of the Rise of the Oral Method in America, as Told in the Writings of the Late Hon. Gardiner G. Hubbard* (Washington, D.C.: W. F. Roberts, 1898), p. 10.

24. Shaw, p. 423.

25. Johann Heinrich Pestalozzi, "Über den Sinn des Gehörs im Hinsicht auf Menschenbildung durch Ton in Sprache," in *Ausgewählten Schriften*, ed. Wilhelm Fitner (Frankfurt am Main: Klett-Cotta im Ullstein Taschenbuch, 1983), p. 251: "Man macht das Kind lesen, ehe es reden kann; man macht es Wörter in ihre Bestandteile auflösen, ehe es weiß, was die Wörter bedeuten; man macht es ganze Sätze sich selbst und dem Lehrer in einer Sprache vorpapageien, die es nicht gelernt hat die gar nicht die Sprache ist, in der es täglich redet." [Children are forced to read before they can speak, forced to break words down into their elements before understanding what they mean, forced to parrot entire sentences to themselves and their teachers in a language they have not learned and which is not even the language they normally speak.]

26. Shaw, p. 433.

27. Martin Heidegger, *Unterwegs zur Sprache*, 7th ed. (Pfullingen: Neske, 1982), p. 161.

28. Maddox, p. 271.

29. Emile Benveniste, *Probleme der allgemeinen Sprachwissenschaft* (Frankfurt am Main: Syndikat, 1974), p. 284.

30. Ibid., p. 286.

31. Herbert Leclerc, "Das 'Frollein vom Amt'—kleine Sizzen zu einem großen Thema," *Archiv für deutsche Postgeschichte* 1 (1977): 147.

32. Karl Sautter, *Geschichte der Deutschen Reichspost, 1871–1945* (Frankfurt am Main: Fischer, 1951), p. 348.

33. Johann Gottlieb Fichte, *Grundlage des Naturrechts* (Hamburg: Meiner,

1960), p. 345. See also Wolf Kittler, *Die Geburt des Partisanen aus dem Geist der Poesie: Heinrich von Kleist und die Strategie der Befreiungskriege* (Freiburg i.B.: Rombach, 1987), pp. 52–55.

34. Quoted in Maddox, p. 266.

35. Quoted in Leclerc, p. 139.

36. Friedrich A. Kittler, *Grammophon Film Typewriter* (Berlin: Brinkmann & Bose, 1986), p. 186.

37. Renate Genth and Joseph Hoppe, *Telephon! Der Draht, an dem wir hängen* (Berlin: Suhrkamp, 1986), p. 135.

38. Leclerc, p. 141.

39. Friedrich Kittler, *Grammophon*, p. 286.

40. Marshall McLuhan, *Die magischen Kanäle: "Understanding Media"* (Frankfurt am Main: Fischer Bücherei, 1970), p. 283.

41. Ibid., p. 286.

42. Sautter, p. 350.

43. Oskar Wagner, "Die Frau im Dienste der Reichs-Post- und Telegraphenverwaltung" (Ph.D. diss., University of Halle, 1913), pp. 18–19.

44. Wagner, p. 237; Sautter, p. 351.

45. Sautter, p. 349.

46. Franz Braun, " 'Die Frau im Staatsdienst,' dargestellt an den Verhältnissen bei der Reichs-Post- und Telegraphen-Verwaltung" (Ph.D. diss., University of Würzburg, 1912), p. 35.

47. Karl Küpfmüller and Paul Storch, "Fernsprechen und fernschreiben," *Europäischer Fernsprechdienst: Zeitschrift für den internationalen Nachrichtenverkehr* 51 (1939): 7. See also Ulfilas Meyer, "Über die Frequenz der Fernsprechströne," *Mitteilungen aus dem Telegraphentechnischen Reichsamt* 9 (1921): 70.

48. Quoted in Wagner, p. 76.

49. Wolfgang Scherer, *Klavier-Spiele: Die Psychotechnik der Klaviere im 18. und 19. Jahrhundert* (Munich: W. Fink, 1989), p. 179.

50. M. D. Fagen, ed., *A History of Engineering and Science in the Bell System*, vol. 1, *The Early Years (1875–1925)* (New York: Bell Telephone Laboratories, 1975), p. 936.

51. Emil Du Bois-Reymond, "Zur Kenntnis des Telephons," *Archiv für Physiologie* 1 (1877): 573, 582.

52. Ludimar Hermann, "Über physiologische Beziehungen des Telephons," *Vierteljahresschrift der Naturforschenden Gesellschaft in Zürich* 23 (1878): 98.

53. H. F. Weber, "Die Induktionsvorgänge im Telephon," *Vierteljahresschrift der Naturforschenden Gesellschaft in Zürich* 23 no. 3 (1878) 265–72; Hermann von Helmholtz, "Telephon und Klangfarbe," *Monatsberichte der Königlich Preußischen Akademie der Wissenschaften zu Berlin* (11 Juli 1878): 488–89.

54. Max Wien, "Die akustischen und elektrischen Constanten des Telephons," *Annalen der Physik* 309 (1901): 457–58.

55. J. Tarchanow, "Das Telephon im Gebiete der thierischen Elektrizität," *St. Petersburger medicinische Wochenschrift* 4, no. 11 (1879): 94.

56. Hermann, p. 98.

57. Hermann Gutzmann, "Untersuchungen über die Grenzen der sprachlichen Perzeptionen," *Zeitschrift für klinische Medizin* 560 (1906): 250, 247.

58. Carl Stumpf, "Über die Tonlage der Konsonanten und die für das Sprachverständnis entscheidende Gegend des Tonreiches," *Sitzungsberichte der Preußischen Akademie der Wissenschaften*, 2nd semivol. (1921): 639.

59. Karl Willy Wagner, "Der Frequenzbereich von Sprache und Musik," *Elektrotechnische Zeitschrift* 45 (1924): 451–54.

60. Clarence John Blake, "Ueber die Verwerthung der Membrana tympani als Phonautograph und Logograph," *Archiv für Augen und Ohrenheilkunde* 5 (1876): 434–39; continued in *Zeitschrift für Ohrenheilkunde* 8 (1879): 5–12; Blake, "Auswahl von Worten zur Prüfung der Hörschärfe in Bezug auf ihren logographischen Werth," *Zeitschrift für Ohrenheilkunde* 11 (1882): 29–31.

61. Bernhard Siegert, "Das Amt des Gehorchens: Hysterie der Telephonistinnen oder Wiederkehr des Ohres, 1874–1913," in *Armaturen der Sinne: Literarische und technische Medien, 1870 bis 1920*, ed. Jochen Hörisch and Michael Wetzel (Munich: Fink, 1990), pp. 83–106.

62. Werner von Siemens, "Über Telephonie," *Monatsberichte der Königlich Preußischen Akademie der Wissenschaften zu Berlin* (21 Januar 1878): 47.

63. Quoted in Fagen, 1:929.

64. Ibid., p. 928.

65. Friedrich-Wilhelm Hagemeyer, "Die Entstehung von Informationskonzepten in der Nachrichtentechnik: Eine Fallstudie zur Theoriebildung in der Technik in Industrie- und Kriegsforschung" (Ph.D. diss., University of Berlin, 1979), p. 141.

66. Harvey Fletcher, "The Nature of Speech and Its Interpretation," *Journal of the Franklin Institute* 193, no. 6 (1922): 744.

67. Franz Kafka, "Hochzeitsvorbereitungen auf dem Lande," in *Hochzeitsvorbereitungen auf dem Lande und andere Prosa aus dem Nachlaß*, ed. Max Brod (Frankfurt am Main: Fischer, 1983), p. 8.

7. Gugerli, Politics on the Topographer's Table

I am deeply indebted to Beat Glaus, Felix Gugerli, Rudolf Jaun, Heinz Lippuner, J. Rafael Martinez, Ulrich Pfister, and, of course, Timothy Lenoir for their critical suggestions and logistical support.

1. See John Brian Harley, "Maps, Knowledge, and Power," in *The Iconography of Landscapes: Essays on the Symbolic Representation, Design, and Use of Past Environments*, ed. Denis Cosgrove and Stephen Daniels (Cambridge, Eng.: Cambridge University Press, 1988), pp. 277–312, esp. pp. 278–80, 283–84. See also David

Buisseret, ed., *Monarchs, Ministers, and Maps: The Emergence of Cartography as a Tool of Government in Early Modern Europe* (Chicago: University of Chicago Press, 1992); and Christian Jacob, *L'Empire des cartes: Approche théorique de la cartographie à travers l'histoire* (Paris: Albin Michel, 1992).

2. Denis Wood, *The Power of Maps* (New York: Gilford Press, 1992), esp. p. 51.

3. Bruno Latour, *Science in Action: How to Follow Scientists and Engineers Through Society* (Cambridge, Mass.: Harvard University Press, 1987), p. 254.

4. John Brian Harley, "Deconstructing the Map," *Cartographica* 26, no. 2 (1989): 1–20; Denis Wood and John Fels, "Designs on Signs / Myth and Meaning in Maps," *Cartographica* 23, no. 3 (1986): 54–103.

5. Cf. Friedrich A. Kittler, *Aufschreibesysteme 1800/1900* (Munich: Wilhelm Fink Verlag, 1987).

6. The Cassinis' "Carte topographique de la France" was based on the first trigonometric survey of an entire country. It utilized vertical projection. See Monique Pelletier, *La carte de Cassini* (Paris: Press de l'École Nationale des Ponts et Chaussées, 1990), and Josef Konvitz, *Cartography in France, 1660–1848: Science, Engineering, and Statecraft* (Chicago: University of Chicago Press, 1987).

7. "Mémoire expliquant sommairement la proposition de faire lever géométriquement les cartes générales et détaillées de toute la suisse composé par le Sr. Micheli du Crest, fait au Château d'Arbourg, le 26 juin 1754," quoted in Johann Heinrich Graf, *Das Leben und Wirken des Physikers und Geodäten Jacques-Barthélémi Micheli-du-Crest aus Genf, Staatsgefangener des alten Bern von 1746–1766* (Bern: K. J. Wyss, 1890), pp. 97–101.

8. Among these contacts were René-Antoine Réaumur, Pierre Louis Moreau de Maupertuis, Albrecht von Haller, and Daniel Bernoulli; see Eduard Fueter, *Grosse Schweizer Forscher* (Zurich: Atlantis-Verlag, 1941), p. 120. Micheli du Crest was a corresponding member of numerous learned societies.

9. See Gardy Fréd, "La carte des environs de Genève, dessinée par J.-B. Micheli du Crest (1730)," in *Geneva, Bulletin du Musée d'Art et d'Histoire de Genève* 2 (1924): 187–92; Jacques Barthélémi Micheli du Crest, *Description de la méthode d'un thermomètre universel* (Paris, 1741); *Recueil des diverses pièces sur le thermomètre* (The Hague, 1756); *Mémoire sur la sphéricité de la terre* (Bern, 1760); *Recueil physique sur le tempère du globe de la terre* (Bern, 1760). For biographical information, see Rudolf Wolf, *Biographien zur Kulturgeschichte der Schweiz* (Zurich: Orell, Füssli, 1858), pp. 229–60; and Graf, *Leben und Wirken*.

10. Jacques Barthélemi Micheli du Crest, *Mémoire sur ce qui s'est passé au sujet des fortifications de Genève* (Geneva, 1728). On the political and cartographical history of fortifications, see Martha D. Pollack, *Military Architecture, Cartography, and the Representation of the Early Modern European City* (Chicago: Newberry Library, 1991); and Henning Eichberg, *Festung, Zentralmacht, und Sozialgeometrie: Kriegsingenieurwesen des 17. Jahrhunderts in den Herzogtümern Bremen und Verden* (Cologne: Böhlau 1989).

11. Jacques Barthélémi Micheli du Crest, *Requêtes, Avertissement placé, Mémoire du Sieur Micheli du Crest, au sujet des sentences rendues contre lui tant au Grand qu'en petit Conseil de Genève, avec les moyens de nullité et recours au Conseil général contre les dites sentences* (Sion, 1735).

12. On the conspiracy led by Samuel Henzi (1701–49) against the Bern patriciate, see Graf, *Leben und Wirken,* and Rudolf Braun, *Das ausgehende Ancien Régime in der Schweiz: Aufriss einer Sozial- und Wirtschaftsgeschichte des 18. Jahrhunderts* (Göttingen: Vandenhoeck & Ruprecht, 1984), pp. 270–71.

13. Wolf, *Biographien,* pp. 229–37.

14. On the relationship between Enlightenment and geographical studies, see Hermann Alfred Schmid, *Die Entzauberung der Welt in der Schweizer Landeskunde: Ein Beitrag zur Geschichte der Aufklärung in der Schweiz* (Basel: Helbing & Lichtenhahn, 1942).

15. Letter of May 15, 1755, to Albrecht von Haller. Quoted in Wolf, *Biographien,* pp. 253–54. Emphasis mine.

16. Svetlana Alpers, *The Art of Describing: Dutch Art in the Seventeenth Century* (Chicago: University of Chicago Press, 1983), pp. 62, 122.

17. Letter of May 15, 1755, to Albrecht von Haller. Quoted in Wolf, *Biographien,* p. 254.

18. Cf. the conception of view in Pascal: "The thing must be seen all at once in a single glance, and not through a progression of reasoning." Blaise Pascal, *Pensées suivies des écrits sur la grâce,* ed. Jacques Chevalier (Paris: Gallimard, 1937), p. 21. On the absolutist view, see Rudolf Braun and David Gugerli, *Macht des Tanzes—Tanz der Mächtigen: Hoffeste und Herrschaftszeremoniell, 1550–1914* (Munich: C. H. Beck, 1993), chap. 2.3.

19. Maps in the ancien régime often had the status of military secrets and, like archival materials, were kept under restriction by the authorities. See Rudolf Wolf, *Geschichte der Vermessungen in der Schweiz* (Zurich: S. Höhr, 1879), p. 71.

20. "Thus, what nature has most magnificently built / Is seen from a mountain with ever-new delight." Albrecht von Haller, *Die Alpen* (1729; Stuttgart: Philipp Reclam Jun., 1978), pp. 15, 323–24. The Chasseral was surveyed by French engineers and later served as the point of departure for designated elevations on Dufour's map. See Louis Puissant, *Nouvelle description géometrique de la France* (Paris, 1832), pp. 407, 527; see also Wolf, *Geschichte,* pp. 175–85.

21. Haller, pp. 15, 339–40.

22. At the same time, Jean-Henri Lambert attempted theoretically to introduce into painting an artificial observational position—still from a central perspective, it is true—that was no longer tied to the constraints of the ground level. "Now, would anyone wish to stand behind the notion that the point which the eye should be occupying is *not* in midair?" Jean-Henri Lambert, *Essai sur la perspective* (Paris: Roger Laurent, 1752).

23. Jacques Barthélémi Micheli du Crest, *Prospect géometrique des Montagnes neigées, dittes Gletscher, telles qu'on les découvre en temps favorable, depuis le château*

d'Aarbourg dans les territoires des Grisons, du Canton d'Ury, et de l'Oberland du Canton Berne (Augsburg, 1755); Micheli du Crest, *Mémoire pour l'explication du Prospect des Montagnes neigées que l'on voit du château d'Aarbourg* (1755).

24. As in his suggestion for a uniform map of Switzerland; see Richard Grob, *Geschichte der schweizerischen Kartographie* (Bern: Kümmerly & Frey, 1941), p. 73.

25. F. L. von Pfyffer, quoted in Wolf, *Geschichte*, p. 108.

26. In 1755, Isaak Iselin published his book *Philosophische und patriotische Träume eines Menschenfreundes*, which was reissued in 1758 and 1762. Franz Urs Balthasar's piece, "Patriotische Träume eines Eydgenossen von einem Mittel, die veraltete Eidgenossenschaft zu verjüngern," appeared in 1758. Micheli du Crest's topographical project (see Grob, p. 73) also belonged in the end to this category of "Helvetic dreams" that emerged around the middle of the eighteenth century and strove for a rejuvenation of "Helvetia" from the space of the bourgeois public sphere. See Ulrich Im Hof, *Die Entstehung einer politischen Öffentlichkeit in der Schweiz: Struktur und Tätigkeit der Helvetischen Gesellschaft* (Frauenfeld: Huber, 1983), pp. 25–26.

27. In 1765 the map was still going into a new edition; it was used for the most diverse illustration purposes, even in popular calendars. Grob, pp. 49–51.

28. Ibid., p. 50.

29. Wolf, *Biographien*.

30. On the Scheuchzer Map, see Wolf, *Geschichte*, pp. 47–56; see also Arthur Dürst, *Johann Jakob Scheuchzer: Nova Helvetiae tabula geographica: Erläutendes Begleitwerk zur Faksimileausgabe der Nova Helvetiae tabula geographica von Johann Jakob Scheuchzer* (Tiguri, 1971).

31. According to the first, more radical version of Haller's poem; in the definitive version the line is softened to "Macht durch der Weisheit Licht die Gruft der Erde heiter." Haller, pp. 17, 365, and 105 (Preface by Adalbert Elschenbroich).

32. On Scheuchzer's biography, see Fueter, pp. 111–12; *Historisch Biographisches Lexikon der Schweiz* (Neuenburg: Administration des Historisch-Biographischen Lexikons der Schweiz, 1931) vol. 6; R. Steiger, *Johann Jakob Scheuchzer* (Zurich, 1927); Wolf, *Biographien*.

33. On his theory of the absolute zero point, Micheli du Crest remarked self-critically: "until one has proved [it] through a certain number of experiences . . . one cannot conclude that it is universal." See Micheli du Crest, *Description*. The quotation is found in Wolf, *Biographien*, pp. 240–41.

34. "Étude de la population du Pays de Vaud," the study produced by Pastor Jean-Louis Muret (1715–96), a statistician and member of the Bern Economic Society, led the Bern Grand Council to ban all publications containing population statistics; Braun, pp. 55–56. On the statistical and political activities (and on the execution) of Johann Heinrich Waser (1742–80), see Rolf Graber, "Der Waser-Handel: Analyse eines sozio-politischen Konflikts in der Alten Eidge-

nossenschaft," *Schweizerische Zeitschrift für Geschichte* 30 (1980): 321–56. See also Christian Simon, "Hintergründe bevölkerungsstatistischer Erhebungen in Schweizer Städteorten des 18. Jahrhunderts: Zur Geschichte des demographischen Interesses," *Schweizerische Zeitschrift für Geschichte* 2 (1984): 186–205.

35. "Moreover, I have sent the Banneret Imhoff a memorandum containing and summarizing the proposal for the geometrical drawing up of general and detailed maps of Switzerland as a whole." Letter to Albrecht Haller, September 5, 1754, quoted in Wolf, *Geschichte*, p. 108, and Graf, *Leben und Wirken*, p. 101.

36. In 1762 Micheli du Crest sent a new version of his memorandum to the Government of Bern, which did not succeed either. Graf, *Leben und Wirken*, p. 109; Grob, p. 73.

37. Haller, pp. 17, 361–72.

38. Reinhart Koselleck, *Kritik und Krise: Eine Studie zur Pathogenese der bürgerlichen Welt* (Frankfurt am Main: Suhrkamp, 1973), p. 41. See also Jürgen Habermas, *Strukturwandel der Öffentlichkeit: Untersuchungen zu einer Kategorie der bürgerlichen Gesellschaft* (Neuwied: Luchterhand, 1962). See also Im Hof.

39. See J. L. Hogreve, *Praktische Anweisung zur topographischen Vermessung eines ganzen Landes* (Leipzig, 1773).

40. Antonio Lafuente and Antonio Mazuecos, *Los caballeros del punto fijo: Ciencia, política, y aventura en la expedición geodésica hispanofrancesa al virreinato del Perú en el siglo XVIII* (Madrid: CSIC, 1987).

41. Daniel Frei, "Mediation," in *Handbuch der Schweizer Geschichte* (Zurich: Berichthaus, 1980), 2:844.

42. As, for example, the border problems of the cantons. Ibid., p. 849.

43. Jean-Charles Biaudet, "Der modernen Schweitz entgegen," in *Handbuch der Schweizer Geschichte* (Zurich: Berichthaus, 1980), 2:892.

44. Ibid., pp. 904, 910.

45. See the Confederal Military Order of 1817. In 1819 a Central School for Officers and Non-Commissioned Officers of the Artillery and Corps of Engineers was set up in Thun, in 1829 a Central School for Infantry and Cavalry Officers. Biaudet, p. 915. For general background, see Georges Rapp, *L'État-major général suisse: Des origines à la Guerre du Sonderbund (1798–1847)* (Basel: Helbing & Lichtenhahn, 1983).

46. Wilhelm Fetscherin, ed., *Repertorium der Abschiede der eidgenössischen Tagsatzungen aus den Jahren 1814–1848* (Bern: K. J. Wyss, 1876), 2:551–564 (§135: Trigonometrische Vermessungen; topographische Arbeiten).

47. On the importance of alliance for the promotion of scientific-technical projects, see Latour, *Science in Action*, pp. 103–44.

48. *Appel au zèle scientifique tendant à obtenir des souscripteurs pour la confection d'une carte topographique détaillée des Alpes de la Suisse* (Lausanne, 1829). On the role of the Naturalist Society in domestic topography, see Wolf, *Geschichte*, pp. 238–40.

49. A resolution of the Diet of Aug. 1, 1822, firmly maintained that all national topographical projects fell within the jurisdiction of the army and were to be concentrated within the army in the confederal General Quartermaster Staff. Cf. Fetscherin, p. 552. See also Hans Rapold, *Strategische Probleme der schweizerischen Landesverteidigung im 19. Jahrhundert* (Frauenfeld: Huber, 1951), pp. 35–42.

50. Johann Heinrich Graf, *Die schweizerische Landesvermessung, 1832–1864: Geschichte der Dufourkarte* (Bern: Eidgenössisches topographisches Bureau, 1896), pp. 19–26.

51. Wolf, *Geschichte*, pp. 240–41.

52. It was decided to use the scales of 1:25,000 and 1:50,000 for surveys of flat land and mountains respectively. Engravings were to be done on a scale of 1:100,000, and this was to be projected according to the modified Flamsteed method. See Graf, *Schweizerische Landesvermessung*, pp. 26, 76–82.

53. Biaudet, pp. 918–19.

54. The "Federal Charter" proposed by the Diet commission at first foundered, then was strongly revised and presented to the cantons on May 17, 1833, but even this version was rejected. Biaudet, pp. 923–27.

55. See *Abschied der ordentlichen eidgenössischen Tagsatzung des Jahres 1833*, §IX, pp. 8–9. The sum appropriated for domestic topography came to 10 percent of the entire military budget. Ibid., pp. 23–25.

56. Bruno Uebel, "Was sind die unerlässlichen Erfordernisse zu einem eidgenössischen Generalquartiermeisterstabs-Offizier?" *Helvetische Militär-Zeitschrift* (1834). Quoted in Rudolf Jaun, *Das eidgenössische Generalstabskorps, 1804–1874: Eine kollektiv-biographische Studie* (Basel: Helbing & Lichtenhahn, 1983), pp. 206–7.

57. Finsler was a businessman and a banker by trade, Wurstemberger had acquired autodidactic knowledge of surveying technology. See ibid., pp. 59, 198.

58. On his topographical training at the École Polytechnique, see Raymond D'Hollander, "Influence de la cartographie française sur Dufour," in *Guillaume-Henri Dufour dans son temps, 1787–1875: Actes du colloque Dufour*, ed. Roger Durand and Daniel Aquillon (Geneva: Société d'histoire et d'archéologie, 1991), pp. 135–51.

59. See also Tom F. Peters, *Transitions in Engineering: Guillaume-Henri Dufour and the Early Nineteenth-Century Cable Suspension Bridges* (Basel: Birkhäuser Verlag, 1987).

60. For biographical information, see Durand and Aquillon. See also Guillaume-Henri Dufour, *Instruction sur le dessin des reconnaissances militaires* (Geneva: Barbezat & Delarue, 1828).

61. The act increasing federal allocations for trigonometric surveying was pushed through in the absence of the delegates of Uri, Unterwalden, Neuenburg, Schwyz, and the City of Basel. Fetscherin, p. 557.

62. The military authorities, who wanted to control weapons smuggling by

occupying the border during the Tirol revolt of 1809, not only were confronted with the fact that most of the cantons mobilized incomplete and poorly equipped operative personnel, but discovered that they did not have a clear enough picture of the boundary lines in the east of Switzerland; entire villages were missing on the existing maps and the shape of Canton Thurgau was completely displaced. The triangulation ordered by Quartermaster Colonel Finsler in order to determine the course of the boundaries could only assume a temporary and improvised nature. Frei, p. 849; Graf, *Sweizerische Landesvermessung,* pp. 5–6.

63. Support for a national cartography project may well have sent out signals similarly with regard to Pope Gregory XVI's Encyclical "Mirari vos" of Aug. 15, 1832, opposing the modern political doctrines of the freedom of conscience and "impudent science."

64. On Dufour's political stance, see Marco Marcacci, "Le député Dufour et l'avènement de la démocratie moderne à Genève," in Durand and Aquillon, pp. 87–98.

65. Guillaume-Henri Dufour, *Notice sur la carte de la Suisse dressée par l'état-major fédéral* (Geneva: J. G. Fick, 1861), p. 6. This polemical pronouncement did not in fact correspond to the established facts, but it more than likely formed an important part of the finance strategy that Dufour pursued vis-à-vis the federal authorities. The preliminary works were already blotted out of memory in the concluding report. Guillaume-Henri Dufour, *Schlussbericht über die topographische Karte der Schweiz* (Bern: Topographisches Bureau, 1864). See also Wolf, *Geschichte,* p. 245.

66. Micheli du Crest had already proposed establishing a groundline in Aarburg.

67. Cf. Bruno Latour, "Give Me a Laboratory and I Will Raise the World," in *Science Observed: Perspectives on the Social Study of Science,* ed. Karin Knorr-Cetina and Michael Mulkay (London: Sage Publications, 1983), pp. 141–70.

68. For the purposes of depiction, the angle *a* of the wedge in Fig. 7.2 is significantly larger than in reality.

69. Graf, *Schweizerische Landesvermessung,* p. 48. For more details on the surveying procedure, which has been simplified here, see Wolf, *Geschichte,* pp. 247–50.

70. "Since I know how easy it is to err, I take every measurement two or three times." Eschmann to Dufour, July 14, 1836, in Graf, *Schweizerische Landesvermessung,* p. 57.

71. "Typewriters do not store individuals, their letters do not communicate a hereafter that perfectly alphabetized readers could then hallucinate as meaning." Friedrich A. Kittler, *Grammophon Film Typewriter* (Berlin: Brinkmann & Bose, 1986), p. 27.

72. Diary entry of Rudolf Wolf, Sept. 20, 1835, in Johann Jakob Burckhardt, "Eine Alpenreise von Rudolf Wolf im Jahre 1835," *Die Alpen* 2 (1989): 98.

73. Johannes Eschmann, *Rapport sur les bases d'Aarberg et celle de Zurich corrigées par de nouvelles expériences.* Quoted in Graf, *Schweizerische Landesvermessung,* p. 58. See also Johannes Eschmann, *Ergebnisse der trigonometrischen Vermessungen in der Schweiz* (Zurich: Orell, Füssli, 1840).

74. For reasons of space, I forgo a description of the measurements of elevation, which were taken with just as much exactitude. See Graf, *Schweizerische Landesvermessung,* pp. 82–85.

75. On the measurements of the mathematic-military society in Zurich, see Wolf, *Geschichte,* pp. 163–68.

76. Eschmann to Dufour, June 14, 1836, in Graf, *Schweizerische Landesvermessung,* pp. 56–57. Emphasis mine.

77. Graf, *Schweizerische Landesvermessung,* p. 61.

78. "Bericht der eidgenössischen Militäraufsichtsbehörde an die H. Tagsatzung über den Zustand der trigonometrischen Arbeiten auf Mitte des Jahres 1835," in *Abschiede der eidgenössischen Tagsatzung des Jahres 1835* (Bern: K. J. Wysz, 1842), supplement Litt. E, p. 6. Emphasis mine.

79. Correspondence notebook of Dufour, no. 3 (July 9, 1836). Quoted in Graf, *Schweizerische Landesvermessung,* p. 91.

80. Latour, *Science in Action,* pp. 223–32. On nineteenth-century attempts at disciplining nature by technical and scientific means, see David Gugerli, *Redeströme. Zur Elektrifizierung der Schweiz 1880–1914* (Zurich: Chronos, 1996), chap. 4.

81. Graf, *Schweizerische Landesvermessung,* pp. 256–58.

82. Ibid., p. 261. On the utilized instruments, see ibid., p. 143.

83. Parallel to the described catalog, a catalog for the original recording on the scale of 1:50,000 existed, which was used in the mountains. Printed as Supplement II in Graf, *Schweizerische Landesvermessung.*

84. Latour, *Science in Action,* pp. 215–57.

85. Cited in Graf, *Schweizerische Landesvermessung,* pp. 141–42.

86. Ibid., p. 144.

87. Ibid., p. 142.

88. It is not astonishing that in the late 1930s a return to names of towns in dialect was demanded. "With the assimilation into written Modern High German, the fathers of these precepts fell victim to the scorn for dialect prevalent in the broadest circles of the people, and to the erroneous opinion that names dressed up this way looked more elegant and beautiful." B. Cueni, "Die Nomenklatur der Landeskarten," in *100 Jahre eidgenössische Landestopographie, 1838–1938* (Bern: Landestopographie, 1983), p. 1.

89. Graf, *Schweizerische Landesvermessung,* pp. 142–43. Emphasis mine.

90. The translation catalog of Folio V (1850) contains translations from German to French, French to German, Italian to German, and German to Rhaeto-Romanic.

91. See Privy Councillor Zollinger's "toast to the fatherland" on the occasion of the Confederal Choral Festival in Zurich, *Neue Zürcher Zeitung*, no. 194 (July 12, 1880).

92. For a linguistic perspective, see Paul Zinsli, *Südwalser Namengut: Die deutschen Orts- und Flurnamen der ennetbirgischen Walsersiedlungen in Bosco-Gurin und im Piemont* (Bern: Stämpfli, 1984).

93. G. Studer and K. J. Durheim, "Unmassgebliche Bermerkungen über die eidgenössische trigonometrische Militärkarte Nr. XVII," *Schweizerischer Beobachter* 41–43 (Apr. 4, 7, and 9, 1846). The article was also reprinted separately and sent to all canton offices. See also K. J. Durheim and G. Studer, *Erwiderung auf den von Herrn Oberst Dufour an den Eidgenössischen Kriegsrath gerichteten Rapport über die Bemerkungen gegen die neue Schweizerkarte* (Bern, 1847).

94. A. J. Buchwalder, "Observations sur la Carte fédérale," *Journal Bernois* 83 and 84 (1846). Quoted in Graf, *Schweizerische Landesvermessung*, p. 154.

95. The complaints concerned Lake Constance and Lake Geneva, on which no national boundaries were marked. Maillardoz to Dufour, Oct. 25, 1845, in Graf, *Schweizerische Landesvermessung*, p. 158.

96. Guillaume-Henri Dufour, *Rapport sur les observations anonymes* (1846).

97. Dufour countered the complaint that "Beistandhorn" should instead be called "Wystätthorn" by noting that his engineer had been thus instructed by a schoolmaster of Turbach, who served as a guide; his engineers had always asked the most educated people of the area. Graf, *Schweizerische Landesvermessung*, p. 160.

98. Eschmann to Dufour, Mar. 14, 1847. Quoted in ibid., p. 156.

99. Eisner Manuel, *Politische Sprache und sozialer Wandel: Eine quantitative und semantische Analyse von Neujahrsleitartikeln in der Schweiz von 1840 bis 1987* (Zurich: Seismo, 1991), p. 92.

100. "Cartography is and always remains a faithful reflection of the state of science and art that has gained acceptance in a certain period, animated by the pulse of economic life during that time. The mode of its appearance is thus configured differently today than in former times." Ulrich Meister, *Der heutige Standpunkt der schweizerischen Kartographie und die Lesbarkeit unserer Karten* (Zurich, 1883). The essay appeared, as is apparent from the text, in the broader context of the Swiss National Exhibition in Zurich in 1883.

101. A list of distinctions appears in Graf, *Schweizerische Landesvermessung*, p. 249.

102. Ibid., p. 247.

103. See also Georg Kreis, *Helvetia—im Wandel der Zeiten: Die Geschichte einer nationalen Repräsentationsfigur* (Zurich: Verlag Neue Zürcher Zeitung, 1991); Guy P. Marchal and Aram Mattioli, eds., *Erfundene Schweiz: Konstruktionen nationaler Identität* (Zurich: Chronos, 1992).

104. Walter Senn-Barbieux, *Das Buch vom General Dufour: Sein Leben und*

Wirken, mit besonderer Berücksichtigung seiner Verdienste um die politische Selbständig-keit der Schweiz, sowie um Wissenschaft, Kunst, und Humanität, unter Benutzung der besten Quellen für das Volk bearbeitet (Glarus: Carl Ziegenhirt, 1878), p. 84.

105. On the Topographical Atlas of Switzerland begun under Hermann Siegfried (1819–79), see Grob, pp. 126–35.

106. "All of these tour, postal, and railway maps, these images of the economic life of the states registered in measurable lines on the map, these statistical, ethnographic, and historical maps are the most telling evidence of how diverse the regions served by cartography are, of how cartography has become an indispensable expedient of the modern world." Meister, p. 2. Special maps were usually produced through a superimposed printing process.

107. Ibid., p. 2.

108. Parliamentary Resolution of Dec. 14, 1853. Published 1873.

109. Meister, p. 12. On the literacy project and the transmission of values since the ancien régime, see Marie Louise von Wartburg, *Alphabetisierung und Lektüre: Untersuchungen am Beispiel einer ländlichen Region im 17. und 18. Jahrhundert* (Bern: Peter Lange, 1981); Rudolf Schenda, *Volk ohne Buch: Studien zur Sozial-geschichte der populären Lesestoffe 1770 bis 1910* (Frankfurt am Main: V. Klostermann, 1970); David Gugerli, *Zwischen Pfrund und Predigt: Die protestantische Pfarrfamilie auf der Zürcher Landschaft im ausgehenden 18. Jahrhundert* (Zurich: Chronos, 1988).

110. Meister, p. 2.

111. Haller, pp. 4, 31; Meister, pp. 6, 67, 83.

112. Meister, p. 2.

113. Haller, pp. 15, 310.

114. Meister, p. 12.

115. Senn, pp. 84–86. Emphasis mine. On mastery of nature and political aesthetics, see Ulrich Jost, "Guillaume-Henri Dufour: L'Esthétique politique et l'appropriation de l'espace," in Durand and Aquillon, pp. 111–21.

116. Meister, p. 3. On Ulrich Meister, see Hans Schmid, *Ulrich Meister: Ein Zürcher Politiker* (Zurich: Verlag der Neuen Zürcher Zeitung, 1925); and Rudolf Jaun, *Das Schweizerische Generalstabskorps, 1875–1945: Eine kollektiv-biographische Studie* (Basel: Helbing & Lichtenhahn, 1991), p. 229.

117. Meister, p. 11.

118. Ibid., p. 10. Emphasis in the original.

119. Stephan Oettermann, *Das Panorama: Die Geschichte eines Massenmediums* (Frankfurt am Main: Syndikat, 1980).

120. Meister, p. 6.

121. Eduard von Orel, "Der Stereoautograph als Mittel zur automatischen Verwertung der Komparatordaten," *Mitteilungen des k.k. milit. geogr. Institutes* 30 (1911): 62; Max Weiss, *Die geschichtliche Entwicklung der Photogrammetrie und die Begründung ihrer Verwendbarkeit für Mess- und Konstruktionszwecke* (Stuttgart: Strecker & Schröder, 1913). For Switzerland, see R. Helbling, "Die Stereoauto-

grammetrische Geländevermessung," *Schweizerische Bauzeitung* 76 (1921). Grob, p. 148.

122. Kittler, *Grammophon*, p. 29.

8. Beer, Writing Darwin's Islands

1. See, in particular, the work of David Lowenthal.

2. Benoit Mandelbrot, *The Fractal Geometry of Nature* (San Francisco: W. H. Freeman, 1982), p. 116.

3. Robert MacArthur and Edward O. Wilson, *Theory of Island Biogeography* (Princeton: Princeton University Press, 1967).

4. See my associated essays, "Discourses of the Island," in *Literature and Science as Modes of Expression*, ed. F. Amrine (Amsterdam: Reidel, 1989), pp. 1–27, and "The Island and the Aeroplane: The Case of Virginia Woolf," in *Nation and Narration*, ed. H. Bhabha (London: Routledge, 1990), pp. 265–90.

5. Robert Fitzroy, ed., *Narrative of the Surveying Voyages of His Majesty's Ship's Adventure and Beagle Between the Years 1826 and 1836, Describing Their Examination of the Southern Shores of South America and the Beagle's Circumnavigation of the Globe*, vol. 3 (London: Henry Colburn, 1839). Subsequent references to the 1839 edition (abbreviated Fitzroy) will appear parenthetically in the text.

6. Charles Darwin, *A Monograph on the Fossil Lepadidaie: Or Pedunculated Cirripedes of Great Britain* (London: Palaeontographic Society Publications, 1851).

7. Frank Sulloway, "Darwin's Thinking: The Vicissitudes of a Crucial Idea," *Studies in the History of Biology* 3 (1979): 23–66; Frank Sulloway, "Darwin's Conversion: The *Beagle* Voyage and Its Aftermath," *Journal of the History of Biology* 15 (1982): 325–96.

8. Howard Gruber, in *Darwin on Man: A Psychological Study of Scientific Creativity* (London: Wildwood House, 1974), pp. 161–63, discusses the related phenomenon of Darwin's awareness of superfecundity before reading Malthus's *Essay on Population* and his earlier repeated access to Malthusian ideas. Gruber describes Darwin's reading Malthus as "at last" stimulating "his conscious recognition of the implications of superfecundity for evolutionary theory" (p. 161). That formulation has its problems, too, because it suggests an already stable concept of "evolutionary theory." For a different view see Sandra Herbert, "Darwin, Malthus, and Selection," *Journal of the History of Biology* 4 (1971):209–17.

9. I discuss these issues further in "Speaking for the Others: Relativism and Authority in Victorian Anthropological Literature," in *Sir Robert Frazer and the Literary Imagination* (London: Macmillan, 1990), pp. 38–60.

10. Charles Darwin, *Journal of Researches into the Natural History and Geology of the Countries Visited During the Voyage of H.M.S. Beagle Round the World*, 2nd ed., corrected, with additions (London: John Murray, 1845), pp. 374–75. Subsequent references (abbreviated *Journal*) will appear parenthetically in the text.

11. Charles Darwin, *Charles Darwin's Diary of the Voyage of H.M.S. "Beagle,"* ed. Nora Barlow (Cambridge: Cambridge University Press, 1933; reprint, New York: Kraus Reprint Co., 1969), p. 335. Subsequent references (abbreviated *Diary*) will appear parenthetically in the text.

12. Charles Darwin, *The Correspondence of Charles Darwin*, ed. F. Burkhardt and S. Smith (Cambridge, Eng.: Cambridge University Press, 1985). See, for example, pp. 471, 485.

13. "I was well repaid by the Cyclopean scene. In my walk I met two large tortoises, each of which must have weighed at least two hundred pounds. One was eating a piece of cactus, and when I approached, it looked at me and then quietly walked away: the other gave a deep hiss and drew in its head. These huge reptiles, surrounded by the black lava, the leafless shrubs, and large cacti, appeared to my fancy like some antediluvian animals" (Fitzroy, p. 456).

14. An example of an apparently neutral comparison is: "I do not think so many faults in Cordillera, as in English coal field—because lowered & raised—so on—but gradually & simply raised" (Charles Darwin, *The Red Notebook of Charles Darwin*, ed. Sandra Herbert [London: British Museum of Natural History, 1980], p. 39).

15. William Shakespeare, *Richard II*, in *The Complete Works* (Oxford: Clarendon Press, 1986), II, i, 40–50, p. 422.

16. Carl Linnaeus, *Reflections on the Study of Nature*, trans. J. E. Smith (London: George Nichol, 1786), p. 2.

17. Janet Browne, *The Secular Ark* (New Haven: Yale University Press, 1983).

18. Charles Darwin, *The Voyage of the Beagle* (London: Everyman, 1979), p. 2.

19. On the treatment of Caribbean peoples in literature see Peter Hulme, *Colonial Encounters: Europe and the Native Caribbean, 1492–1797* (London: Methuen, 1986).

20. Notebook of "Books to Be Read," 1838, Box 119, Darwin Papers, University Library, Cambridge, Eng.

21. Jonathan Swift, *Gulliver's Travels and Other Writings* (New York: Bantam Books, 1971), p. 275.

22. Darwin's *Diary* (p. 311) describes Panuncillo's inhabitants' views of the English. May 12, 1835: "To this day they hand down the atrocious actions of the Buccaniers; one of them took the Virgin Mary out the Church & returned the ensuing year for St. Joseph, saying it was a pity the poor lady should not have a husband. I heard Mr. Caldcleugh say, that sitting by an old lady at a dinner in Coquimbo, she remarked how wonderfully strange it was that she should live to dine in the same room with an Englishman.—Twice as a girl, at the cry of 'Los Inglese' every soul, carrying what valuables they could, had taken to the mountains."

23. Alfred Russel Wallace, *Island Life: Or, the Phenomena and Causes of Insular Faunas and Floras, Including a Revision and Attempted Solution of the Problem of Geological Climates* (London: Macmillan, 1880), p. 7 n. 1.

24. For further discussion of the case of Jemmy Button see my pamphlet *Can the Native Return?* (London: University of London, 1989) and essay "Four Bodies on the *Beagle*: Touch, Sight, and Writing in a Darwin Letter," in *Textuality and Sexuality*, ed. M. Worton and J. Still (Manchester, Eng.: Manchester University Press, 1993).

25. Charles Darwin, *The Voyage of the Beagle*, ed. Michael Neve and Janet Browne (London: Penguin, 1989).

26. In 1845 the word "lakes" is substituted for "seas." In 1839 the quoted sentence about the world as habitable appears in a footnote (Fitzroy, p. 77).

27. Darwin, *Diary*, 1839 version, p. 22. Darwin made no changes to this passage, or to that below describing the unjust game laws in St. Helena, in the later edition. Passages in the *Diary* are often openly ironic, for example, on the journey from "Patagones to Buenos Ayres," 1833: "If this warfare is successful, that is if all the Indians are butchered, a grand extent of country will be gained for the production of cattle. . . . The country will be in the hands of white Gaucho savages instead of copper coloured Indians. The former being a little superior in civilization, as they are inferior in every moral virtue" (pp. 172–73).

28. Darwin, *Diary*, 1839 version, p. 520. Darwin added an excoriating passage on the evils of slavery to the conclusion of the 1845 edition. In the added passage, as so often, the wish to believe in his society as good is pitted against his knowledge of past and present evil: "It makes one's blood boil, yet heart tremble, to think that we Englishmen and our American descendants, with their boastful cry of liberty, have been and are so guilty; but it is a consolation to reflect, that we at least have made a greater sacrifice, than ever made by any nation, to expiate our sin" (Darwin, *Diary*, 1845 version, p. 500).

29. Darwin, *Correspondence* 3:380.

30. Charles Darwin, *The Origin of the Species by means of natural selection; or The preservation of favored races in the struggle for life*, ed. J. Burrow (Harmondsworth: Penguin, 1968), p. 131.

31. Gillian Beer, "Darwin's Reading and the Fictions of Development," in *The Darwinian Heritage*, ed. D. Kohn (Princeton: Princeton University Press, 1985), pp. 543–88.

32. Adrian Desmond and James Moore, *Darwin* (London: Michael Joseph, 1991). Subsequent page references to this edition appear parenthetically in the text.

33. Thomas Macaulay, *History of England* (London: Macmillan, 1913), 1:13.

34. Discussing Conybeare's *Introduction to Geology* in the *Red Notebook*, concerning beds of wood deposited in different basins, Darwin comments: "If such

can happen in troubled England, the more minute equalities of elevation may well be preserved in Patagonia. The English fact is astonishing" (p. 46).

35. Darwin, *Correspondence*, 3:20.

9. Prodger, Darwin's *Expression of the Emotions*

1. Charles Bell, *The Anatomy and Philosophy of Expression as Connected with the Fine Arts* (London: G. Bell, 1912), p. 194.

2. Darwin's uncle, Sir Thomas Wedgwood, had devised a method of creating light-sensitive emulsions using silver nitrate in 1802. Although Wedgwood had attempted to use his process to make photographic pictures, he was unable to prevent the images he created from fading upon subsequent exposure to light.

3. Charles Darwin, *The Autobiography of Charles Darwin*, ed. Nora Barlow (New York: W. W. Norton, 1993), p. 68. Today Herschel is perhaps best known for his contributions to the science of astronomy, although he was also an influential natural scientist, inventor, philosopher, and draftsman.

4. Ibid., p. 61.

5. Ibid. Darwin identifies Sebastiano del Piombo's early sixteenth-century painting "The Raising of Lazarus" as one of his favorites, as it excited in him a "sense of sublimity."

6. William Henry Fox Talbot, "Some Account of the Art of Photogenic Drawing," in *Photography in Print*, ed. Vicki Goldberg (Albuquerque: University of New Mexico Press, 1990), p. 39.

7. "The Edinburgh Review, January 1843," in Goldberg, p. 52.

8. Gillian Beer, *Darwin's Plots: Evolutionary Narrative in Darwin, George Eliot, and Nineteenth-Century Fiction* (London: Ark, 1985), p. 100.

9. Edward Manier, *The Young Darwin and His Cultural Circle* (Boston: D. Reidel, 1978), p. 184.

10. Darwin, *Autobiography*, p. 131.

11. Ibid.

12. Charles Darwin, *The Descent of Man, and Selection in Relation to Sex*, ed. John T. Bonner and Robert T. May (Princeton: Princeton University Press, 1981), p. 191.

13. Charles Darwin, *On the Origin of Species*, ed. Ernst Mayr (Cambridge, Mass.: Harvard University Press, 1964), p. 235.

14. Charles Darwin, *The Expression of the Emotions in Man and Animals*, ed. Konrad Lorenz (Chicago: University of Chicago Press, 1965), p. 367.

15. Ibid., p. 1.

16. Pseudo-Aristotle, "Physiognomica," in Aristotle, *Minor Works*, trans. W. S. Hett (London: W. Heinemann, 1936), 4:808b.

17. Michael W. Kwakkelstein, "Leonardo da Vinci's Grotesque Heads and

the Breaking of the Physiognomic Mould," *Journal of the Warburg and Courtauld Institutes* 54 (1991): 127.

18. I have listed here some of the notable artists who appear to have been at least an indirect influence on Darwin's documentary enterprise. A wide array of nineteenth-century Romantic artists shared Darwin's preoccupation with expression. I have been unable to ascertain whether or not Darwin was exposed to the work of continental European artists such as Géricault and Goya, whose interests in expression, and insanity in particular, are a striking parallel to Darwin's work. This similarity lends itself to the extremely interesting question of the relationship between Romantic interest in individual experience and the popularity of Darwin's theory of evolution.

19. Brewster Rogerson, "The Art of Painting the Passions," *Journal of the History of Ideas* 14, no. 1 (Jan. 1953): 73.

20. René Descartes, *The Passions of the Soul*, trans. Stephen Voss (Indianapolis: Hackett, 1989), p. 24.

21. Charles Le Brun, *A Method to Learn to Design the Passions*, ed. David S. Rodes, trans. John Williams (Los Angeles: William Andrews Clark Memorial Library, 1980), p. i.

22. John Graham, *Lavater's Essays on Physiognomy: A Study in the History of Ideas* (Bern: Peter Lang, 1979), p. 61.

23. Bell, pp. 2–3.

24. Charles Darwin, "Notebook C" 243, in *Notebooks on Transmutation of Species*, ed. Gavin de Beer (London: British Museum of Natural History, 1960–67), p. 315.

25. Bell, p. 77.

26. William Paley, *Natural Theology; or, Evidences of the Existence and Attributes of the Deity, Collected from the Appearances of Nature* (Philadelphia: Printed for J. Morgan by H. Maxwell, 1802).

27. Graham, p. 62.

28. Edgar Yoxall Jones, *O. G. Rejlander: Father of Art Photography, 1813–1875* (London: David & Charles, 1973), p. 10.

29. Adrian Desmond and James Moore, *Darwin: The Life of a Tormented Evolutionist* (New York: Warner Books, 1991), p. 594.

30. Darwin, *Expression of the Emotions*, p. 181.

31. Malcolm Rogers, *Camera Portraits* (New York: Oxford University Press, 1990), p. 84.

32. Darwin, *Autobiography*, p. 77.

33. Ibid., p. 101.

34. Darwin Papers, Cambridge University Library, vol. 53.1, folio 85, item 164, and vol. 53.1, folio 89, item 168.

35. The other photographers were the psychiatrist James Crichton Browne

(1840–1938), the medical doctor/oceanographer George Charles Wallich (1815–99), and the painter/photographer Adolph Diedrich Kindermann (1823–92).

36. R. Andrew Cuthbertson, "The Highly Original Dr. Duchenne," in G. B. Duchenne de Boulogne, *The Mechanism of Human Facial Expression*, ed. R. Andrew Cuthbertson (Cambridge, Eng.: Cambridge University Press, 1990), p. 227.

37. Desmond and Moore, p. 32.

38. Sander L. Gilman, "Darwin Sees the Insane," *Journal of the History of the Behavioral Sciences* 15 (1979): 258.

39. Darwin Marable, "Photography and Behaviour in the Nineteenth Century," *History of Photography* 9, no. 2 (Apr.–June 1985): 141.

40. G. B. Duchenne de Boulogne, "Recherches électro-physiologiques, pathologiques, et thérapeutiques," a series of papers addressed to the Académie des Sciences, Paris, May 21, 1849.

41. Duchenne de Boulogne, *Mechanism*, p. 1.

42. Ibid., p. 7.

43. Ibid., p. 34.

44. Cuthbertson, p. 230.

45. Duchenne de Boulogne, *Mechanism*, p. 101.

46. Darwin, *Expression of the Emotions*, pp. 13–14.

47. Lorraine Daston and Peter Galison, "The Image of Objectivity," *Representations* 40 (Fall 1992): 81–128.

48. Jones, p. 9.

49. Ibid., p. 15.

50. Oscar G. Rejlander, "On Photographic Composition; With a Description of the 'Two Ways of Life,'" *Journal of the Royal Photographic Society* (Apr. 21, 1858): 191.

51. Yoxall Jones, p. 13.

52. Ibid., p. 38.

53. Ibid.

54. Darwin Papers, Cambridge University Library, vol. 53.1, folio 54, item 96, and vol. 53.1, folio 63, item 132.

55. Darwin, *Expression of the Emotions*, p. 248.

56. Frank Spencer, "Some Notes on the Attempt to Apply Photography to Anthropometry During the Second Half of the Nineteenth Century," in *Anthropology and Photography: 1860–1920*, by Elizabeth Edwards (New Haven: Yale University Press, 1972), p. 99.

57. Ibid., p. 100.

58. Ibid.

59. Roslyn Poignant, "Surveying the Field of View: The Making of the RAI Photographic Collection," in Edwards, p. 56.

60. Muybridge began his studies of horses in motion at Stanford University in 1872, the same year Darwin's *Expression of the Emotions* was published. Muybridge did not begin to publish his results formally until 1878, some six years later. By the time Muybridge visited Marey and the painter Jean-Louis-Ernest Meissonier in 1881, *The Expression of the Emotions* would have been widely available. The photometric backdrops Muybridge employed in his motion sequences are direct descendants of Huxley's method for producing "objective" photographic documents.

61. Etienne-Jules Marey, as quoted in Francois Forster-Hahn, "Marey, Muybridge, and Meissonier: The Study of Movement in Science and Art," in *Eadweard Muybridge: The Stanford Years, 1872–1882*, ed. Anita Mozley (Stanford, Calif.: Stanford University Museum of Art, 1972), p. 98.

10. Schaffer, Leviathan of Parsonstown

1. Steven Shapin, "Pump and Circumstance: Robert Boyle's Literary Technology," *Social Studies of Science* 14 (1984): 481–520.

2. Steven Shapin and Simon Schaffer, *Leviathan and the Air Pump* (Princeton, N.J.: Princeton University Press, 1985), pp. 76–79.

3. James Paradis, "Montaigne, Boyle, and the Essay of Experience," in *One Culture*, ed. George Levine (Madison: University of Wisconsin Press, 1987), pp. 59–91; Peter Dear, "Jesuit Mathematical Science and the Reconstitution of Experience in the Early Seventeenth Century," *Studies in History and Philosophy of Science* 18 (1987): 133–75; Robert Iliffe, "Is He Like Other Men? The Meaning of the *Principia Mathematica*," in *Culture and Society in the Stuart Restoration*, ed. Gerald MacLean (Cambridge: Cambridge University Press, 1995), pp. 159–76; F. L. Holmes, "Argument and Narrative in Scientific Writing," in *The Literary Structure of Scientific Argument*, ed. Peter Dear (Philadelphia: University of Pennsylvania Press, 1991), pp. 164–81.

4. Robert Darnton, *The Literary Underground of the Old Regime* (Cambridge, Mass.: Harvard University Press, 1982); Roger Chartier, *The Cultural Uses of Print in Early Modern France* (Princeton: Princeton University Press, 1987); Daniel Roche, *Les Républicains des Lettres* (Paris: Fayard, 1988).

5. Shapin and Schaffer, pp. 143–54; Steven Shapin, *A Social History of Truth* (Chicago: University of Chicago Press, 1994), p. 381.

6. On chains of representations in technical and social networks, see Michael Lynch and Steve Woolgar, "Sociological Orientations to Representational Practice in Science," in *Representation in Scientific Practice*, ed. Michael Lynch and Steve Woolgar (Cambridge, Mass.: MIT Press, 1990), pp. 1–18; Bruno Latour, "Visualization and Cognition," *Knowledge and Society* 6 (1986): 1–40. In *Open Fields: Science in Cultural Encounter* (Oxford: Oxford University Press, 1996), p. 159, Gillian Beer writes that "not what is said, but the agreement as to constraints on

its reception, will stabilize scientific discourse. . . . The enclosing within a community is a necessary condition for assuring stable signification."

7. Ian Hacking, *Representing and Intervening* (Cambridge: Cambridge University Press, 1983), pp. 130–46; Bruno Latour, "Opening One Eye While Closing the Other," in *Picturing Power*, ed. Gordon Fyfe and John Law (London: Routledge, 1988), pp. 15–38, on pp. 15–16; Michael Lynch, "Discipline and the Material Form of Image: An Analysis of Scientific Visibility," *Social Studies of Science* 15 (1985): 37–66.

8. Martin Rudwick, "The Emergence of a Visual Language for Geological Science, 1760–1840," *History of Science* 14 (1976): 149–95; John Law and Michael Lynch, "Lists, Field Guides, and the Descriptive Organization of Seeing: Birdwatching as an Exemplary Observational Activity," in Lynch and Woolgar, *Representation*, pp. 267–99; Michael Lynch and Samuel Edgerton, "Aesthetics and Digital Image Processing," in Fyfe and Law, pp. 184–220. On aesthetic judgment in these sciences, see Nicholas Jardine, *The Scenes of Inquiry* (Oxford: Clarendon, 1991), pp. 204–24.

9. For example, William Smyth, *A Cycle of Celestial Objects*, 2 vols. (London: John Parker, 1844), 1:3: "The department of Urania is entitled to the strict regard of the educated classes, as furnishing the most exquisite proof of what the human intellect is capable"; and John Herschel, "Address of the President," *Report of the Fifteenth Meeting of the British Association for the Advancement of Science* (London: John Murray, 1846), pp. xxvii–xliv, on p. xxxiii: "Astronomy can be advanced by a combined system of observation and calculation carried on uninterruptedly; where, in way of experiment, man has no control, and whose only handle is the continual observation of Nature as it develops itself under our eyes." For astronomy as the positive science see Jacques Merleau-Ponty, *La Science de l'Univers à l'Age du Positivisme* (Paris: Vrin, 1983), chap. 3; Simon Schaffer, "Where Experiments End: Tabletop Trials in Victorian Astronomy," in *Scientific Practice*, ed. J. Z. Buchwald (Chicago: University of Chicago Press, 1995), pp. 257–99, esp. pp. 260–63.

10. Harriet Martineau, *A History of the Thirty Years' Peace*, 4 vols., revised edition (London: George Bell, 1877–78), 4:442–43, 451.

11. On artisan skill and secrecy see John Rule, "The Property of Skill in the Period of Manufacture," in *The Historical Meanings of Work*, ed. Patrick Joyce (Cambridge: Cambridge University Press, 1987), pp. 99–118; and Clive Behagg, "Secrecy, Ritual, and Folk Violence: The Opacity of the Workplace in the First Half of the Nineteenth Century," in *Popular Culture and Custom in Nineteenth-Century England*, ed. R. D. Storch (London: Croom Helm, 1982), pp. 154–79. On London astronomers and political economy see William Ashworth, "The Calculating Eye: Baily, Herschel, Babbage and the Business of Astronomy," *British Journal for the History of Science* 27 (1994): 409–42.

12. Agnes Clerke, *A Popular History of Astronomy in the Nineteenth Century* (London: Black, 1908), p. 115.

13. On patents and discoveries see Simon Schaffer, "Scientific Discoveries and the End of Natural Philosophy," *Social Studies of Science* 16 (1986): 387–420; Geof Bowker, "What's in a Patent?" in *Shaping Technology / Building Society*, ed. Wiebe Bijker and John Law (Cambridge, Mass.: MIT Press, 1992), pp. 53–74; Christine MacLeod, "Concepts of Innovation and the Patent Controversy in Victorian Britain," in *Technological Change*, ed. Robert Fox (Amsterdam: Harwood, 1996), pp. 137–54; Paul Lucier, "Court and Controversy: Patenting Science in the Nineteenth Century," *British Journal for the History of Science* 29 (1996): 139–54.

14. Iwan Rhys Morus, "Manufacturing Nature: Science, Technology, and Victorian Consumer Culture," *British Journal for the History of Science* 29 (1996): 403–34, esp. p. 418.

15. Henry C. King, *The History of the Telescope* (New York: Charles Griffin, 1955), pp. 120–44, 198–204.

16. J. A. Bennett, "On the Power of Penetrating into Space: The Telescopes of William Herschel," *Journal for the History of Astronomy* 7 (1976): 75–108; Simon Schaffer, "Herschel in Bedlam: Natural History and Stellar Astronomy," *British Journal for the History of Science* 13 (1980): 211–39.

17. Maskelyne to Méchain, 1803, in Schaffer, "Herschel in Bedlam," p. 216. On the novelty of stellar astronomy see Mari Williams, "Was There Such a Thing as Stellar Astronomy in the Eighteenth Century?" *History of Science* 21 (1983): 369–85.

18. Bennett, "Power," p. 88; Mary Cornwallis Herschel, *Memoir and Correspondence of Caroline Herschel* (New York: Appleton, 1876), p. 73.

19. Herschel to Watson, Jan. 7, 1782, in J. L. E. Dreyer, ed., *Collected Scientific Papers of William Herschel*, 2 vols. (London: Royal Society, 1912), I:xxxiii.

20. Bennett, "Power," p. 86.

21. Dreyer, I:xxvi.

22. Edwards to Herschel, July 4, 1783, in A. J. Turner, *Science and Music in Eighteenth Century Bath* (Bath: University of Bath, 1977), p. 105.

23. Dreyer, I:xlviii–1; Thomas Dick, *The Practical Astronomer* (Edinburgh, 1845), p. 310.

24. Charles Parsons, ed., *Scientific Papers of William Parsons, Third Earl of Rosse* (London: Lund, Humphries, 1926), pp. 13, 8.

25. On Herschel's long-term commitment to the existence of true nebulosity see Michael Hoskin, *Stellar Astronomy: Historical Studies* (Chalfont St Giles: Science History, 1982), pp. 125–36. On the authority of his unique instruments see J. A. Bennett, "A Viol of Water or a Wedge of Glass," in *The Uses of Experiment*, ed. David Gooding, Trevor Pinch, and Simon Schaffer (Cambridge: Cam-

bridge University Press, 1989), pp. 105–14, esp. pp. 108–10. On the role of calibration in breaking the vicious circle set up when there is no agreed criterion of ascertaining correct outcomes of a trial or observation see H. M. Collins, *Changing Order* (Beverly Hills: Sage, 1985), pp. 100–106.

26. Dreyer, 1:90.

27. Ibid., 2:148.

28. For the naming of Uranus see Simon Schaffer, "Uranus and the Establishment of Herschel's Astronomy," *Journal for the History of Astronomy*, 12 (1981): 11–26; Robert Smith, "The Impact on Astronomy of the Discovery of Uranus," in *Uranus and the Outer Planets*, ed. Garry Hunt (Cambridge: Cambridge University Press, 1982), pp. 81–90. On "asteroids" see Herschel to Watson, April 25, 1802, in Constance Lubbock, *The Herschel Chronicle* (Cambridge: Cambridge University Press, 1933), p. 270.

29. Dreyer, 1:44–45.

30. Ibid., p. 49.

31. Ibid., p. 58.

32. Herschel to Maskelyne, Jan. 30, 1782, Royal Astronomical Society, Herschel MSS, W 1/1 pp. 29–30; Herschel to Caroline Herschel, June 3, 1782, in Turner, p. 107.

33. Dreyer, 1:59.

34. Dreyer, 2:592–93, 608–11.

35. Ibid., p. 609.

36. [Henry Brougham], *Edinburgh Review* 1 (1803): 426–31.

37. See Olivia Smith, *The Politics of Language, 1791–1819* (Oxford: Clarendon, 1984).

38. Mary Cornwallis Herschel, p. 309; Fanny Burney, *Diary*, ed. L. Gibbs (London: Everyman, 1940), pp. 160–61.

39. Dreyer, 2:654–56.

40. Hoskin, *Stellar Astronomy*, pp. 126–27.

41. Ibid., pp. 128–29.

42. George Airy, "Address on Presenting the Honorary Medal to Sir J. F. W. Herschel," *Memoirs of the Royal Astronomical Society* 9 (1836): 303–12.

43. John Herschel, "An Account of the Actual State of the Great Nebula in Orion Compared with Those of Former Astronomers," *Memoirs of the Astronomical Society of London* 2 (1826): 487–95, and "Observations of Nebulae and Clusters of Stars," *Philosophical Transactions* 123 (1833): 359–505, on pp. 501–502. See Michael Hoskin, "John Herschel's Cosmology," *Journal of the History of Astronomy* 18 (1987): 1–32.

44. David Evans, Terence J. Deeming, Betty Hall Evans, and Stephen Goldfarb, eds., *Herschel at the Cape* (Austin: University of Texas Press, 1969), p. 50; Hoskin, "Herschel's Cosmology," pp. 18–19.

45. Simon Schaffer, "The Nebular Hypothesis and the Science of Progress,"

in *History, Humanity and Evolution*, ed. J. R. Moore (Cambridge: Cambridge University Press, 1989), pp. 131–64.

46. [John Pringle Nichol], "On Some Recent Discoveries in Astronomy," *Tait's Edinburgh Magazine* 4 (Oct. 1833): 57–64, esp. p. 62; and "State of Discovery and Speculation Concerning the Nebulae," *London and Westminster Review* 3 (July 1836): 390–409, esp. pp. 394–95, 398n.

47. Nichol, "State of Discovery," p. 405n. On the political context of the attack on transmutation see Adrian Desmond, *The Politics of Evolution* (Chicago: University of Chicago Press, 1989), chap. 5.

48. Nichol to Herschel, Nov. 4, 1838, Royal Society Herschel MSS HS 13.131.

49. [George Gilfillan], "Popular Lecturers—Professor Nichol," *Tait's Edinburgh Magazine* 15 (Mar. 1848): 145–53, esp. p. 152.

50. John Pringle Nichol, *Architecture of the Heavens* (Edinburgh: William Tait, 1837), p. 127.

51. John Stuart Mill, *System of Logic* (1843), in *Collected Works*, vols. 7–8, ed. J. M. Robson (London: Routledge, 1973–74), 7:507–8; [Robert Chambers], *Vestiges of the Natural History of Creation* (London: John Churchill, 1844), pp. 1–26; [David Brewster], "Review of M. Comte's *Course of Positive Philosophy*," *Edinburgh Review* 67 (1838): 271–308. See Stephen G. Brush, *Nebulous Earth* (Cambridge: Cambridge University Press, 1996), pp. 42–44, 69–72.

52. On Mill's problems see Schaffer, "Nebular Hypothesis," p. 148; on Chambers see Marilyn Bailey Ogilvie, "Robert Chambers and the Nebular Hypothesis," *British Journal for the History of Science* 8 (1975): 214–32; on Brewster's switch see Paul Baxter, "Brewster, Evangelism and the Disruption of the Church of Scotland," in *Martyr of Science: Sir David Brewster 1781–1866*, ed. A. D. Morrison-Low and J. R. R. Christie (Edinburgh: Royal Scottish Museum, 1984), pp. 45–50, esp. pp. 47–49.

53. John Herschel, "Address," pp. xxix, xxxviii. On Herschel and Mill see S. S. Schweber, "Auguste Comte and the Nebular Hypothesis," in *In the Presence of the Past*, ed. R. T. Bienvenu and M. Feingold (Dordrecht: Kluwer, 1990), pp. 131–91; on astronomy, business, and speculation see Ashworth, p. 424.

54. John Herschel, "Address," p. xxxvii.

55. George Airy, "Address on Herschel," p. 311.

56. Parsons, pp. 190–91, 206; Nichol to Herschel, Nov. 4, 1838, Royal Society MSS Herschel HS 13.131.

57. William Martin, *The Exposure of Dr. Nichol, the Impostor and Mock Astronomer* (Newcastle: Pattison and Ross, 1839), p. 2.

58. On "journalism" and the sciences in the 1830s and 1840s see Desmond, p. 14; Susan Sheets-Pyenson, "Popular Scientific Periodicals in Paris and London: The Emergence of a Low Scientific Culture, 1820–1875," *Annals of Science* 42 (1985): 549–72; James Secord, "Extraordinary Experiment: Electricity and

the Creation of Life in Victorian England," in Gooding, Pinch, and Schaffer, pp. 337–83. On "plebification" see Samuel Taylor Coleridge, *On the Constitution of the Church and State* (London: William Pickering, 1830; reprint London: Routledge, 1976), pp. 68–70: "you wish for *general* illumination. . . . You begin, therefore, with the attempt to *popularize* science: but you will only effect its *plebification*."

59. "Extraordinary Nebula," *Pictorial Times* 6 (July 12, 1845): 26.

60. Lynch and Edgerton, p. 218 n. 22.

61. For Rosse as mechanic and chemist see "Electrical Soirée," *Literary Gazette* 27 (1843): 352; and for his moral standing see "The Earl of Rosse's Great Telescope, at Parsonstown," *Illustrated London News* (Sep. 9, 1843): 165. On Robinson see J. A. Bennett, *Church, State and Astronomy in Ireland* (Belfast: Armagh Observatory, 1990), pp. 59–138. Robinson attacks Nichol in Robinson to Rosse, April 7, 1841, Rosse MSS K 5.4.

62. Thomas Romney Robinson, "An Account of a Large Reflecting Telescope Lately Constructed by Lord Oxmantown," *Proceedings of the Royal Irish Academy* 2 (1843–44): 2–12 (read Nov. 9, 1840), esp. p. 3.

63. Thomas Romney Robinson, "On Lord Rosse's Telescopes," *Proceedings of the Royal Irish Academy* 3 (1845–47): 114–33 (read April 25, 1842 and April 14, 1845), esp. pp. 116–17.

64. [Thomas Romney Robinson], "Speculum," in *A Cyclopaedia of the Physical Sciences*, 2nd ed., ed. John Pringle Nichol (Glasgow: Richard Griffin, 1860), pp. 779–90, esp. p. 781.

65. Robinson, "Lord Rosse's Telescopes," p. 119; [Anon.], "The Earl of Rosse's Telescopes," *Dublin Review* 18 (Mar. 1845): 1–43, esp. p. 19; George Airy, "On the Large Reflecting Telescopes of the Earl of Rosse and Mr Lassell," *Monthly Notices of the Royal Astronomical Society* 9 (1848–49): 110–21 (read Nov. 1848), esp. pp. 113, 121.

66. R. P. Graves, *Life of Sir William Rowan Hamilton*, 3 vols. (Dublin: Dublin University Press, 1882–89), 2:121, 620–35.

67. K. M. Lyell, *Life, Letters and Journals of Sir Charles Lyell*, 2 vols. (London: John Murray, 1881), 2:76; W. Airy, ed., *Autobiography of Sir George Biddell Airy* (Cambridge: Cambridge University Press, 1896), p. 198. "Astigmatism" was a (human) optical defect that Airy himself had first identified.

68. W. Airy, pp. 221–22 (Sept. 1854).

69. Michael Hoskin, "Rosse, Robinson and the Resolution of the Nebulae," *Journal for the History of Astronomy* 21 (1990): 331–44.

70. Parsons, p. 102.

71. Rosse to Scoresby, Apr. 6, 1846, Scoresby MSS, Whitby Philosophical Society; Hunter Christie to Rosse, enclosing Herschel's report on Lord Oxmantown's paper, Birr MSS B 6.1, Astronomical Scrapbook, July 17, 1840.

72. Birr MSS B 6.1, Oct. 29, 1840; Hoskin, "Rosse, Robinson," p. 334.

South's later remark is reprinted in *Astronomische Nachrichten* no. 536 (May 1845): cols. 113–15.

73. Birr MSS B 6.1, Nov. 5, 1840; Hoskin, "Rosse, Robinson," p. 335.

74. South to Rosse, Nov. 23, 1840, Birr MSS K 1.3; Robinson, "Account," pp. 5, 10; Hoskin, "Rosse, Robinson," p. 337.

75. Robinson, "Lord Rosse's Telescopes," p. 130; Hoskin, "Rosse, Robinson," p. 339.

76. Hoskin, "Rosse, Robinson," p. 340; Herschel to Rosse, Mar. 9, 1845, Birr MSS K 2.2; John Herschel, "Address," p. xxxvii.

77. John Pringle Nichol, *Thoughts on Some Important Points Relating to the System of the World*, 2nd ed. (Edinburgh: Johnstone, 1848), p. 109.

78. Ibid., pp. 110–13.

79. See Ogilvie; on Chambers and Nichol see James Secord, "Behind the Veil: Robert Chambers and *Vestiges*," in J. R. Moore, pp. 165–94, esp. pp. 174–77.

80. Nichol, *Thoughts*, pp. iv, 132; Nichol to William Thomson, Feb. 1, 1846, Cambridge University Library MSS 7342 N28; Schaffer, "Nebular Hypothesis," p. 155.

81. [David Brewster], "Review of *Vestiges of the Natural History of Creation*," *North British Review* 3 (1845): 470–515, esp. p. 481; Baxter, p. 48.

82. Whewell to Rosse, Sep. 3, 1853, Birr MSS K 17.18. See John Brooke, "Natural Theology and the Plurality of Worlds: Observations on the Brewster-Whewell Debate," *Annals of Science* 34 (1977): 221–86; Michael Crowe, *The Extraterrestrial Life Debate* (Cambridge: Cambridge University Press, 1986), pp. 312, 330.

83. Bennett, *Church, State and Astronomy*, p. 107; [David Brewster], "The Earl of Rosse's Reflecting Telescopes," *North British Review* 2 (Nov. 1844): 175–212, esp. p. 177.

84. John Pringle Nichol, "Address Delivered at the Soirée of the Stirling School of Arts," in *Importance of Literature to Men of Business* (Glasgow: Griffin, 1852), pp. 212–52, esp. p. 241.

85. Nichol, *Thoughts*, p. ix.

86. "Recent Astronomy," *British Quarterly Review* 6 (1847): 1–40.

87. "Earl of Rosse's Telescopes," p. 41. On other treatments of the Leviathan see Jonathan Smith, "De Quincey's Revisions to 'The System of the Heavens,'" *Victorian Periodicals Review* 26 (1993): 203–12, esp. p. 204.

88. William Scoresby, "The Earl of Rosse's Telescopes and Their Revelations in the Sidereal Heavens," *Edinburgh Philosophical Journal* 54 (1852–53): 113–18; Tom Stamp and Cordelia Stamp, *William Scoresby: Arctic Scientist* (Whitby: Caedmon, 1975), p. 145.

89. Hoskin, "Rosse, Robinson," p. 342.

90. Humboldt to Arago, July 7, 1848 and Nov. 9, 1849, in *Correspondance*

d'Alexandre de Humboldt avec François Arago, ed. E. T. Hamy (Paris: Guilmoto, 1909), pp. 286–89, 304–5.

91. King, pp. 217–24.

92. B. Z. Jones, ed., "Diary of the Two Bonds, 1846–1849," *Harvard Library Bulletin* 16 (1968): 49–71, esp. pp. 56–57; Bond to Everett, Sep. 22, 1847, in Ronald Numbers, *Creation by Natural Law: Laplace's Nebular Hypothesis in American Thought* (Seattle: University of Washington Press, 1977), p. 35.

93. Jones, pp. 61, 64. Bond claimed resolution in his "Description of the Nebula About the Star θ Orionis," *Memoirs of the American Academy of Sciences* 3 (1848): 87–96. His claim was recognized in John Herschel, *Outlines of Astronomy,* 4th ed. (London: Longman, 1851), p. 609, then denied in Clerke, p. 119.

94. George Airy, "Address," *British Association Report* (1851): xxxix–liii, esp. p. xli.

95. Hoskin, *Stellar Astronomy,* p. 150.

96. John Pringle Nichol, "Nebulae," in Nichol, *Cyclopaedia,* pp. 603–4, 612–13.

97. Herbert Spencer, *Essays, Scientific, Political and Speculative,* 3 vols. (London: Williams and Norgate, 1891), 1:111, 114; David Duncan, *Life and Letters of Herbert Spencer* (London: Methuen, 1908), pp. 425–29, 591. See Brush, pp. 50–53, 72–74.

98. William Huggins, *Scientific Papers* (London: Wesley, 1909), pp. 105–8.

99. Ibid., p. 106. On early astrospectroscopy see Karl Hufbauer, *Exploring the Sun: Solar Science Since Galileo* (Baltimore: Johns Hopkins University Press, 1991), pp. 57–59.

100. William Huggins, "On the Physical and Chemical Constitution of the Fixed Stars and Nebulae," in *Royal Institution Library of Science: Astronomy,* ed. Bernard Lovell (Dordrecht: Reidel, 1970), 1:42–50, esp. pp. 49–50.

101. Huggins, *Scientific Papers,* p. 118.

102. Parsons, pp. 190, 200.

103. Ibid., pp. 203–6.

104. Struve to Rosse, July 28, 1869, Birr MSS, L 6.1.

105. Ludwik Fleck, *Genesis and Development of a Scientific Fact* (Chicago: University of Chicago Press, 1979), pp. 111–24.

106. Simon Schaffer, "Babbage's Intelligence: Calculating Engines and the Factory System," *Critical Inquiry* 21 (1994): 203–27. Rosse lectured the Royal Society on Babbage in 1854; see Parsons, p. 78.

11. Pang, Development of Astrophotography

My thanks to Robert Kohler, Rikki Kuklick, and Tim Lenoir for their advice and encouragement, and to Dorothy Schaumberg, director of the Mary Lea Shane Archives of the Lick Observatory, whose help made this article possible.

The following abbreviations are used: *ApJ, Astrophysical Journal*; MLS, Mary Lea Shane Archives of the Lick Observatory, University of California, Santa Cruz; *PASP, Publications of the Astronomical Society of the Pacific.*

1. On the discourse of astronomical photography in the 1880s and the construction of incommensurability in photography and drawing, see Alex Soojung-Kim Pang, "The Industrialization of Vision in Victorian Astronomy," in *Cultural Babbage: Technology, Time, and Invention*, ed. Francis Spufford and Jenny Uglow (London: Faber, 1996).

2. Edward S. Holden, "Photography the Servant of Astronomy," *Overland Monthly* n.s. 8 (Nov. 1886): 459–70, quote on p. 468.

3. "Astronomical Photography," *Edinburgh Review* (Jan. 1888), reprinted in *Siderial Messenger* 7 (Apr. 1888): 138–91, quotes from pp. 138, 149.

4. Owen Gingerich, ed., *The General History of Astronomy*, vol. 4, *Astrophysics and Twentieth-Century Astronomy to 1950: Part A* (Cambridge, Eng.: Cambridge University Press, 1984), part I; John Lankford, "Amateurs and Astrophysics: A Neglected Aspect in the Development of a Scientific Specialty," *Social Studies of Science* 11 (1981): 275–303; Simon Schaffer, "Astronomers Mark Time: Discipline and the Personal Equation," *Science in Context* 2 (1988): 115–45. For the American case, see Marc Rothenberg, "Organization and Control: Professionals and Amateurs in American Astronomy, 1899–1918," *Social Studies of Science* 11 (1981): 305–25.

5. Charles Piazzi Smyth, "On Astronomical Drawing," *Memoirs of the Royal Astronomical Society* 15 (1846): 71–82; John Herschel, *Outlines of Astronomy*, 4th ed. (Philadelphia, 1860), p. 512. See also Alex Soojung-Kim Pang, "Victorian Observing Practices, Printing Technology, and Representations of the Solar Corona, Part 1: The 1860s and 1870s," *Journal for the History of Astronomy* 25 (Nov. 1994): 249–74.

6. Lorraine Daston and Peter Galison, "The Image of Objectivity," *Representations* 40 (Fall 1992): 81–128.

7. Lynn White, Jr., "Natural Science and Naturalistic Art in the Middle Ages," *American Historical Review* 52 (1947): 421–35, quote on p. 424.

8. Important exceptions to this generalization are Ann Shelby Blum, *Picturing Nature: American Nineteenth-Century Zoological Illustration* (Princeton: Princeton University Press, 1993); Michael Aaron Dennis, "Graphic Understanding: Instruments and Interpretation in Robert Hooke's *Micrographia*," *Science in Context* 3 (1989): 306–64; Bernard Smith, *Imagining the Pacific: In the Wake of Cook* (New Haven: Yale University Press, 1992).

9. Bernard Smith, *Imagining*; Bernard Smith, *European Vision and the South Pacific* (New Haven: Yale University Press, 1985); Samuel Edgerton, "Galileo, Florentine 'Disegno,' and the 'Strange Spottednesse' of the Moon," *Art Journal* 44 (1984): 225–32; Samuel Edgerton, *The Heritage of Giotto's Geometry: Art and Science on the Eve of the Scientific Revolution* (Ithaca: Cornell University Press, 1991).

10. Samuel Edgerton and Michael Lynch, "Aesthetics and Digital Image Processing: Representational Craft in Contemporary Astronomy," in *Picturing Power: Visual Depiction and Social Relations*, ed. Gordon Fyfe and John Law (London: Routledge, 1988), pp. 184–219, quotes on p. 212, emphasis added.

11. Wilhelm Cronenberg, *Half-Tone on the American Basis*, trans. William Gamble (London: Country Press, 1896), p. 49.

12. H. Jenkins, *A Manual of Photoengraving: Containing Practical Instructions for Producing Photoengraved Plates in Relief-Line and Half-Tone* (Chicago: Inland Printer Co., 1902), pp. 55–57.

13. Alfred Elson, "The Reproductive Processes of the Graphic Arts," in *Lectures on Printing* (Cambridge, Mass.: Harvard University Press, 1912), p. 30. The best screens, made by Philadelphia manufacturer Max Levy, were made by etching lines into glass with a ruling engine and rubbing ink into the grooves, then cementing two plates together: see *Dictionary of American Biography* (1928 ed.), s.v. "Levy, Max."

14. Cronenberg, pp. 112–15; Jenkins, chap. 9.

15. Jenkins, pp. 86–89; Elson, p. 30.

16. Jenkins, chap. 11.

17. "Some Highlights in the History of Photogravure & Color Company," undated ms., Division of Graphic Arts, National Museum of American History, Smithsonian Institution. My thanks to Helena Wright for a copy of this document.

18. Advertisements in Ernest Edwards, *The Heliotype Process* (Boston: James R. Osgood & Co., 1876), pp. 13–16; William Hogarth, *The Works of William Hogarth Reproduced by the Heliotype Process from the Original Engravings* (Boston: James R. Osgood, 1876); Paolo Toschi, *Toschi's Engravings from Frescos by Correggio and Parmegiano* (Boston: James R. Osgood, 1875).

19. On Edwards's work for Muybridge, see Robert Haas, *Muybridge: Man in Motion* (Berkeley: University of California Press, 1976), pp. 154–55, 180; Gordon Hendricks, *Eadweard Muybridge: The Father of the Motion Picture* (New York: Grossman, 1975), chap. 8; Estelle Jussim, *Visual Communication and the Graphic Arts: Photographic Technologies in the Nineteenth Century* (New York: Bowker, 1983), p. 229.

20. Jussim, pp. 133, 140.

21. Ernest Edwards, "The Art of Making Photo-Gravures," *Anthony's Photographic Bulletin* 17 (1887): 430.

22. Ibid.   23. Ibid., p. 431.

24. Cronenberg, p. 123.   25. Ibid., p. 76.

26. J. P. Ourdan, *The Art of Retouching* (New York: E. and H. T. Anthony, 1880).

27. Jenkins, p. 49; Cronenberg, pp. 41–44; C. E. Wright to William Wallace Campbell, Feb. 14, 1902, MLS, box 65, file "Miscellaneous W 1900–1904." On

judgment of Wright's work, see George Ellery Hale to James Keeler, May 9, 1900, MLS, box 22, file "G. E. Hale, 1900."

28. Jenkins, p. 110.

29. Cronenberg, pp. 151–54.

30. Edwards, "Art of Making Photo-Gravures," pp. 430–31.

31. E. E. Barnard to Campbell, Feb. 9, 1909, MLS, file "E. E. Barnard, 1905–1912."

32. Cronenberg, p. 77.

33. Carl Nemethy, "Photo-Etching and Printing," *American Art Printer* (Nov. 1891): 99–100.

34. Holden to Edwards, Dec. 7, 1896, MLS, Director's Copybook 55.

35. Campbell to Barnes-Crosby, Mar. 12, 1912, MLS, file "Barnes-Crosby, 1910–1933."

36. Robert Aitken to Barnes-Crosby, Feb. 12, 1927, MLS, file "Barnes-Crosby, 1910–1933."

37. Barnard to Campbell, Jan. 26, 1909, MLS, file "E. E. Barnard, 1905–1912." On tacit knowledge in scientific practice, see Harry Collins, *Changing Order: Replication and Induction in Scientific Practice* (Beverly Hills: Sage, 1985).

38. Campbell to Royal Engraving, Apr. 12, 1906, MLS, file "Wright & Co./Royal Engraving."

39. Wright to Campbell, Apr. 4, 1906, Apr. 26, 1906, MLS, file "Wright & Co./Royal Engraving"; Campbell to Wright, May 1, 1906, MLS, file "Wright & Co./Royal Engraving"; Wright to Campbell, May 3, 1906, MLS, file "Wright & Co./Royal Engraving."

40. Barnard to Campbell, Aug. 5, 1908; Campbell to Barnard, Jan. 13, 1909, MLS, file "E. E. Barnard, 1905–1912."

41. Campbell to Chicago Photogravure Co., MLS, file "Chicago Photogravure Co."

42. Barnard to Campbell, May 15, 1911, MLS, file "E. E. Barnard, 1905–1912." This reliance on particular specialists improved the quality of work, but it could also cause delays. When Campbell complained to the Photogravure & Color Company about delays in their work, he was told that their etcher had been sick and "we did not dare to put another man . . . [to] work on the 'Star' plates." Karl Arvidson to Campbell, Jan. 25, 1905, MLS, file "Photogravure & Color Co."

43. Campbell to Arvidson, Nov. 5, 1906, MLS, file "Photogravure & Color Co."

44. J. L. Shilling to Campbell, Apr. 13, 1904, MLS, box 7, file "Miscellaneous Bin–Biz, 1874–1904."

45. Barnard to Campbell, Feb. 9, 1909, MLS, file "E. E. Barnard, 1905–1912."

46. Barnard to Campbell, Feb. 21, 1909, MLS, file "E. E. Barnard, 1905–1912."

47. Campbell to Barnard, Feb. 27, 1909, MLS, file "E. E. Barnard, 1905–1912."

48. J. L. Shilling to Campbell, Apr. 13, 1904, MLS, box 7, file "Miscellaneous Bin–Biz, 1874–1904."

49. Barnes-Crosby to Campbell, Apr. 7, 1910, MLS, file "Barnes-Crosby, 1910–1933."

50. Campbell to Barnes-Crosby, Apr. 12, 1910, MLS, file "Barnes-Crosby, 1910–1933."

51. Keeler to Holden, Dec. 13, 1898, MLS, Director's Copybook 63.

52. Campbell to Barnard, Feb. 15, 1907, Feb. 3, 1909, MLS, file "E. E. Barnard, 1905–1912"; Campbell to Arvidson, Jan. 29, 1909, MLS, file "Photogravure & Color Co."

53. Hale to Keeler, June 12, 1899, MLS, box 22, file "George Ellery Hale 1899."

54. Keeler to Edwards, Jan. 11, 1900, MLS, Director's Copybook 67.

55. Keeler to Hale, Mar. 7, 1900, MLS, Director's Copybook 67.

56. Hale to Keeler, Apr. 12, 1900, MLS, box 22, file "G. E. Hale, 1900."

57. Hale to Keeler, May 9, 1900, MLS, box 22, file "G. E. Hale, 1900." Chicago halftone printers Binner-Wells agreed: a halftone plate "will stand any number of impressions," they told Campbell, because it was the silver coating that wore down rather than the actual image: J. L. Shilling to Campbell, Nov. 26, 1904, MLS, box 7, file "Miscellaneous Bin–Biz, 1874–1904."

58. Keeler to Hale, Apr. 20, 1900, MLS, Director's Copybook 67.

59. Keeler to Edwards, Apr. 10, 1900, MLS, Director's Copybook 67.

60. Campbell to C. L. Wright & Co. (New York), Binner Engraving Co. (Chicago), John Andrews and Son (Boston), Photogravure & Color Co. (New York), W. S. Scudder (Houghton, Mifflin & Co., Boston), Feb. 6, 1902, MLS, Director's Copybook 75; Campbell to Binner Engraving, Mar. 20, 1902, MLS, Director's Copybook 76. Keeler had planned to send a few test subjects to C. L. Wright & Co. Keeler to Hale, Apr. 20, 1900, MLS, Director's Copybook 67.

61. C. E. Wright to Campbell, Feb. 14, 1902, MLS, box 65, file "Miscellaneous W 1900–1904."

62. C. E. Wright to Campbell, Feb. 14, 1902, Apr. 16, 1902, MLS, box 65, "Wright & Co./Royal Engraving."

63. Campbell to Binner, June 13, 1902, June 16, 1902, June 20, 1902, July 14, 1902, July 16, 1902, MLS, Director's Copybook 77.

64. Campbell to Henry Phipps, May 4, 1904; Campbell to William Thaw, May 13, 1904; Campbell to E. J. Molera, May 19, 1904; Campbell to Benjamin Ide Wheeler, May 25, 1904; Campbell to John Brashear, June 8, 1904, MLS, Director's Copybook 84; E. J. Molera to Campbell, May 22, 1904; Campbell to Ernest A. Denicke, May 22, 1904, MLS, box 39, file "Molera, E. J. 1900–1904."

65. Campbell to Photogravure & Color, Apr. 9, 1904, MLS, Director's Copybook 83.

66. Campbell to Photogravure & Color, May 2, 1904, May 24, 1904, MLS, Director's Copybook 83.

67. Campbell to C. E. Wright, Apr. 8, 1902, MLS, Director's Copybook 76.

68. J. L. Shilling to Campbell, Apr. 13, 1904, MLS, box 7, file "Miscellaneous Bin–Biz, 1874–1904."

69. Campbell to E. E. Barnard, Feb. 18, 1908, MLS, file "E. E. Barnard, 1905–1912." See also Holden to J. K. Rees, Dec. 18, 1896, MLS, Director's Copybook 55.

70. Holden to Edwards, Aug. 3, 1896, MLS, file "Photogravure & Color Co."

71. Edwin Frost to Campbell, Jan. 27, 1904, MLS, box 18, file "Edwin Frost, 1900–1904"; see also Campbell to Chicago Photogravure Co., Feb. 13, 1915; Campbell to Alex Copelin, Apr. 21, 1915, MLS, file "Chicago Photogravure Co."

72. Campbell to Arvidson, Sept. 16, 1907, Jan. 30, 1908, MLS, file "Photogravure & Color Co."

73. Campbell to Barnes-Crosby, July 9, 1918, MLS, file "Barnes-Crosby, 1910–1933."

74. Keeler to Edwards, Sept. 21, 1899, MLS, Director's Letterbook 65; Campbell to Wright, June 19, 1906, July 27, 1906, MLS, file "Wright & Co./ Royal Engraving"; Campbell to Arvidson, Jan. 18, 1905, May 10, 1907, June 22, 1907, June 27, 1907, July 15, 1907, Sept. 16, 1907, Apr. 29, 1908, July 18, 1908, MLS, file "Photogravure & Color Co."

75. Campbell to Barnard, Dec. 28, 1911, MLS, file "E. E. Barnard, 1905–1912." Campbell also suspected that printers who were given large orders for astronomical engravings regularly "ran in" quick commercial work under their bigger projects, rather than turn them away: Campbell to Barnard, Aug. 14, 1908, MLS, file "E. E. Barnard, 1905–1912."

76. Wright to Campbell, Apr. 4, 1906, Apr. 26, 1906, MLS, file "Wright & Co./Royal Engraving"; Edwards to Holden, May 30, 1896, MLS, file "Photogravure & Color Co."

77. Perrine to Barnard, Oct. 10, 1911, MLS, file "E. E. Barnard, 1905–1912."

78. Arvidson to Campbell, July 18, 1904; Arvidson refers to only one etcher being assigned to this project in Arvidson to Campbell, Jan. 20, 1905, MLS, file "Photogravure & Color Company."

79. C. D. Perrine, "Report of Work Done, July 1, 1904 to July 1, 1906," handwritten ms., n.d., MLS, file "C. D. Perrine, 1905–1912," p. 7.

80. Perrine to Campbell, Sept. 28, 1904, MLS, box 44, file "C. D. Perrine, 1900–1904."

81. Perrine to Campbell, Sept. 30, 1904, MLS, box 44, file "C. D. Perrine, 1900–1904."

82. Barnard to Campbell, Feb. 25, 1907, MLS, file "E. E. Barnard, 1905–1912."

83. Barnard to Campbell, Jan. 27, 1907, MLS, file "E. E. Barnard, 1905–1912."

84. Campbell to Barnes-Crosby, Feb. 6, 1921, MLS, file "Barnes-Crosby, 1910–1933."

85. Alfred Brothers, *Photography: Its History, Processes, Apparatus, and Materials*, 2nd ed. (London: Charles Griffin & Co., 1899), p. 295.

86. Campbell to C. E. Wright, Mar. 7, 1906, MLS, file "Wright & Co./ Royal Engraving."

87. Aitken to Barnes-Crosby, July 20, 1911, MLS, file "Barnes-Crosby, 1910–1933"; Aitken to Oakland National Engraving Company, Nov. 28, 1928, MLS, file "Oakland National Engraving Co., 1927–1933." See also Campbell to San Francisco Photo-Engraving Co., June 21, 1917, Wright to San Francisco Photo-Engraving Co., July 15, 1915, MLS, file "San Francisco Photo-Engraving Co."

88. Campbell to Oakland National Engraving Company, Mar. 22, 1928, MLS, file "Oakland National Engraving Co."

89. Heber D. Curtis to Sunset Publishing House, Jan. 29, 1917, MLS, file "Sunset Publishing House, 1915–1932."

90. Campbell to Barnes-Crosby, Feb. 6, 1921, MLS, file "Barnes-Crosby, 1910–1933."

91. Holden to Edwards, Oct. 23, 1896, Nov. 14, 1896, MLS, file "Photogravure & Color Co."

92. Campbell to Barnes-Crosby, Nov. 25, 1913, MLS, file "Barnes-Crosby, 1910–1933."

93. Wright to Barnes-Crosby, Mar. 12, 1927, MLS, file "Barnes-Crosby, 1910–1933."

94. Campbell to Sunset Publishing, Nov. 23, 1916, MLS, file "Sunset Publishing House, 1915–1932."

95. Campbell to Arvidson, Feb. 7, 1905, Feb. 28, 1905, Dec. 5, 1906, Feb. 15, 1907, May 10, 1907, MLS, file "Photogravure & Color Co."

96. Campbell to Barnes-Crosby, Apr. 12, 1910, MLS, file "Barnes-Crosby, 1910–1933."

97. Campbell to Binner-Wells, Jan. 9, 1915; Campbell to Chicago Photogravure Co., Jan. 9, 1915; Campbell to Charles Elliott Co., Jan. 9, 1915, and to Chicago Photogravure Co., Feb. 13, 1915, MLS, file "Chicago Photogravure Co."; Aitken to Oakland National Engraving Co., Aug. 17, 1931, Nov. 10, 1932, MLS, file "Oakland National Engraving Co., 1927–1933." See also Alex Soojung-Kim Pang, "Victorian Observing Practices, Printing Technology, and Representations of the Solar Corona, Part 2: The Age of Photomechanical Reproduction," *Journal for the History of Astronomy* 26 (1995): 63–75.

98. Campbell to Barnes-Crosby, Jan. 30, 1912, Mar. 12, 1912, MLS, file "Barnes-Crosby, 1910–1933."

99. Campbell to Barnes-Crosby, June 7, 1915, MLS, file "Barnes-Crosby, 1910–1933."

100. Campbell to Barnes-Crosby, Aug. 15, 1911, June 25, 1914; Wright to Barnes-Crosby, Feb. 11, 1918, MLS, file "Barnes-Crosby, 1910–1933"; Aitken to Oakland National Engraving, Dec. 30, 1929, Apr. 23, 1932, MLS, file "Oakland National Engraving Co., 1927–1933."

101. Campbell to Barnes-Crosby, Aug. 15, 1911, Aug. 28, 1911, MLS, file "Barnes-Crosby, 1910–1933."

102. Alfred Brothers, p. 195.

103. Barnes-Crosby to Aitken, Sept. 5, 1911, MLS, file "Barnes-Crosby, 1910–1933."

104. Barnes-Crosby to Campbell, July 1, 1915, July 23, 1915, Feb. 23, 1918, MLS, file "Barnes-Crosby, 1910–1933." On Barnes-Crosby, see "Mr. Gamble Visits a Chicago Photoengraving Company," *Inland Printer* (Sept. 1906): 876.

105. The terms come from Aitken to Barnes-Crosby, Aug. 28, 1911, MLS, file "Barnes-Crosby, 1910–1933."

106. Campbell to Arvidson, May 2, 1904, MLS, file "Photogravure & Color Co."

107. Campbell to San Francisco Photo-Engraving Co., Feb. 22, 1918, MLS, file "San Francisco Photo-Engraving Co."

108. Campbell to San Francisco Photo-Engraving Co., July 10, 1918, MLS, file "San Francisco Photo-Engraving Co."

109. Campbell to Sunset Publishing House, Dec. 9, 1916, MLS, file "Sunset Publishing House, 1915–1932."

110. Campbell to Arvidson, Apr. 9, 1904, MLS, file "Photogravure & Color Co." Campbell says of another proof by the same company, "No. 2 shows the extent of the nebulosity not so well [but] I like its snappy character." Campbell to Arvidson, May 2, 1904.

111. Campbell to Arvidson, May 2, 1904, MLS, file "Photogravure & Color Co."

112. William Wright to Barnes-Crosby, Feb. 11, 1918, MLS, file "Barnes-Crosby, 1910–1933."

113. Campbell to C. E. Wright, Mar. 7, 1906, MLS, file "Wright & Co./Royal Engraving"; Campbell to Arvidson, Mar. 7, 1906, MLS, file "Photogravure & Color Co."

114. James Keeler, "The Ring Nebula in Lyra," *ApJ* 10 (1899): 193–201; "The Annular Nebula H IV-13 in Cygnus," *ApJ* 10 (1899): 266–68; "The Spiral Nebula H I-55 Pegasi," *ApJ* 11 (1900): 1–5; "Photograph of the Trifid Nebula in Sagittarius," *PASP* 12 (1900): 89–90.

115. Holden to Edwards, Dec. 7, 1896, MLS, file "Photogravure & Color Co."

116. James Keeler, "Reviews," *ApJ* 5 (1897): 150–51.

117. Campbell to Sunset Publishing House, Dec. 9, 1916, MLS, file "Sunset Publishing House, 1915–1932."

118. Campbell to Barnes-Crosby, June 7, 1915, MLS, file "Barnes-Crosby, 1910–1933."

119. Robert G. Aitken to Barnes-Crosby, Feb. 23, 1927, MLS, file "Barnes-Crosby, 1910–1933"; Aitken to Oakland National Engraving Co., Apr. 23, 1932, MLS, file "Oakland National Engraving Co., 1927–1933."

120. Karl Arvidson to Campbell, July 18, 1904, MLS, file "Photogravure & Color Company."

121. Campbell to Sunset Publishing House, Nov. 23, 1916, MLS, file "Sunset Publishing House, 1915–1932"; William H. Wright to Campbell, Mar. 16, 1918, MLS, file "Wright, W. H. 1918–1919"; see also Edward Pickering to James McKeen Cattell, May 17, 1915, MLS, file "Cattell, J. M. 1905–1936."

122. Campbell to Barnes-Crosby, Jan. 16, 1916, MLS, file "Barnes-Crosby, 1910–1933."

123. Campbell to Barnes-Crosby, Feb. 21, 1916, MLS, file "Barnes-Crosby, 1910–1933"; Aitken to Oakland National Engraving Co., Dec. 30, 1929, MLS, file "Oakland National Engraving Co., 1927–1933."

124. Perrine, handwritten ms., Nov. 9, 1908, MLS, file "C. D. Perrine, 1905–1912."

125. It had also been criticized by astronomers for decades: see Charles Piazzi Smyth's condemnation of such special effects, pp. 71–84.

126. Campbell to Arvidson, Aug. 29, 1908, MLS, file "Photogravure & Color Co."

127. Campbell to Arvidson, Jan. 29, 1909, MLS, file "Photogravure & Color Co."

128. Barnard to Campbell, Jan. 26, 1909, MLS, file "E. E. Barnard, 1905–1912."

129. Barnes-Crosby to Campbell, Oct. 24, 1917, MLS, file "Barnes-Crosby, 1910–1933."

130. Edwards to Holden, June 19, 1896, MLS, file "Photogravure & Color Co."

131. Campbell to Barnard, Feb. 3, 1909, MLS, file "E. E. Barnard, 1905–1912."

132. Barnard to Campbell, Jan. 26, 1909, MLS, file "E. E. Barnard, 1905–1912."

12. Brain, Standards and Semiotics

The author gratefully acknowledges the fellowship support of the UCLA Paris Program in Critical Theory, the Deutsche Akademische Austauschdienst, and the Renaissance Trust for the research and writing of this paper. The author thanks Jean-Paul Gaudillière, Michael Hagner, Arne Hessenbruch, Nicholas Jardine, Timothy Lenoir, Simon Schaffer, and Richard Staley, and the participants of the Berlin Rathenau Sommerakademie 1994 for helpful comments on previous recensions.

1. The literature on the relationship between language and twentieth-century thought is vast and usually related to one of the different intellectual traditions. A comprehensive view is provided by Ernst Cassirer, *Philosophy of Symbolic Forms*, 3 vols. (New Haven: Yale University Press, 1953–57); Ian Hacking, *Why Does Language Matter to Philosophy?* (Cambridge, Eng.: Cambridge University Press, 1975); Richard Rorty, *Philosophy and the Mirror of Nature* (Oxford: Blackwell, 1980). Katherine Hayles considers the relation between field theories and structural linguistics in *The Cosmic Web: Scientific Field Models and Literary Strategies in the Twentieth Century* (Ithaca, N.Y.: Cornell University Press, 1984); in *Chaos Bound: Orderly Disorder in Contemporary Literature and Science* (Ithaca, N.Y.: Cornell University Press, 1989), Hayles considers the relations between poststructuralism and nonlinear dynamics.

2. Michel Foucault, *The Order of Things: An Archaeology of the Human Sciences* (London: Routledge, 1970), pp. 296–300. Compare the remarks of Ian Hacking, "Night Thoughts on Philology," *History of the Present* 4 (1987): 3–10.

3. See Gyorgy Markus, "Why Is There No Hermeneutics of Natural Sciences? Some Preliminary Theses," *Science in Context* 1 (1987): 2–51; Steven Shapin, "Pump and Circumstance: Robert Boyle's Literary Technology," *Social Studies of Science* 14 (1984): 481–520; Geoffrey Cantor, "The Rhetoric of Science," in *The Uses of Experiment: Studies in the Natural Sciences*, ed. David Gooding, Trevor Pinch, and Simon Schaffer (Cambridge, Eng.: Cambridge University Press, 1989), pp. 159–80; Mi Gyung Kim, "The Layers of Chemical Language, I: The Constitution of Bodies v. Structure of Matter," *History of Science* 30 (1992): 69–96; and Kim, "The Layers of Chemical Language, II: Stabilizing Atoms and Molecules in the Practice of Organic Chemistry," *History of Science* 30 (1992): 397–437; see also Martin Jay, "Should Intellectual History Take a Linguistic Turn? Reflections on the Habermas-Gadamer Debate," in *Rethinking Intellectual History: Reappraisals and New Perspectives*, ed. Steven Kaplan and Dominick LaCapra (Ithaca, N.Y.: Cornell University Press, 1982), pp. 86–110.

4. For a historiographically alert treatment of nineteenth-century linguistics see Olga Amsterdamska, *Schools of Thought: The Development of Linguistics from Bopp to Saussure* (Dordrecht: D. Reidel, 1987).

5. For the historical context of nineteenth-century laboratories, see Frank A. J. L. James, ed., *The Development of the Laboratory: Essays on the Place of Experiment in Industrial Civilization* (Basingstoke, Eng.: Macmillan, 1989); and Andrew Cunningham and Perry Williams, eds., *The Laboratory Revolution in Medicine* (Cambridge, Eng.: Cambridge University Press, 1992). See also the essays in Richard Staley, ed., *Empires of Physics: A Guide to the Exhibition* (Cambridge, Eng.: Whipple Museum Publications, 1993).

6. Michel Bréal, "Les lois phoniques: À propos de la création du laboratoire

de phonétique expérimentale au Collège de France," *Mémoires de la Société de linguistique de Paris* 10 (1898): 11.

7. The description of the modern laboratory is from Bruno Latour and Steve Woolgar, *Laboratory Life* (Beverly Hills, Calif.: Sage, 1979). See also Hans-Jörg Rheinberger, *Experiment, Differenz, Schrift* (Marburg: Basiliskenpresse, 1992). On the development of a modern science of signification see Hans Aarsleff, "Bréal, 'la sémantique,' and Saussure," in *From Locke to Saussure: Essays on the Study of Language and Intellectual History* (Minneapolis: University of Minnesota Press, 1982), pp. 382–400.

8. The discontinuity is recounted in Aarsleff, *From Locke to Saussure*. One of the few works by a historian of linguistics to grant importance to the phonetics laboratory is Sylvain Auroux, "La catégorie du *parler* et la linguistique," *Romantisme* (1976): 170–73.

9. Friedrich A. Kittler, *Discourse Networks 1800/1900*, trans. Michael Metteer with Chris Cullens (Stanford, Calif.: Stanford University Press, 1990).

10. Timothy Mitchell, *Colonising Egypt* (Cambridge, Eng.: Cambridge University Press, 1988). On the role of technology in nineteenth-century empire-building, see Daniel R. Headrick, *The Tools of Empire: Technology and European Imperialism in the Nineteenth Century* (Oxford: Oxford University Press, 1981).

11. Ian Hacking's claim that "almost the only item that you cannot wrest from another people by barter or victory is language," overlooks the more effective strategy of standardizing technologies. See his excellent article, "How, Why, When, and Where Did Language Go Public?" *Common Knowledge* 2 (Fall 1992): 77. I am grateful to Skuli Sigurdsson for calling my attention to this article. See also Jacques Attali and Yves Stourdze, "The Birth of the Telephone and Economic Crisis: The Slow Death of Monologue in French Society," in *The Social Impact of the Telephone*, ed. Ithiel Pool (Cambridge, Mass.: MIT Press, 1977), pp. 97–111. On the role of communications media in the formation of a standard Italian language, see T. De Mauro, *Storia linguistica dell'Italia unita* (Bari: Laterza, 1970).

12. Eugen Weber, *Peasants into Frenchmen: The Modernization of Rural France, 1870–1914* (London: Chatto & Windus, 1979); Stephen Kern, *The Culture of Time and Space, 1880–1918* (Cambridge, Mass.: Harvard University Press, 1983). On scientific research in the French provinces during the same period, see Mary Jo Nye, *Science in the Provinces: Scientific Communities and Provincial Leadership in France, 1860–1930* (Berkeley and Los Angeles: University of California Press, 1986).

13. Aarsleff, "Bréal," p. 393.

14. The linguist Roy Harris claims Saussure's chapter on *La valeur linguistique* is "arguably, the most important single chapter in the whole of the *Cours*." See Roy Harris, *Reading Saussure: A Critical Commentary on the "Cours de linguistique générale"* (London: Gerald Duckworth, 1987), p. 118.

15. Etienne-Jules Marey, *La méthode graphique en sciences expérimentales* (Paris: G. Masson, 1878), pp. 390–98.

16. On Marey, see François Dagognet, *Etienne-Jules Marey: La Passion de la trace* (Paris: Hazan, 1987); and Martha Braun, *Picturing Time: The Work of Etienne-Jules Marey* (Chicago: University of Chicago Press, 1993).

17. On German physiologists and graphic methods, see Kathryn Olesko and Frederic L. Holmes, "Experiment, Quantification, and Discovery: Helmholtz's Early Physiological Researches, 1843–50," in *Hermann von Helmholtz and the Foundations of Nineteenth-Century Science*, ed. David Cahan (Berkeley and Los Angeles: University of California Press, 1993), pp. 50–108; Soraya de Chadarevian, "Graphical Method and Discipline: Self-Recording Instruments in Nineteenth-Century Physiology," *Studies in the History and Philosophy of Science* 24, no. 2 (1993): 267–91; Timothy Lenoir, "Helmholtz and the Materialities of Communication," *Osiris* 9, special issue (1994): 185–207.

18. See Gustav Le Bon, "La méthode graphique dans l'Exposition," in *Études sur l'Exposition de 1878: Annales et archives de l'industrie au XIX siècle*, ed. Paul Lacroix, 8 vols. (Paris: Lacroix, 1881), vol. 7; and G. Stanley Hall, "The Graphic Method," *The Nation* 29 (1879): 238–39.

19. Auroux, p. 169.

20. On the decline of monologue and the rise of the new circuitous communicational economy of the telephone at this very moment, see Attali and Stourdze.

21. Michel Bréal, *Essai de sémantique*, 5th ed. (Paris: Hachette, 1911), pp. 255–56. On the impact of the telegraph on the commodification of information in the nineteenth-century United States see JoAnne Yates, *Control Through Communication: The Rise of System in American Management* (Baltimore: Johns Hopkins University Press, 1989), pp. 21–36; Menahem Blondheim, *News Over the Wires: The Telegraph and the Flow of Public Information in America, 1844–1897* (Cambridge, Mass.: Harvard University Press, 1994).

22. On the calibration of international electrotechnical standards, see Simon Schaffer, "Late Victorian Metrology and Its Instrumentation: A Manufactory of Ohms," in *Invisible Connections: Instruments, Institutions, and Science*, ed. Susan Cozzens and Robert Bud (Bellingham, Wash.: SPIE Optical Engineering Press, 1992), pp. 23–56; and David Cahan, *An Institute for an Empire: The Physikalisch-Technische Reichsanstalt, 1871–1918* (Cambridge, Eng.: Cambridge University Press, 1979).

23. Simon Schaffer, "The Laboratory Measurement of Modern Values" (Cambridge University, n.d.), takes up the role of the laboratory as a mediator of important themes of cultural modernism.

24. On the centrality of philology in the constitution of the German research university, see William Clark, "From the Medieval *Universitas Scholarium* to the German Research University: A Sociogenesis of the German Academic" (Ph.D. diss., University of California, Los Angeles, 1986). On nineteenth-century Ger-

man philology, see Amsterdamska, esp. chap. 3, "Linguistics at the German University," pp. 63–89.

25. Renan wrote, "They say that the victor at Sadowa was the [Prussian] school teacher. Not at all, the victor in that battle was German science." Cited in Antoine Prost, *L'Histoire de l'enseignement en France* (Paris: A. Colin, 1968).

26. For a study of the contributors and content of the principal organ of the Société, see Piet Desmet, "La *Revue de linguistique et de philologie comparée* (1867–1916)—Organe de la linguistique naturaliste en France," *Beiträge zur Geschichte der Sprachwissenschaft* 4, no. 1 (1994): 49–80.

27. For a study of Bernard's epistemology of relevance to the present comparison with Bréal, see P. Q. Hirst, *Durkheim, Bernard, and Epistemology* (London: Routledge & Kegan Paul, 1975).

28. On Bernard's discipline-building efforts, see William Coleman, "The Cognitive Basis of the Discipline: Claude Bernard on Physiology," *Isis* 76 (1985): 49–70.

29. For an account of his experiences, see Michel Bréal, *Excursion pédagogique: Un voyage scolaire en Allemagne* (Paris: Hatchette, 1882). A brief biographical sketch of Bréal is contained in the translator's preface to Michel Bréal, *The Beginnings of Semantics: Essays, Lectures, and Reviews*, ed. and trans. George Wolf (London: Duckworth, 1991), pp. 3–4.

30. François Bopp, *Grammaire comparée des langues indo-europeenes comprenant le sanscrit, le zend, l'armenien, le grec, le latin, le lithuanien, l'ancien slave, le gothique et l'allemand*, trans. Michel Bréal, 2 vols. (Paris: Imprimerie Nationale, 1966).

31. Michel Bréal, ed., *Collection philologique: Recueil de travaux originaux ou traduits relatifs à la philologie et à l'histoire littéraire, avec un avant-propos de M. Michel Bréal—Premier fascicule. La Théorie de Darwin;—De l'importance du langage pour l'histoire naturelle de l'homme, par A. Schleicher* (Paris: A. Franck, 1867).

32. Dietmar Rothermund, *The German Intellectual Quest for India* (New Delhi: Manihar, 1986).

33. Michel Bréal, "Introduction," to Bopp, p. xxiv.

34. Ibid., pp. viii–x.

35. On phonetic self-experiment, see Kittler, pp. 25–69. On Romantic self-experiment more generally, see Simon Schaffer, "Self-Evidence," *Critical Inquiry* 18 (1992), esp. pp. 359–62; and Michael Dettelbach, "Romanticism and Administration: Mining, Galvanism, and Oversight in Alexander von Humboldt's Global Physics" (Ph.D. diss., Cambridge University, 1992), pp. 82–94.

36. Bréal, "Introduction," pp. x; see also Ruth Römer, *Sprachwissenschaft und Rassenideologie in Deutschland* (Munich: Fink, 1985).

37. Bréal, "Introduction," p. viii.

38. Michel Bréal, "The Science of Language" (1879), reprinted in *Beginnings of Semantics*, p. 128.

39. Bréal frequently used this fad as an example against phonetic laws. See

Michel Bréal, "Le langage et les nationalités," *Revue des deux mondes* 6 (1891): 622–23; and Bréal, "Lois phoniques," p. 3.

40. See Hans Aarsleff, "Bréal vs. Schleicher: Reorientations in Linguistics During the Latter Half of the Nineteenth Century," in *From Locke to Saussure*, pp. 293–335. On the relation of Schleicher's work to that of his friend Haeckel, see E. F. K. Koerner, "Schleichers Einfluß auf Haeckel: Schlaglichter auf die wechselseitige Abhängigkeit zwischen linguistischen und biologischen Theorien im 19. Jahrhundert," *Zeitschrift für vergleichende Sprachforschung* 95 (1981): 3–21.

41. August Schleicher, *Darwinism Tested by the Science of Language*, trans. A. V. W. Bikkers (London: 1869), pp. 20–21. Translation of *Die Darwinsche Theorie und die Sprachwissenschaft* (1863).

42. Bréal consistently opposed all vestiges of organicism. In a later work, after having compared words to telegraphic signals, Bréal continued, "to say that language is an organism is to obscure things and to throw a seed of error into minds. One could just as well say that writing is an organism. I have avoided comparisons with botany, physiology, geology, with the same care that others have shown in seeking them out. As a consequence, my exposition is more abstract, but I think I can say it is truer." See Bréal, *Essai de sémantique*, p. 256. In this connection, Aarsleff points out that both Bréal and Saussure drew upon architecture to illustrate structure, form, and interrelationships. Aarsleff, "Bréal," p. 397. On the interaction between architecture and language in French cultural forms, see Philippe Hamon, *Expositions: Literature and Architecture in Nineteenth-Century France*, trans. Katia Sainson-Frank and Lisa Maguire (Berkeley and Los Angeles: University of California Press, 1993).

43. Gaston Paris, review of Bréal, ed., *Collection philologique*, in *Revue critique d'histoire et de littérature* 42 (Oct. 17, 1868): 242.

44. Excerpts from Duruy's initiative have been published in Michel de Certeau, Dominique Julia, and Jacques Revel, *Une politique de la langue: La révolution française et le patois* (Paris: Gallimard, 1975), pp. 270–72. Eugen Weber records and analyzes the results in *Peasants into Frenchmen*, pp. 67–94, 498–501.

45. In addition to the excellent analysis of these events furnished by De Certeau, Julia, and Revel, see in their appendix (pp. 300–17) Gregoire's report to the national convention, "Sur la necessité et les moyens d'anéantir les patois et d'universaliser l'usage de la langue française."

46. Gabriel Bergouinioux, "La science du langage en France de 1870 à 1885: Du marché civil au marché étatique," *Langue française* 63 (1984): 14. Antoine Meillet claimed that Saussure's system was largely worked out during these years: "It is well known that Saussure's thought was formed very early. The doctrines he explicitly taught in his course in general linguistics were those that already distinguished his teaching at the École des Hautes Études when I was a student." See Antoine Meillet, review of Saussure, in *Bulletin de la Société de linguistique de Paris* 20 (1916): 32–36.

47. Martin Bernal, *Black Athena: The Afroasiatic Roots of Classical Civilization,* Vol. 1, *The Fabrication of Ancient Greece, 1785–1985* (London: Vintage, 1987), pp. 224–33.

48. Bergouinioux, pp. 25–26; George Weisz, "La réforme de l'enseignement supérieur sous la IIIe République," in Alain Drouard, *Analyse comparative des processus de changement et des mouvements de réforme de l'enseignement supérieur français* (Paris: Editions du Centre national de la recherche scientifique, 1978).

49. Fustel de Coulanges, cited in Bergouinioux, p. 36.

50. Ernest Renan, *Nouvelle lettre à M. Strauss* (1871), cited in H. Peyre, *Renan* (Paris: Presses Universitaires de France, 1969), p. 406. On the background to Renan's views, see also Lia Formigari, "Théories du langage et théories du pouvoir en France, 1800–1848," *Historiographica Linguistica* 12, no. 1/2 (1985): 63–83; G. Pflug, "Ernest Renan und die deutsche Philologie," in *Philologie und Hermeneutik im 19. Jahrhundert,* ed. Mayotte Bollack and Heinz Wismann (Göttingen: Vandenhoeck & Ruprecht, 1983), pp. 156–77; Maurice Olender, *The Languages of Paradise: Race, Religion, and Philology in the Nineteenth Century,* trans. Arthur Goldhammer (Cambridge, Mass.: Harvard University Press, 1992).

51. Gaston Paris, *Leçon d'ouverture du cours de grammaire historique de la langue française professé à la Sorbonne en 1867* (Paris: Franck, 1867).

52. Gaston Paris, *Mélanges linguistiques* (Paris: Société Amicale de Gaston Paris, 1905), pp. 28–31.

53. On the moral authority of self-registering instruments, see Schaffer, "Self-Evidence," p. 362.

54. Jean-Marc-Gaspard Itard, *The Wild Boy of Aveyron* (New York: Century, 1932); J. M. de Gerando, *De l'éducation des sourds-muets de naissance* (Paris: Mequignon, 1827).

55. The legacy of Itard's labors for deaf-mute education is taken up in Harlan Lane, *The Wild Boy of Aveyron* (London: George Allen & Unwin, 1977).

56. Leon Vaïsse, "Discours du Président," *Bulletin de la Société de linguistique de Paris* 1–5 (1869–85): 151.

57. Ibid., p. 153.

58. Vaïsse was one of the most outspoken defenders of oral methods in teaching deaf-mutes. After a heated exchange at a conference in 1878, one of his allies took the podium and shouted "Vive la parole!" See Lane, p. 252.

59. Vaïsse, "Discours," p. 153.

60. For an account of how thermodynamics upset the notion of a rigid separation of animate and inanimate nature and thereby created a new framework for thinking about communicative codes, see François Jacob, *The Logic of Life: A History of Heredity,* trans. Betty E. Spillman (New York: Pantheon Books, 1973), p. 194.

61. L. Vaïsse, "Notation des sons du langages," *Bulletin de la Société de linguistique de Paris* 1–5 (1869–85): 155.

62. On the Parisian medical debates surrounding Marey's demonstration, see Jules Gavarret, *Sur la théorie des mouvements du coeur* (Paris: J.-B. Bailliere & fils, 1864). Marey's most general account of his cardiography was given in *Physiologie médicale de la circulation du sang basée sur l'étude graphique des mouvements du coeur et du pouls artériel, avec application aux maladies de l'appareil circulatoire* (Paris: Delahaye, 1863). On Marey's work in the context of cardiology, see Robert G. Frank, Jr., "The Telltale Heart: Physiological Instruments, Graphic Methods, and Clinical Hopes, 1854–1914," in *The Investigative Enterprise: Experimental Physiology in Nineteenth-Century Medicine,* ed. William Coleman and Frederic L. Holmes (Berkeley and Los Angeles: University of California Press, 1988), pp. 211–90.

63. Marey, *Méthode graphique,* p. vi.

64. Leon Vaïsse, "Notes pour servir à l'histoire des machines parlantes," *Mémoires de la Société de linguistique de Paris* 3 (1878): 257–68.

65. See Adolphe Regnier, *Études sur la grammaire védique: Pratiçakya du Rig-Véda* (Paris: Imprimerie Impériale, 1859).

66. W. S. Allen later remarked upon the mingling of prescription with description in the aphorisms of the *Pratisakhya*: "Their avowed purpose is to preserve the oral tradition of the sacred texts: to this end the direst penalties are threatened for mispronunciation, including descent to the hell of Kumbhipaka; the competent pupil, on the other hand, is encouraged by verses such as that which closes the *Tattiriya-Pratisakhya*—'He who knows the distinctions of tone and length may go and sit with the professors.'" See W. S. Allen, *Phonetics in Ancient India* (London: Geoffrey Cumberlege, 1953), p. 6.

67. For a twentieth-century discussion of the *yamas,* see Ibid., pp. 75–78.

68. The term *phoneme* was first used in conjunction with these experiments. First used in a paper delivered to the Société de linguistique by the philologist A. Dufriche-Desgenettes as an alternative to the inelegant *son du langage,* it was taken up by Louis Havet, who is credited by historians of phonetics with authoring the term and disseminating it through the auspices of the Association phonétique internationale, which he helped inaugurate in 1886. See Jiri Kramsky, *The Phoneme: Introduction to the History and Theories of a Concept* (Munich: Wilhelm Fink, 1974), p. 21; David Jones, *The History and Meaning of the Term "Phoneme"* (London: International Phonetic Association, 1957).

69. Charles Rosapelly, "Inscription de mouvements phonétiques," *Travaux du laboratoire de M. Marey* (Paris: G. Masson, 1875), p. 113. For other attempts to apply the measure of machines to different aspects of Indian culture in the period, see Michael Adas, *Machines as the Measure of Men: Science, Technology, and Ideologies of Western Dominance* (Ithaca, N.Y.: Cornell University Press, 1989).

70. Rosapelly, p. 110.

71. For a fascinating account of this principle in the work of Wilhelm Wundt, Adolf Fick, and especially Hermann Helmholtz, see Timothy Lenoir, "The Eye as Mathematician: Clinical Practice, Instrumentation, and Helmholtz's

Construction of an Empiricist Theory of Vision," in *Hermann von Helmholtz*, ed. Cahan, pp. 109–53.

72. Rosapelly, p. 112.

73. Timothy Lenoir, "Helmholtz and the Materialities of Communication."

74. Auroux, p. 169.

75. Etienne-Jules Marey, "Inscription des phénomènes phonétiques d'après les travaux de divers auteurs," *Journal des savants* 20 (Oct. 1897): 564. A twentieth-century discussion of the *yamas* can be found in Allen, pp. 75–78.

76. Rosapelly, pp. 129–30.

77. Wilbur A. Benware, *The Study of Indo-European Vocalism in the Nineteenth Century* (Amsterdam: John Benjamins, 1974), pp. 13–14. See also Ferdinand de Saussure, "Mémoire sur le système primitif des voyelles dans les langues indo-europeenes" (Ph.D. diss., University of Leipzig, 1878).

78. For a comprehensive survey of acoustic instruments in the mid-1860s, including those used to inscribe the voice, see Franz-Joseph Pisko, *Die neureren Apparate der Akustik* (Vienna: Carl Gerhold's Sohn, 1865).

79. Leon Scott de Martinville, *La fixation graphique de la voix* (Paris: Société d'encouragement, 1857), p. 1. On the growth of stenographic instruments in the United States, see Yates, pp. 36–45.

80. Leon Scott de Martinville, *Histoire de la sténographie depuis les temps anciens jusqu'à nos jours* (Paris: C. Tondeur, 1849).

81. Franziskus Cornelius Donders, "Zur Klangfarbe der Vocale," *Annalen der Physik und Chemie* 73 (1868).

82. Rudolph Koenig, *Quelques expériences d'acoustique* (Paris: A. Blanchard, 1882). Koenig won gold medals for his apparatus at both the International Exhibition of 1862 in London and at the Philadelphia Centennial Exposition of 1876. For further description, see Koenig, *Catalogue des appareils d'acoustique* (Paris, 1882); and Gerard L'E. Turner, *Nineteenth-Century Scientific Instruments* (London/Berkeley and Los Angeles: Sotheby Publications/University of California Press, 1983), pp. 142–46.

83. See Ludimar Hermann, "Phonophotographische Untersuchungen," *Archiv für die gesammte Physiologie* 45 (1889): 582; Hermann, "Über das Verhalten der Vokale am neuen Edison'schen Phonographen," *Archiv für die gesammte Physiologie* 47 (1890): 44; Ph. Wagner, "Über die Verwendung des Grützner-Marey'schen Apparats u.d. Phonographen zu phonetischen Untersuchungen," *Phonetische Studien* 4 (1891): 68.

84. H. Marichelle, *La parole d'après le tracé du phonographe* (Paris: Delagrave, 1897).

85. Georges Demeny, "Analyse des mouvements de la parole par la chronophotographie," *Comptes rendus de l'Académie des Sciences de Paris* 113 (1891): 216–17.

86. Braun, pp. 180–82.

87. Marey, p. 582.

88. Yates, p. 45.

89. Louis Couturat, *Les nouvelles langues internationales* (Paris: s.n., 1907); Otto Jespersen, *Phonetische Grundfragen* (Leipzig: B. G. Teubner, 1904); John G. McKendrick, "Experimental Phonetics," *Nature* 65 (1901): 1678.

90. See the review of a course at the École des Hautes Études entitled "The Method of Dialectological Information," in *Revue des patois gallo-romans* 2 (1887): 20.

91. Eduard Koschwitz, "La phonétique expérimentale et la philologie Franco-Provençal," *Revue des patois gallo-romans* 4 (1891): 215.

92. The rapid disappearance of the patois can be attributed in part to the previous attempts to eradicate them, as well as to the zealous program of public education in the Third Republic. Already in 1887, the Abbé Rousselot wrote, "The patois are disappearing and it is already late to recover them." *Revue des patois gallo-romans* 1 (1887): 3.

93. Lucien Adam, *Les patois lorrains* (Paris: Maisonneuve, 1881); and Lucien Adam, *Les classifications, l'objet, la méthode, les conclusions de la linguistique* (Paris: Maisonneuve, 1882).

94. Koschwitz, "Phonétique expérimentale," p. 227.

95. Hubert Pernot, "L'Abbé Rousselot (1846–1924)," *Revue de l'Alliance française* 22 (1925): 135.

96. Jean-Pierre Rousselot, *Les modifications phonétiques du langage étudié dans le patois d'une famille de Cellefrouin* (Paris: H. Welter, 1891); and Rousselot, *La méthode graphique appliquée à la phonétique* (Paris: Macon, 1890).

97. Rousselot, *Modifications*, p. 2.

98. Ibid., p. 23.

99. In his treatise on Hebrew grammar, Spinoza expressed it thus: "vowels are not letters, but their 'soul,' and letters without vowels, or consonants, are the 'bodies without soul.'" Cited in De Certeau, Julia, and Revel, pp. 97–98.

100. Ibid., p. 114.

101. Rousselot, *Modifications*, pp. 7–8. In the wake of Rousselot's enormous international success among linguists and ethnographers Verdin sold many examples of the kit, especially in Germany, for 1,724 francs. See Charles Verdin, *Catalogue des instruments de précision pour la physiologie et la médecine* (Paris: s.n., 1890); Eduard Koschwitz, "Der Registrierapparat," *Archiv für das Studium der neueren Sprachen* 88 (1892): 244–53.

102. Rousselot, *Modifications*, pp. 8–19.

103. Ibid., pp. 23–60.

104. Two celebrated examples of such works from the period are Rudolfo Livi, *Antropometria militare, risultati ottenuti dallo spoglio dei fogli sanitarii dei militari delle classi 1859–63,* 2 vols. (Rome: Presso il Giornale medico del regio esercito, 1896); John Beddoe, *The Races of Britain, A Contribution to the Anthropology of Western Europe* (Bristol: Arrowsmith, 1885).

105. Rousselot, *Modifications*, p. 147.

106. Ibid., p. 1.

107. Gillian Rose, *Feminism and Geography: The Limits of Geographical Knowledge* (Cambridge, Eng.: Polity Press, 1993).

108. Luce Irigary, "Sexual Difference," reprinted in *French Feminist Thought: A Reader*, by Toril Moi (Oxford: Blackwell, 1987), p. 122.

109. Rousselot, *Modifications*, p. 97.

110. The Abbé Gorse gave a similar account, in an anecdote recounted by Eugen Weber: " 'Parlez français,' says the master to the pupil. 'Monsieur, je parle comme je save et comme je poude.' The villager forgets a little of his mother tongue at school, and learns only a parody of French." See E. Weber, p. 67.

111. Rousselot, *Modifications*, p. 98.

112. Anonymous review of Rousselot, *Les modifications phonétiques du langage étudié dans le patois d'une famille de Cellefrouin*, and Paul Passy, *Étude sur les changements phonétiques et leurs caractères généraux*, in *Literaturblatt für germanische und romanische Philologie* 9 (1892): 314.

113. Hermann Breymann, *Die phonetische Literatur von 1876–1895* (Leipzig: Deichert, 1897), pp. 129–35. A list of reviews of Rousselot's work appears on p. 55.

114. Paul Azoulay, "L'ère nouvelle des sons et des bruits: Musées et archives phonographiques," *Bulletins et Mémoires de la Société d'anthropologie de Paris* 1 (1900): 172–78. This article was also published in *Revue scientifique* (1900): 712–15.

115. Quoted in McKendrick, p. 1678.

116. Azoulay, "L'ère nouvelle," p. 178.

117. Ibid., p. 176.

118. For anthropometric studies at the previous (1889) Paris exhibition, see J. Deniker and L. Laloy, "Les races exotiques à l'Exposition Universelle de 1889," *L'Anthropologie* 1 (1890): 257–94, 513–47.

119. Paul Azoulay, "Sur la constitution d'une musée phonographique," *Bulletin et Mémoires de la Société d'anthropologie de Paris* 1 (1900): 226.

120. Ibid., p. 222.

121. On the role of international exhibitions in the creation of international standards institutions, see Robert Brain, *Going to the Fair* (Cambridge, Eng.: Whipple Museum Publications, 1993).

122. Azoulay, "Sur la constitution," pp. 222–23.

123. On the Bertillon system of judicial identification, see Whipple Museum of the History of Science, *1900: The New Age. A Guide to the Exhibition* (Cambridge, Eng.: Whipple Museum Publications, 1994), pp. 75–78.

124. William H. Schneider, "Colonies at the 1900 World's Fair," *History Today* 31 (1981): 31–36.

125. For another attempt to place Saussure in an intellectual context, see Jacques Derrida, "The Linguistic Circle of Geneva," *Critical Inquiry* 8 (Summer 1992): 675–91.

126. Ferdinand de Saussure, *Course in General Linguistics*, trans. Roy Harris (London: Duckworth, 1983). All references to Saussure's text are to this edition.

127. Ibid., p. 33.

128. Ibid., p. 118.

129. Ibid., p. 120.

130. Rousselot made this argument in his "La méthode graphique appliquée à la recherche des transformations inconscientes du langage," *Comptes rendus du Congrès scientifique internationale des Catholiques* (1891), sect. S, pp. 109–12. See also Michael Hagner's very fine accounts of Broca's and Wernicke's work on the cerebral localization of language functions, an important component of Saussure's concept: Michael Hagner, "Vom Stottern des Menschen zum Stocken der Maschine," in *Im Zug der Schrift*, ed. Norbert Haas, Rainer Nägele, and Hans-Jörg Rheinberger (Munich: Wilhelm Fink, 1994), pp. 26–31; and Hagner, "Hirnbilder: Cerebrale Representätionen im 19. und 20. Jahrhundert" (paper presented at the 1994 Berlin Summer Academy).

131. See Harris, pp. 204–18.

132. Saussure, *Course*, pp. 110–20. See also Samuel Weber, "Saussure and the Apparition of Language: The Critical Perspective," *MLN* 91 (1976): 913–38.

133. Saussure, *Course*, p. 116.   134. Ibid., p. 13.

135. Ibid.   136. Ibid.

137. Emile Durkheim, *The Rules of Sociological Method*, trans. W. D. Halls (London: Macmillan, 1982), p. 51.

138. French engineers since the 1830s had spoken of graphic inscriptions in such terms, referring to them as *monnaie mécanique*. See J.-V Poncelet, *Cours de mécanique appliquée aux machines* (Paris: Gauthier-Villars, 1874), p. 2.

139. Saussure, *Course*, p. 110.

140. Bréal, "Science of Language," p. 133.

141. Saussure, *Course*, p. 115.   142. Ibid.

143. Ibid., p. 117.   144. Ibid., p. 115.

145. Harry Collins, *Artificial Experts: Social Knowledge and Intelligent Machines* (Cambridge, Mass.: MIT Press, 1990), pp. 22–29.

146. Saussure, *Course*, p. 18.

147. Ibid., p. 10.

148. See Derrida, "Linguistic Circle"; and *Of Grammatology*, trans. Gayatri Spivak (Baltimore: Johns Hopkins University Press, 1974).

13. Rheinberger, Graphematic Spaces

This essay was presented at the Summer Academy of the Rathenau Program on History of Science on "Writing Science," Berlin, July 1991. The essay partly relies on material published in Hans-Jörg Rheinberger, "Experiment, Difference, and Writing: I. Tracing Protein Synthesis," and "II. The Laboratory Production of Transfer RNA," *Studies in the History and Philosophy of Science* 23 (1992): 305–31, 389–422. I thank an anonymous reviewer for challenging but encouraging comments.

1. Ian Hacking, *Representing and Intervening* (Cambridge, Eng.: Cambridge University Press, 1983), pp. 149–50.

2. Peter Galison, *How Experiments End* (Chicago: University of Chicago Press, 1987).

3. Peter Galison, "History, Philosophy, and the Central Metaphor," *Science in Context* 2 (1988): 197–212, quoted from p. 211.

4. Andrew Pickering, ed., *Science as Practice and Culture* (Chicago: University of Chicago Press, 1992). See also, from varying perspectives, Karin Knorr Cetina, *The Manufacture of Knowledge* (Oxford: Pergamon Press, 1981); Harry M. Collins, *Changing Order* (London: Sage Publications, 1985); Steven Shapin and Simon Schaffer, *Leviathan and the Air-Pump: Hobbes, Boyle, and the Experimental Life* (Princeton: Princeton University Press, 1985); Allan Franklin, *The Neglect of Experiment* (Cambridge, Eng.: Cambridge University Press, 1986); Bruno Latour and Steve Woolgar, *Laboratory Life* (Princeton: Princeton University Press, 1986); David Gooding, Trevor Pinch, and Simon Schaffer, eds., *The Uses of Experiment* (Cambridge, Eng.: Cambridge University Press, 1989); David Gooding, *Experiment and the Making of Meaning* (Dordrecht: Kluwer, 1990); Homer E. Le Grand, ed., *Experimental Inquiries* (Dordrecht: Kluwer, 1990); Allan Franklin, *Experiment, Right or Wrong* (Cambridge, Eng.: Cambridge University Press, 1990).

5. Thomas Kuhn, *The Trouble with the Historical Philosophy of Science* (Cambridge, Mass.: Harvard University Press, 1992), p. 90.

6. Bruno Latour, "One More Turn After the Social Turn: Easing Science Studies into the Non-Modern World," in *The Social Dimensions of Science*, ed. Ernan McMullin (Notre Dame: Notre Dame University Press, 1992), pp. 272–94.

7. Bruno Latour, *Science in Action* (Cambridge, Mass.: Harvard University Press, 1987), pp. 132–34.

8. Jacques Derrida, "Structure, Sign, and Play in the Discourse of the Human Sciences," in *Writing and Difference* (Chicago: University of Chicago Press, 1978), pp. 278–93.

9. Gaston Bachelard, *The New Scientific Spirit* (1934; Boston: Beacon Press, 1984), p. 13.

10. Gaston Bachelard, *L'Activité rationaliste de la physique contemporaine* (Paris: Presses Universitaires de France, 1951), p. 84.

11. Bachelard, *New Scientific Spirit*, p. 6; the translation has been altered slightly.

12. Ludwig Wittgenstein, *Philosophical Investigations* (Oxford: Basil Blackwell, 1953), § 7, § 654.

13. The notion of "experimental system," as it is introduced here, primarily

focuses on research in biomedicine, biochemistry, biology, and molecular biology, and draws upon the use made of it in the everyday language of the scientist. See, among many references, François Jacob, *The Statue Within* (New York: Basic Books, 1988), p. 234. Only very recently have historians of science started to become aware of the historiographic potential of the idea of the experimental system. Robert Kohler, in dealing with Drosophila, Neurospora, and the rise of biochemical genetics, speaks of "systems of production." See Robert Kohler, "Systems of Production: Drosophila, Neurospora, and Biochemical Genetics," *Historical Studies in the Physical and Biological Sciences* 22 (1991): 87–130; Robert E. Kohler, *Lords of the Fly* (Chicago: University of Chicago Press, 1994). David Turnbull and Terry Stokes use the notion of "manipulable systems" in their analysis of malaria research at the Walter and Eliza Hall Institute of Medical Research in Melbourne. See David Turnbull and Terry Stokes, "Manipulable Systems and Laboratory Strategies in a Biomedical Institute," in Le Grand, pp. 167–92.

14. There is a simple reason for this restriction: We cannot claim that all scientific activity is locally constrained without accepting that this holds for our own activity, too.

15. Mahlon Hoagland, *Toward the Habit of Truth* (New York: W. W. Norton, 1990), p. xvii. Hoagland speaks of the "selection of a good system" as one key to success on the "itinerary into the unknown" (p. xvi).

16. George Kubler has labeled his thoughts about the temporal forms of artistic production as "Remarks on the History of Things." Under the aspect of formal sequences of works of art, experiments, tools, and technical constructs, Kubler notes: "The value of any rapprochement between the history of art and the history of science is to display the common traits of invention, change, and obsolescence that the material works of artists and scientists both share in time." See George Kubler, *The Shape of Time* (New Haven: Yale University Press, 1962), p. 10.

17. Bruno Latour, "The Force and the Reason of Experiment," in Le Grand, p. 66, emphasis added.

18. See, e.g., Karl Popper, *The Logic of Scientific Discovery* (New York: Harper & Row, 1968), p. 107.

19. Ludwik Fleck, *Genesis and Development of a Scientific Fact* (Chicago: University of Chicago Press, 1979), p. 96.

20. Ibid., p. 86.

21. Jacob, *Statue Within*, p. 9.

22. Bachelard, *New Scientific Spirit*, p. 139.

23. Paul C. Zamecnik, "The Use of Labeled Amino Acids in the Study of the Protein Metabolism of Normal and Malignant Tissues: A Review," *Cancer Research* 10 (1950): 659–67, quoted from p. 659.

24. Robert Loftfield, "Preparation of C^{14}-Labeled Hydrogen Cyanide, Alanine, and Glycine," *Nucleonics* 1 (1947): 54–57.

25. For a retrospective review, see Robert Loftfield, "The Biosynthesis of Protein," *Progress in Biophysics and Biophysical Chemistry* 8 (1957): 348–86.

26. Zamecnik, p. 663.

27. W. F. Loomis and Fritz Lipmann, "Reversible Inhibition of the Coupling Between Phosphorylation and Oxidation," *Journal of Biological Chemistry* 173 (1948): 807–8.

28. Ivan D. Frantz, Jr., Paul C. Zamecnik, John W. Reese, and Mary L. Stephenson, "The Effect of Dinitrophenol on the Incorporation of Alanine Labeled with Radioactive Carbon into the Proteins of Slices of Normal and Malignant Rat Liver," *Journal of Biological Chemistry* 174 (1948): 773–74.

29. Jacob, *Statue Within*, p. 274.

30. Ibid., p. 255.

31. François Jacob, *The Possible and the Actual* (Seattle: University of Washington Press, 1982). The French title is *Le jeu des possibles*.

32. Jacques Derrida, *Of Grammatology*, trans. Gayatri Spivak (Baltimore: Johns Hopkins University Press, 1974), pp. 23–24, and elsewhere.

33. Ibid., p. 24.

34. Fleck, p. 96.

35. Derrida, *Grammatology*, p. 23.

36. Jacques Derrida, "Une 'folie' doit veiller sur la pensée: Interview with François Ewald," *Magazine littéraire* (Mars 1991): 18–30, see pp. 26–27, my translation.

37. For a contemporary overview, see Albert Claude, "Studies on Cells: Morphology, Chemical Constitution, and Distribution of Biochemical Function," *Harvey Lectures* 43 (1950): 121–64.

38. Philip Siekevitz, "Uptake of Radioactive Alanine in Vitro into the Proteins of Rat Liver Fractions," *Journal of Biological Chemistry* 195 (1952): 549–65.

39. "Our position is that representations and objects are inextricably interconnected; that objects can only be 'known through representation.'" Michael Lynch and Steve Woolgar, eds., *Representation in Scientific Practice* (Cambridge, Mass.: MIT Press, 1990), p. 13.

40. The recipe came from a neighboring lab of the Huntington Memorial Hospital and had been developed for quite a different purpose. Nancy L. R. Bucher, "The Formation of Radioactive Cholesterol and Fatty Acids from C^{14}-Labeled Acetate by Rat Liver Homogenates," *Journal of the American Chemical Society* 75 (1953): 498.

41. The instrument was introduced into the system in 1953.

42. There is a growing body of literature on the problem of representation in science. For an overview, see Lynch and Woolgar. See also George Levine, ed., *Realism and Representation* (Madison: University of Wisconsin Press, 1993).

43. "We cannot see reality in its positivity. We can only feel it through iso-

morphic constraints operating upon competing local representations." N. Katherine Hayles, "Constrained Constructivism: Locating Scientific Inquiry in the Theater of Representation," in Levine, pp. 27–43, quotation on p. 33.

44. Latour and Woolgar, p. 51; see also Latour, *Science in Action*, pp. 64–70.

45. Marvin Lamborg and Paul C. Zamecnik, "Amino Acid Incorporation into Protein by Extracts of E. Coli," *Biochimica et Biophysica Acta* 42 (1960): 206–11.

46. Hacking, p. 136.

47. Jean Baudrillard, *Simulations* (New York: Semiotext(e), 1983), pp. 31–32.

48. Harry Collins has described this situation lucidly. But because he still wants to break the circle of the "experimenter's regress," he gives "society" the power to do so. See Collins.

49. Compare notes 40 and 41. The first fractionations by ultracentrifugation from this laboratory were reported in Elizabeth B. Keller and Paul C. Zamecnik, "Anaerobic Incorporation of C^{14}-Amino Acids into Protein in Cell-Free Liver Preparations," *Federation Proceedings* 13 (1954): 239–40. Compare also Paul C. Zamecnik and Elizabeth B. Keller, "Relation Between Phosphate Energy Donors and Incorporation of Labeled Amino Acids into Protein," *Journal of Biological Chemistry* 209 (1954): 337–54.

50. In 1952, Strittmatter and Ball from the Harvard Medical School had found that deoxycholate rapidly clarified a microsome suspension—it solubilized aggregated material, presumably the lipoproteins. Cornelius F. Strittmatter and Eric G. Ball, "A Hemochromogen Component of Liver Microsomes," *Proceedings of the National Academy of Sciences of the United States of America* 38 (1952): 19–25. Initially, deoxycholate had been used for disrupting bacterial cells and for extracting macromolecular components. Avery, MacLeod, and McCarty had used it for isolating their "transforming agent."

51. George E. Palade, "A Small Particulate Component of the Cytoplasm," *Journal of Biophysical and Biochemical Cytology* 1 (1955): 59–68.

52. Mary L. Petermann, Nancy A. Mizen, and Mary G. Hamilton, "The Macromolecular Particles of Normal and Regenerating Rat Liver," *Cancer Research* 13 (1953): 372–75; Mary L. Petermann, Mary G. Hamilton, and Nancy A. Mizen, "Electrophoretic Analysis of the Macromolecular Nucleoprotein Particles of Mammalian Cytoplasm," *Cancer Research* 14 (1954): 360–66.

53. In paraphrasing the adequation theory of truth, Latour has called this procedure, in his wonderful Kitchen Latin, *adaequatio laboratorii et laboratorii*. Bruno Latour, *The Pasteurization of France* (Cambridge, Mass.: Harvard University Press, 1988), p. 227.

54. Derrida, *Grammatology*, p. 145.

55. Brian Rotman, *Signifying Nothing: The Semiotics of Zero* (New York: St. Martin's Press, 1987), p. 102.

14. Doyle, Artificial Life

1. Marcello Barbieri, in *The Semantic Theory of Evolution* (New York: Harwood Academic Publishers, 1985) points out that "synthetic biology," in terms of producing lifelike objects, actually goes back at least to 1907: "Stephane Le Duc's 'The Mechanism of Life' featured a group of mushrooms, a colony of algae and a cell undergoing mitosis. In fact they were all inorganic artefacts that Le Duc had created in saturated solutions of potash with dyes, phosphates, chlorides and other salts" (p. 89).

2. Christopher G. Langton, "Artificial Life," in *Artificial Life: The Proceedings of an Interdisciplinary Workshop on the Synthesis and Simulation of Living Systems*, ed. Christopher G. Langton (Redwood City, Calif.: Addison-Wesley, 1989), p. 1.

3. F. Jacob and J. Monod, "Genetic Regulatory Mechanisms in the Synthesis of Proteins," *Journal of Molecular Biology* 3 (1960): 354.

4. Langton, "Artificial Life," p. 2. Of course, the idea that the diversity of life somehow constitutes a "single example" is itself historically constituted. See Michel Foucault, *The Order of Things* (London: Routledge, 1970), p. 128.

5. Friedrich Nietzsche, *Twilight of the Idols* (Harmondsworth: Penguin, 1990), p. 30.

6. Langton, *Artificial Life*, p. xiii.

7. Foucault, p. 269.

8. Ibid., p. 161.

9. C. H. Waddington, ed., *Towards a Theoretical Biology* (Chicago: Aldine Publishing Co., 1972), 4:289.

10. See Niels K. Jerne, "The Generative Grammar of the Immune System," *Scandinavian Journal of Immunology* 38, no. 1 (1993): 2–9.

11. See Richard M. Doyle, *On Beyond Living: Rhetorical Transformations of the Life Sciences* (Stanford, Calif.: Stanford University Press, 1997).

12. Christopher Longuet-Higgins, "The Seat of the Soul," in Waddington, 2:236–41.

13. Andrew Hodges, *Alan Turing: The Enigma* (New York: Simon & Schuster, 1983), pp. 102–3.

14. Longuet-Higgins, p. 241. Of course, what becomes important for our purposes is the overlooking of the platform for the tape; as in molecular biology's persistent "overlooking" or forgetting of the body, accounts of Turing machines often forget that tapes must be read by some agent, mechanical or otherwise.

15. Ibid., p. 236.

16. H. Atlan and M. Koppel, "The Cellular Computer DNA—Program or Data," *Bulletin of Mathematical Biology* 52, no. 3 (1990): 335.

17. Jacob and Monod, p. 354.

18. Atlan and Koppel, p. 346.

19. See Doyle.

20. On the one-way relationship of the analogy between machines and organisms, historian of science George Canguilhem writes: "The relationship between machine and organism has generally been studied in only one way. Nearly always, the organism has been explained on the basis of a preconceived idea of the structure and functioning of the machine; but only rarely have the structure and function of the organism been used to make the construction of the machine itself more understandable." George Canguilhem, "Machine and Organism," in *Incorporations*, ed. Jonathan Crary and Sanford Kwinter (New York: Zone, 1992), pp. 45–65, quotation from p. 45.

21. Claude Shannon, "The Band Wagon," *IRE Transactions* IT-2 (1956): 2–3.

22. By contrast, the history of automata—as opposed to computer life—is full of these reversals in which the model displaces the "original" as the object of study. Vaucanson, for example, had plans to produce "an automatic figure whose motions will be an imitation of all animal operations, such as the circulation of the blood, respiration, digestion, the movement of muscles, tendons, nerves and so forth. He claims that by using this automaton we shall be able to carry out experiments on animal functions, and to draw conclusions from them which will allow us to recognize the different states of human health, in order to remedy his ills." Of course, no claims were made that this automata lived—this would have to wait for the modern notion of life outlined above. See Jean-Claude Beaune, "The Classical Age of Automata: An Impressionistic Survey from the Sixteenth to the Nineteenth Century," in *Fragments for a History of the Human Body, Part One*, ed. Michael Feher (New York: Urzone, 1989), pp. 431–80.

23. One of Freud's paradigmatic examples of the uncanny (*unheimlich*) is the inability to distinguish between living and nonliving entities. See Sigmund Freud, "The Uncanny," vol. 17 in *The Standard Edition of the Complete Psychological Works of Sigmund Freud*, trans. Strachey (New York: Norton, 1976).

24. Karl Sigmund, *Games of Life: Explorations in Ecology, Evolution, and Behaviour* (Oxford: Oxford University Press, 1993), p. 10.

25. Ibid.

26. Anthropologist Stefan Helmreich, who has studied the community of A-life researchers at the Santa Fe Institute, confirms this observation (personal communication). See his "Replicating Reproduction in Artificial Life: or, The Essence of Life in the Age of Virtual Electronic Reproduction," in *Reproducing Reproduction*, ed. Sarah Franklin and Helena Ragone (forthcoming).

27. Brian Rotman, "Toward a Semiotics of Mathematics," *Semiotica* 72 (1988): 15.

28. Beaune, p. 435.

29. David Lavery, *Late for the Sky: The Mentality of the Space Age* (Carbondale: Southern Illinois University Press, 1992).

30. Langton, "Artificial Life," p. 2.

31. Ibid., p. 20.

32. Remarkably, Freeman Dyson's plans for an Artificial-Life mission to Mars relies on the same Icarus imagery, this time with Icarus planning on the big meltdown: "Dyson turned his imagination to the cosmos and proposed a self-reproducing automaton sent to the snow-covered Saturnian moon Enceladus. In his vision, this particular machine would draw on the distant sun's energy to create factories that produced a long stream of solar-powered sailboats, each carrying a block of ice. The sailboats would head toward Mars, and the fiery ride into the Martian atmosphere would melt the ice blocks." From Steven Levy, *Artificial Life: The Quest for a New Creation* (New York: Pantheon Books, 1992), p. 33.

33. Langton, "Artificial Life," p. 2.

34. Ibid.

35. As quoted in Ed Regis, *The Great Mambo Chicken and the Transhuman Condition: Science Slightly Over the Edge* (Reading, Mass.: Addison-Wesley, 1990), p. 192.

36. Donna Haraway invests microelectronics with this capability: "Micro-electronics is the technical basis of simulacra; that is, copies without originals." See Donna Haraway, "A Cyborg Manifesto: Science, Technology, and Socialist-Feminism in the Late Twentieth Century," in *Simians, Cyborgs, and Women: The Reinvention of Nature* (New York: Routledge, 1991), p. 165. Here I want to highlight the linguistic artifacts necessary to produce the effects of such simulation, what Beaune refers to as the "language of the technostructure." Although microelectronics are themselves "written" artifacts, it is also true that they are limited by their rhetorical softwares, textual artifacts that make possible the explication of the simulation and produce the effect of "originality" or "life." These softwares themselves reach their limit at both the limits of the hardware and the limits of wetwares, the threshold at which the rhetorical software becomes inarticulate, disjointed, unable to explicate anything but its own inadequacy. The problematic of "definition" in artificial life is one such threshold.

37. Howard Pattee sums this up well when he writes "The fact that human thought can be simulated by computation is treated as evidence in support of the Physical Symbol System Theory. But since virtually everything can be simulated by a computer, it is not really evidence for the theory at all." Similarly, we have no evidence from A-life for a general theory concerning living systems; the simulations of life tell us more about the capabilities of computers (and their operators) than about the formal attributes of all living systems. See H. H. Pattee, "Simulations, Realizations, and Theories of Life" in Langton, ed., *Artificial Life*, p. 67.

38. Jean Baudrillard, *Simulations* (New York: Semiotext(e), 1983), p. 146.

39. Richard Dawkins, *The Blind Watchmaker* (London: Penguin Books, 1991), p. 206. A further twist on this logic of the simulacrum occurred in central Pennsylvania, at a store that offered for sale "Genuine Xerox Copies."

40. Baudrillard, pp. 5–6. Note that the limit of this argument arrives the moment one attempts to fake death, and medicine becomes the arbiter of "true"

death, a decision that seems to take on a higher degree of arbitrariness in the age of postvitality.

41. Ibid., p. 3.

42. Ibid., p. 7.

43. Ibid., p. 9.

44. Waddington, 2:75.

45. Levy, p. 95.

46. Ibid., p. 58.

47. Lavery, p. 112.

48. Dawkins, p. 111.

49. Tom Ray, Tierra abstract, 1992. Available by anonymous ftp at: ftp://alife.santafe.edu/pub/SOFTWARE/Tierra/doc.

50. Atlan and Koppel, p. 346.

51. See, for example, D. H. Adams, "Self-Organization and Living Systems: Is DNA an Artificial Intelligence?" *Medical Hypotheses* 29, no. 4 (1989): 223–30, in which Adams writes: "If in fact the nature of life is embodied in a unique molecular electronic structure of DNA, this would suggest that any extraterrestrial life would probably resemble the forms occurring on this plant [*sic*]." Here we can see the way the notion that DNA is formally a "program" or "artificial intelligence" argues for the universality of form, if not substance, in the processes of life.

52. Beaune, p. 437.

53. Hans Moravec, *Mind Children: The Future of Robot and Human Intelligence* (Cambridge, Mass.: Harvard University Press, 1988), p. 4.

54. Michel Serres, *The Parasite* (Baltimore: Johns Hopkins University Press, 1982), p. 37.

55. From a grant proposal available online at ftp://alife.santafe.edu/pub/SOFTWARE/Tierra/doc/proposal.tex.

56. Thus A-life researchers do not simply "construct" A-life creatures; A-life organisms are real entities that use up energy and space and are not simply the result of human will. Like viruses, however, they require a host, and A-lifers are such hosts, and as usual, the host is transformed in the process. Thus, following traditional usage in which viruses are only alive when they have colonized a cell, A-life creatures are never in themselves alive; it is only through the ecology of wetwares, softwares, and hardwares that such emergent phenomena occur. Thus, when A-life organisms are alive, they are not, strictly speaking, artificial, as they include corporeal traces of organic elements, humans. When they are not networked with humans, they are not alive, but they are artificial. In this sense the A-life creature is *beyond living*.

57. From a grant proposal available online at ftp://alife.santafe.edu/pub/SOFTWARE/Tierra/doc/proposal.tex.

58. Nietzsche, p. 75.

15. Bloom, Science and Writing

1. E. L. Atkinson, "The Finding of the Dead," in *Scott's Last Expedition: The Personal Journal of Captain R. F. Scott, R.N., C.V.O., on His Last Journey to the*

South Pole, by R. F. Scott, with a biographical introduction by J. M. Barrie (New York: Dodd, Mead, 1923), p. 467.

2. For critical approaches on masculinities from a feminist viewpoint, see: Donna Haraway, "Teddy Bear Patriarchy: Taxidermy in the Garden of Eden, New York City, 1908–1936," *Social Text* 11 (Winter 1984–85): 20–64; Susan Jeffords, " 'Things Worth Dying For': Gender and the Ideology of Collectivity in Vietnam Representation," *Cultural Critique* 8 (Winter 1987–88): 79–104; Eve Kosofsky Sedgewick, *Between Men: English Literature and Male Homosocial Desire* (New York: Columbia University Press, 1985); and Klaus Theweleit, *Male Fantasies* (Minneapolis: University of Minnesota Press, 1987).

3. R. F. Scott, "Message to the Public," in *Scott's Last Expedition*, p. 477.

4. Atkinson, "The Finding of the Dead," p. 469.

5. The following is a partial listing of later editions published under the same title: 1914 (London), 1923 (London and New York), 1957 (Boston), 1964 (London and New York).

6. Roland Huntford, *The Last Place on Earth* (New York: Atheneum, 1983), p. 524. Originally published as *Scott and Amundsen* (London: Hodder & Stoughton, 1979).

7. Ibid., p. 527.

8. Ibid., p. 528.

9. Ibid.

10. Vilhjalmur Stefansson's and other U.S. reactions to Robert Falcon Scott's death can be found in an album of newspaper clippings entitled *Expedition to the South Pole, 1910–1913* in the collection of the Annex of the New York Public Library, 521 West 43 Street, New York, N.Y.

11. Ibid.

12. Ibid.

13. For an examination of how changes in technology and culture between 1880 and World War I created new modes of understanding time and space, see Stephen Kern, *The Culture of Time and Space, 1880–1918* (Cambridge, Mass.: Harvard University Press, 1983).

14. Robert Peary, *The North Pole: Its Discovery in 1909 Under the Auspices of the Peary Arctic Club* (New York: Stokes, 1910; reprint, Mineola, N.Y.: Dover, 1986), p. 201.

15. Ibid., pp. xxi–xxxii.

16. Ibid., p. xxxii.

17. Ibid.

18. On the history of British polar exploration, see Frank Rasky, *Explorers of the North: The North Pole or Bust* (Toronto: McGraw-Hill Ryerson, 1977), pp. 7–128. On the British historical background, see Correlli Barnett, *The Collapse of British Power* (London: Methuen, 1972); Eric Hobsbawm, "Waving Flags: Nations and Nationalism," in *The Age of Empire, 1875–1914* (New York: Pantheon

Books, 1987), pp. 142–64; V. G. Kiernan, *The Lords of Humankind: Black Man, Yellow Man, and White Man in an Age of Empire* (Boston: Little, Brown, 1969). On the history of the British Royal Navy, see William Clowes, *The Royal Navy: A History from the Earliest Times to the Present* (New York: AMS Press, 1966). For details on Scott's career in the Royal Navy prior to his final antarctic expedition and his rivalry with Shackleton, see Roland Huntford, *Shackleton* (New York: Fawcett, 1985); L. B. Quartermain, *South to the Pole* (Oxford: Oxford University Press, 1967).

19. Huntford, *Last Place*, p. 117.

20. Clements Markham, *The Lands of Silence: A History of Arctic and Antarctic Exploration* (Cambridge, Eng.: Cambridge University Press, 1921), p. 174.

21. Huntford, *Last Place*, p. 126.   22. Ibid., p. 127.

23. Ibid.   24. Markham, p. 174.

25. Huntford, *Last Place*, p. 137.   26. Ibid., p. 161.

27. Ibid.

28. Robert Falcon Scott, *The Voyage of the Discovery* (London: John Murray, 1913), 1:334.

29. Ibid., 1:343.

30. Immediately following Scott's death the British government raised a generous memorial fund that went toward the expenses of compiling and publishing Scott's letters, erecting a memorial statue to honor Scott in Trafalgar Square, and providing for the families of Scott and the men who died with him in Antarctica. See the following newspaper clippings from 1913 (see note 10, this chapter) for details concerning the Scott fund: *New York Times,* Feb. 26; *New York Sun,* Feb. 25 and Mar. 1. For an understanding of how the Scott story continues to be considered a part of England's national heritage see: L. P. Kirwan, *A History of Polar Exploration* (London: Penguin, 1962), p. 292. For an account of the significance of the Scott legend outside of England, see Huntford, *Last Place*, p. 528.

31. Scott, *Scott's Last Expedition*, p. 470.

32. Ibid.

33. Ibid.

34. Ibid., p. 471.

35. Mary Pratt, "Imperial Stylistics, 1860–1980," in *Imperial Eyes: Travel Writing and Transculturation* (London: Routledge, 1992), pp. 201–27, identifies some of the major strategies by which the British explorer Sir Richard Burton creates positions of dominance and distance in his 1860 narrative on his discovery of Lake Tanganyika. For Pratt, Burton's strategy of what she refers to as "making the home viewpoint appear normative" is one of the ways in which British exploration narratives of the period create value for their achievements in the non-West. Dean MacCannell, in *The Tourist: A New Theory of the Leisure Class* (New York: Schocken Books, 1976), makes a similar argument about the work-

ings of tourism. For MacCannell, the tourist always places himself at the center of the imaginary construction of the universe he creates. Tourists' homes are the reference points by which they interpret the attractions they visit (p. 56).

36. For an examination of other British military heroes who expressed the imagination of British empire, see Martin Green, *Dreams of Adventure, Deeds of Empire* (New York: Basic Books, 1979), pp. 3–27. For a comprehensive examination of how adventure played a key role in the history of Western European capitalistic societies, see Michael Nerlich, *Ideology of Adventure: Studies in Modern Consciousness, 1100–1750* (Minneapolis, University of Minnesota Press, 1987), vol. 2.

37. Quotations from Evans's, Bowers's, and Oates's original notes appear in Roland Huntford, *Scott and Amundsen* (London: Hodder & Stoughton, 1979), pp. 486, 489, 509 (reprinted in 1983 as *The Last Place on Earth*). Huntford's quotations from Oates's diary are not publicly available. Bowers's final letters to his mother and sister are at the Scott Polar Research Institute, Cambridge, Eng. E. A. Wilson's antarctic diaries were published as *Terra Nova* (London: Blandford Press, 1972).

38. See Huntford, *Scott and Amundsen*, p. 606. According to Huntford, the authority for Vitamin C deficiency as a cause of P. O. Evans's death is Dr. A. F. Rogers of the Department of Physiology of the University of Bristol, who has presented his findings in "The Death of Chief Petty Officer Evans," *Practitioner* (Apr. 1974): 570–80.

39. Scott, *Scott's Last Expedition*, p. 473.

40. Huntford, *Last Place*, p. 628.

41. Kirwan, p. 292.

42. Scott, *Scott's Last Expedition*, p. 473.

43. On the genre of tragedy, see Raymond Williams, *Writing in Society* (London: Verso, 1983).

44. *The Last Place on Earth* cost Britain's Central Television 7 million pounds to produce. It was screened on British television in February 1985. The television series appeared in the United States in October 1985 and July 1988 on PBS, presented by WGBH-TV, Boston, and funded by the Mobil Corporation.

45. John Wyver, "Hero Caught in an Icy Blast," London *Times* (Feb. 11, 1985): 12.

46. Ibid.

47. Ibid.

48. Andy Metcalf and Martin Humphries, eds., *The Sexuality of Men* (London: Pluto, 1985), p. 13.

49. Quoted in J. Hoberman, "Vietnam: The Remake," in *Dia Art Foundation's Remaking History: Discussions in Contemporary Culture* (Seattle: Bay Press, 1989), p. 76.

50. Ibid., p. 195.

51. See Dov Zakheim, "Is the Vietnam Syndrome Dead? Happily, It's Buried

in the Gulf," *New York Times* (Mar. 4, 1991); John Carlos Rowe, "The 'Vietnam Effect' in the Persian Gulf War," *Cultural Critique* 19, special issue (Fall 1991): 121–50.

52. Abdul Jan Mohamed and Donna Przybylowicz, "Introduction: The Economy of Moral Capital in the Gulf War," *Cultural Critique* 19, special issue (Fall 1991): 9.

53. Robert Stam, "Mobilizing Fictions: The Gulf War, the Media, and the Recruitment of the Spectator," *Public Culture* 4, no. 2 (Spring 1992): 117. This observation originally appeared in Jason de Parle, "Keeping the News in Step: Are the Pentagon's Gulf War Rules Here to Stay?" *New York Times* (May 1, 1991): 9A.

54. See the Socialist Review Bay Area Collective's article, "Warring Stories: Reading and Contesting the New World Order," *Socialist Review* 21, no. 1 (Jan.–Mar. 1991): 11–26.

55. See Tom Wicker, "An Unknown Casualty," *New York Times* (Mar. 20, 1991), 29A.

56. For a more elaborate analysis of how the discourse of colonialism operated during the Gulf War, see Les Levidow, "The Gulf War as Paranoid Rationality" (paper presented at the Dia Center for the Arts Ideologies of Technology Conference, Apr. 12, 1992); Ella Shohat, "The Media's War," *Social Text* 9, no. 28 (1991): 135–41; Edward Said, "The Arab Portrayed," *Arab World* 14 (1968): 5–7; Edward Said, "Islam Through Western Eyes," *Nation* (Apr. 26, 1980): 488–92.

57. See Stam, pp. 120–21.

58. Wicker, p. 29A.

16. Gumbrecht, Perception Versus Experience

Most of the arguments in this text regarding cultural and media history originated in a graduate seminar on "Practices and Theories of Recording 1820–1940," which my friend Tim Lenoir and I held in the History of Science program at Stanford University in the fall of 1991. I presented a preliminary sketch of the history of epistemology that converges with these arguments in a lecture entitled "The Depth of Hermeneutic Life Form and the Facility of System Theory" at an *Autorenkolloquium* (Colloquium on an author) on the work of Niklas Luhmann, held on the occasion of his retirement at the Center for Interdisciplinary Research at the University of Bielefeld in February 1993.

1. The association of aesthetic knowledge with the silencing of discourse stems from Luiz Costa Lima, *Limites da voz* (Rio de Janeiro: Rocco, 1993), 1:136. An English edition is *The Limits of Voice: Montaigne, Schlegel, Kafka* (Stanford, Calif.: Stanford University Press, 1996).

2. I analyze the genealogy and significance of this incompatibility for the epistemological situation of the present in *The Non-Hermeneutic* (forthcoming).

My hypothesis of the incompatibility between sensuality/perception and concepts/experience points to the impossibility of anything like an "adequate translation" of sensual perceptions into concepts. I agree, of course, with potential objections that first, this fact has belonged to the premises of Western philosophy for centuries, and second, it should not be directed against the necessity of trying to search for individual and collective orientation and agreement *in language*. The point of departure for my argument is therefore the suspicion that a broad philosophical consensus has underestimated (if not repressed) the significance of the fact that "adequate translation" of perceptions into experience is impossible.

3. This argument was central to a seminar Lyotard gave during the summer semester of 1988 at the Graduierten-Kolleg in Siegen. For the historical and systematic background, see Jean-François Lyotard, *Le Différend* (Paris: Editions de Minuit, 1983), pp. 189ff; see also Lyotard, *Leçons sur l'analytique du sublime* (Paris: Galilee, 1991), pp. 61ff.

4. Ferdinand Fellmann, *Symbolischer Pragmatismus: Hermeneutik nach Dilthey* (Hamburg: Rowohlt Taschenbuch Verlag, 1991), p. 65ff.

5. Among the large number of publications in Germany that have responded to the growing importance of the observer-concept in Luhmann's work, the essays by Niklas Luhmann, Humberto Maturana, Mikioi Namiki, Volker Redder, and Francisco Varela are especially illuminating; see Niklas Luhmann, et al., *Beobachter: Konvergenz der Erkenntnistheorien?* (Munich: Fink, 1990).

6. As interpretation in this concrete sense, the practice of the Inquisition from the late fifteenth century was closely related to the genesis of early modern subjectivity. See Claudia Krülls-Hepermann, *Die Unwahrscheinlichkeit neuzeitlicher Subjektivität: Spanische Schäferromane des späten 16. und des frühen 17. Jahrhundert* (Frankfurt: 1990), pp. 4–55.

7. See Hans Ulrich Gumbrecht, "Stimme als Form: Zur Topik lyrischer Selbstinszenierung im vierzehnten und fünfzehnten Jahrhundert," in *Musique naturelle: Interpretationen zur französischen Lyrik des Spätmittelalters*, ed. Wolf-Dieter Stempel (Munich: W. Fink, 1994), fn30.

8. This observation plays a central role in the cultural-historical sketches by Niklas Luhmann. See, for example, "Das Kunstwerk und die Selbstreproduktion der Kunst," in *Stil: Geschichten und Funktionen eines kulturwissenschaftlichen Diskurselements*, ed. Hans Ulrich Gumbrecht and K. Ludwig Pfeiffer (Frankfurt: 1986), pp. 621–72, esp. pp. 633–34.

9. See, for example, Michel Foucault, *Les mots et les choses* (Paris: Gallimard, 1966), pp. 60ff.

10. The brilliant standard work on this development is Hayden White, *Metahistory: The Historical Imagination in Nineteenth-Century Europe* (Baltimore: Johns Hopkins University Press, 1973).

11. See Foucault, p. 229 (on the crisis of representation of the world), and pp. 262–64 (on the concepts of "life," "labor," and "language").

12. See Charles Grivel, "Die Identitätsakte bei Balzac: Prolegomena zu einer allgemeinen Theorie des Gesichts"; Hans Ulrich Gumbrecht and Jürgen E. Müller, "Sinnbildung als Sicherung der Lebenswelt. Ein Beitrag zur funktionsgeschichtlichen Situierung der realistischen Literatur am Beispiel von Balzacs Erzählung *La Bourse*"; and Rainer Warning, "Chaos und Kosmos: Kontingenzbewältigung in der *Comédie humaine*," all in *Honoré de Balzac*, ed. Hans Ulrich Gumbrecht, Karlheinz Stierle, and Rainer Warning (Munich: Fink, 1980), pp. 93–142, 339–90, 9–56.

13. This thesis was presented by Tim Lenoir in our jointly conducted seminar on the history of recording technologies (Autumn 1991).

14. On the legal consequences of this (supposed) substitution of the camera for the human observer, see Gerhard Plumpe, *Der tote Blick: Zum Diskurs der Photographie in der Zeit des Realismus* (Munich: W. Fink, 1990).

15. On the history of "Realism" as a concept of literary and art criticism, see Helmut Pfeiffer, *Roman und historischer Kontext: Strukturen und Funktionen des französischen Romans um 1857* (Munich: W. Fink, 1984), pp. 100–25.

16. On the motif of the "new way of seeing," see Hans Robert Jauss, *Ästhetische Erfahrung und literarische Hermeneutik* (Frankfurt: Suhrkamp, 1982), pp. 125–28.

17. This corresponds to the "reality-concept of the world as resisting experience" illustrated in Hans Blumenberg, "Wirklichkeitsbegriff und Möglichkeit des Romans," in *Nachahmung und Illusion: Poetik und Hermeneutik*, ed. Hans Robert Jauss (Munich: W. Fink, 1964), 1:9–27, esp. pp. 24–25.

18. On the philosophical-historical context of Flaubert's work, see Franz Koppe, *Literarische Versachlichung: Zum Dilemma der neueren Literatur zwischen Mythos und Szientismus. Paradigmen: Voltaire, Flaubert, Robbe-Grillet* (Munich: W. Fink, 1977), pp. 53–62.

19. From the 49th chapter of *Le rouge et le noir*, quoted in Hugo Friedrich, *Drei Klassiker des französischen Romans: Stendhal, Balzac, Flaubert* (Frankfurt: V. Klostermann, 1961), pp. 15–16.

20. Fellmann, fn 5.

21. Wlad Godzich discusses the consequences of the shift from language to (animated) images as the dominant medium of communication from a philosophical perspective in "Language, Images, and the Postmodern Predicament," in *Materialities of Communication,* ed. Hans Ulrich Gumbrecht and K. Ludwig Pfeiffer, trans. William Whobrey (Stanford, Calif.: Stanford University Press, 1994), pp. 355–70.

22. The concept of "time-objects in the specific sense" comes from Edmund Husserl, *Zur Phänomenologie des inneren Zeitbewußtseins (1893–1917)*, vol. 10 of *Husserliana* (The Hague: Nijhoff, 1960), p. 23. On the definition of "rhythm," see Hans Ulrich Gumbrecht, "Rhythm and Meaning," in Gumbrecht and Pfeiffer, *Materialities*, pp. 170–82; on the historical constellation of the topic of "rhythm"

around the turn of the century, see Michael Golston, " 'Im Anfang war der Rhythmus': Rhythmic Incubations in Discourses of Mind, Body, and Race from 1850–1944," *Stanford Humanities Review* special issue (1996).

23. On the historical site and criticism of Bergson's polemic against the *illusion cinématographique*, see Gilles Deleuze, *Cinéma 1: L'image-mouvement* (Paris: Editions de Minuit, 1983), p. 9.

24. Niklas Luhmann, "Modes of Communication and Society," in *Essays on Self-Reference* (New York: Columbia University Press, 1990), pp. 99–106, esp. p. 102.

25. Especially in the second chapter of Henri-Louis Bergson, *Essai sur les données immédiates de la conscience* (Paris: F. Alcan, 1889). Husserl, in *Zur Phänomenologie*, repeatedly thematized the *Bild-Bewußtsein* (Consciousness of images). In contrast, above all, to Freud and Mead, however, Husserl's main interest seemed to remain focused on contents and structures of consciousness that can be accounted for conceptually.

26. *The Interpretation of Dreams* came out in Vienna in 1900, but "in reality already in November 1899." Cf. "Vorbemerkung der Redaktion" in the Fischer edition (Frankfurt: Taschenbuchverlag, 1961), p. 5. That Freud's method— despite its concentration on the preconceptual levels of consciousness—was informed by the structures of the hermeneutic field becomes clear through the surface/depth topology in the motto of his book: *Flectere si nequeo superos, acheronta movebo*.

27. See George Herbert Mead, "Die Philosophie der Sozialität," in *Philosophie der Sozialität: Aufsätze zur Erkenntnisanthropologie* (Frankfurt: Suhrkamp, 1969), pp. 229–324, esp. pp. 306–7.

28. The intense interest in preconceptual levels of consciousness, on the one hand, and the return to language and conceptuality (in full consciousness of their "inadequacy in relation to reality"), on the other, can be portrayed as two divergent types of the reaction to the "second demystification" that set in during the second half of the nineteenth century as a demystification of the Enlightenment project of demystification. See Hans Ulrich Gumbrecht, "Déconstruction deconstructed: Transformationen französischer Logozentrismuskritik in der amerikanischen Literaturwissenschaft," *Philosophische Rundschau* 33 (1986): 1–35.

29. Regarding our thesis of the incompatibility between motion (animated images) and static conceptuality, it is worth noting that Alfred Schütz—reaching back to Bergson's concept of *durée* and Husserl's concept of "time-objects in the specific sense"—distinguished between *Handeln* as "polythetic" execution and *Handlung* as a unity of meaning that can be attributed "monothetically" to *Handeln* in retention and protention. Cf. Alfred Schütz, *Der sinnhafte Aufbau der sozialen Welt; Eine Einleitung in die verstehende Soziologie* (Vienna: J. Springer, 1932), pp. 43–72 ("Die Konstitution des sinnhaften Erlebnisses in der je eigenen Dauer").

30. Martin Heidegger, *Being and Time*, trans. John Macquarrie and Edward Robinson (New York: Harper & Brothers, 1962); see especially paragraph 31f.

31. Ibid., paragraph 74, p. 437.

32. In a graduate seminar on the "Cultural and Technological Incubations of Fascism" conducted jointly with Jeffrey Schnapp and me, Tim Lenoir showed how the reduction of human existence articulated by the concept of "human material" in the first decades of the twentieth century made possible the integration of the body into new structures of industrial and military systems.

33. It is this assertion alone that I do *not* agree with in Godzich, "Language, Images," fn 22.

34. The conception of reality described here was among the themes of a colloquium that Niklas Luhmann conducted at Stanford University in March 1993. See also Niklas Luhmann, "Gleichzeitigkeit und Synchronisation," in *Soziologische Aufklärung 5: Konstruktivistische Perspektiven* (Opladen: Westdeuscher Verlag, 1990), pp. 95–130, esp. pp. 98–99.

Index

In this index an "f" after a number indicates a separate reference on the next page, and an "ff" indicates separate references on the next two pages. A continuous discussion over two or more pages is indicated by a span of page numbers, e.g., "57–59." *Passim* is used for a cluster of references in close but not consecutive sequence.

Aarsleff, Hans, 419n42, 451
Aboriginal peoples, Europeans and, 132, 138
Absentation, presentation and, 297
Academy of Sciences (France), 48, 50
Adam, Lucien, 270
Adams, D. H., 433n51
Addison, Joseph, 30
Aesthetics, 228, 356; astrophotography and, 240–48
Against Method (Feyerabend), 2
Agassiz, Alexander, 231
Agency, 59, 63, 68
Agent, 15, 67f; subject and, 65f, 69
Agriculture, women and, 80
Airy, George, 199f, 204, 210f, 219, 404n67
Album des Photographies Pathologiques, illustrations for, 162
Aldrovandis, Ulisse, 20–21
Alfred Brothers, 241, 243
A-life. *See* Artificial Life
Allen, W. S., 421n66
Alphabetic graphism, 11
Alphabet of Nature, 254
Alphabets, 41–45 *passim*
"Alps, The" (Haller), 95

Alterations: intervention and, 18; improvements and, 240–48
American Mathematical Society, 39, 52
"Am I Not a Man and a Brother" (Wedgwood), 131
Amphiareo, Vespasiano, 375n11
Amundsen, Roald, 328, 342
Anatomy and Philosophy of Expression as Connected with Fine Arts, The (Bell), 140, 153f, 162
Andromeda nebula, 235, 244
Animal Locomotion (Muybridge), illustrations for, 231
Anoine-Réaumur, René, 384n8
Appenzeller Zeitung, 99
Aquinas, Thomas, 73
Arbib, Michael, 312
Architectural inscriptions, 41
Architecture of the Heavens, The (Nichol), 201, 207
Arendt, Hannah, 346
Aristotle, 5, 22, 73, 150
Arithmetica, logistica and, 68
Art, 226; science and, 207
Artificial Life, 304, 311f, 324–27 *passim*, 432n32, 433n56; metaphors of, 315f,

322–23; plausibility of, 317–18, 319–20; rhetoric of, 305–6, 317; vision of, 313–18
"Artificial Life" (Langton), 314
Art of Computer Programming, The (Knuth), 50–51
Arvidson, Karl, 240, 245, 247
Astronomers, 226, 233–34, 236–40, 240ff, 245ff
Astronomical Society of London, 186, 199, 204
Astronomy, 185, 187, 198, 202, 223–24; nebular, 196–97, 198, 203f, 221
Astrophotography, 17, 223–48
Astrophysical Journal, photographs in, 233, 237f
Atkinson, Surgeon, 329
Atlan, Henri, 309f, 323f
Aub, Joseph, 288
Auroux, Sylvain, 252, 261
Auxerre, William of, 72
Azoulay, Paul, 277, 279

Babbage, Charles, 76, 222
Bachelard, Gaston, 286
Bacon, Francis, 23–24, 30
Ball, Robert, 220f, 429n50
Balthasar, Franz Urs, 386n26
Balzac, Honoré de, 357
Banks, Joseph, 190
Barbari, Jacopo de, 150
Barnard, E. E., 233, 235f, 240, 247
Barrie, J. M., 331
Baselines (surveying), 102–9 *passim*
Bateson, Gregory, 60
Baudrillard, Jean, 3, 298, 318–20
Beagle, HMS: Darwin aboard, 120, 125–27, 130, 134, 137f, 146, 151, 157; voyage of, 121, 128–29, 141
Beaune, Jean-Claude, 314, 324, 432n36
Bebel, August, 79f
Behavior: evolution and, 147f; instinct and, 149, 154; lifelike, 317, 319
Being and Time (Heidegger), 363
Bell, Alexander Graham, 81–85 *passim*, 89
Bell, Alexander Melville, 83, 259
Bell, Charles, 140f, 152–55 *passim*, 162, 164

Bell, David Charles, 83
Bell, Eliza, 84
Bell, Melville, 81, 83
Bell Patent Association, 82
Bell Telephone Company, 82
Benveniste, Emile, 64, 84f
Bergson, Henri, 361f, 440n29
Bernard, Claude, 253
Bernoulli, Daniel, 75, 384n8
Bernshteyn polynomials, 51
Bertillon, Alphonse, 279
Bézier cubics, 52
Bildungsroman, 134, 136
Billettes, Gilles Filleau Des, 43
Biochemistry, 288, 292, 298, 427n13
Birth of the Clinic, The (Foucault), 325
Blake, Clarence John, 89
Blake, William, 151f
Blitzer, Wolf, 348
Body, 151, 356–60
Bologna Stone, 21
Bond, W. C., 218, 220
Bopp, Franz, 253ff
Boston Telephone Dispatch Company, 85
Bottom-up, metaphor of, 315
Bower, Henry Robertson, 331, 339f, 436n37
Boyle, Robert, 21, 27, 29–32 *passim*, 36, 61f, 182f
Bréal, Michel, 251, 253–58 *passim*, 262, 270, 280, 282, 419n42; linguistics and, 250–53 *passim*
Breath, graphic measures of, 272
Brewster, David, 189, 203, 215, 217
Breymann, Hermann, 277
Bride, The (Bell), 83
Bridgeman, Francis Charles, 342
British Association for the Advancement of Science, 203, 209–14 *passim*, 218
Broad Romic (Sweet), 83
Brougham, Henry, 197f
Browne, James Crichton, 162, 397n35
Browne, Janet, 127, 130
Browne, William, 162
Brunel, Marc Isambard, 185, 207
Brunk, A. B., 235
Buckler, Robert, 345

Burney, Fanny, 198
Burton, Richard, 435n35
Bush, George, 347
Button, Jemmy, 129, 137

CA. *See* Cellular automata
Cairns-Smith, A. G., 325
Calculating engines, 222
Cameron, Julia Margaret, 142
Campbell, Wallace, 233, 235–47 *passim*,
    409n42, 410n57, 411n75
Canguilhem, George, 431n20
Carlyle, Thomas, 127, 136
"Carte des environs de Genève" (Micheli
    du Crest), 96
Cartography, 92–95 *passim*, 93, 94–95,
    391n100, 392n106; ethical influence
    of, 116; national, 114; politics and, 97–
    101; Swiss, 116, 117–18
Cassini, César-François, 93
Cassini, Gian Domenico, 21, 32
Cassini, Jacques, 93
Cassirer, Ernst, 249
Celestial light, 32f
Celestial mechanics, 194
Cellefrouin, 273–76 *passim*
Cellular automata (CA), 315, 321
Chaîne acoustique, 281
Chaîne parlée, 281
Chambers, Robert, 203, 206, 215
Champ fleury (Tory), 376n19
Chance, 144, 154
Charcot, Jean Martin, 161
Charles Island, 127, 138
Chartier, Roger, 11–12
Chartism (Carlyle), 136
Chasseral, 94, 385n20
Chatham Island, 123
Chicago Photogravure Company, 235
Chiromancy, 150
Chomsky, Noam, 308
Chronophotographs, 268
Civil history, 22
Clerke, Agnes, 186
"Climb up to Gemmi" (Füssli), 96
Code, 63–67 *passim*, 312; communicative,
    420n60; mathematical, 314

Codescript, 306
Colbert, Minister: typography and, 43
Collins, Harry, 283, 429n48
Colonization, 120
Comets, 194, 233, 242–46 *passim*
Communication, 10, 12, 70, 87, 89, 355
Computer modern (typeface), 52ff
Condorcet, Marquis de, 11
Conservative Revolution, 363
Consonants, theory of, 89
Constellations, celestial light from, 32f
Constructed alphabet, 44
Constructivism, 363
Conway, John Horton, 311f, 316, 326
Cooper, James Davis, 160
Copper plates, 45, 110, 113, 230ff
Courbet, Gustave, 358
Cours de linguistique générale (Saussure),
    251, 280
Couturat, Louis, 269
Crab nebula, 212
Crandal, Irving B., 89f
Creuzer, Georg Friedrich, 254
Crick, Francis, 304, 309
Crossley reflector, 237, 244
Crusoe, Robinson, 127f
Cultural production, 2f
Cultural studies, 2, 227
Culture wars, 3f
Cursus publicus, 78
Cybernetics, 309, 311
Cyclopaedia of the Physical Sciences, 218

Daguerrotype process, 141
D'Alembert, Jean, 48, 355
Darwin, Charles, 18, 119f, 122, 126f, 130,
    134, 136–41, 152, 154, 162, 180,
    395nn27, 28, 399n60; evolution and,
    125, 132, 143, 147, 149, 158, 370n42,
    397n18; expressions and emotions
    and, 148, 154, 177, 397n18; on habit-
    ual behavior, 147–48; illustrations and,
    156–60 *passim*, 169f; language and,
    16–17; narrative strategies of, 144,
    166, 179; photography and, 144ff,
    155f, 178–81 *passim*; reading by, 127–
    28; Rejlander and, 170, 176–79 *passim*

Darwin, William, 146
"Darwin's Conversion: The *Beagle* Voyage and Its Aftermath" (Sulloway), 121
Darwin's Plots (Beer), 16
"Darwin's Thinking: The Vicissitudes of a Crucial Idea" (Sulloway), 121
Daston, Lorraine, 167, 224, 248
Da Vinci, Leonardo, 43, 150
Dawkins, Richard, 319–27 *passim*
Deaf-mutes, 82f, 258, 260, 420nn55, 58
Dear, Peter, 22
De causis mirabilium (Oresme), 22
Deconstruction, 12–18 *passim*, 291, 302, 369n36
De la parole considérée au double point de vue de la physiologie et de la grammaire (Vaïsse), 258
Deleuze, Gilles, 361
Della Porta, G. B., 150
Del Piombo, Sebastiano, 396n5
Demeny, Georges, 268
Demonology, 26f
DeMorgan, Augustus, 76
Depôt de la Guerre de France, Dufour and, 109–10
Derrida, Jacques, 3, 6–12 *passim*, 15, 18f, 55f, 286, 302, 304; analyses and, 71; deconstruction and, 4–5, 13f, 291; difference and, 291f
Descartes, René, 23, 151f, 155
Descent of Man, and Selection in Relation to Sex, The (Darwin), 148; *Expression* and, 146–47
Desmond, Adrian, 133, 139
Desprez: electric signal by, 273; galvanometer by, 262
De Tourtoulon, Charles, Baron, 271
Dhyaana, 260
Diamond, Hugh Welch, 162–63
Diary (Darwin), 136, 395nn27, 28; quotations from, 124–27 *passim*
Dickens, Charles, 137
Dictionnaire des idées reçues, 359
Diderot, 48, 258, 355
Die Lehrlinge zu Sais (Novalis), 134
Différance, 9, 14, 291f, 369n36
Differential reproduction, 287, 292
Dilthey, Wilhelm, 352, 360

Dime, The (Stevin), 59
Diophantus, 74
Discourse Networks, 1800 / 1900 (Kittler), 9
Discourses at the Royal Academy (Reynolds), 142
DNA, 1, 14, 308, 310, 322f, 326
Domestic services, women and, 80
Donders, Franziskus, 264, 267
Doolittle, Eliza, 83f
Double stars, 193–96 *passim*
Down House, island state of, 139
Dragon's blood, 230, 232
Dream work, 8–9
Du Bois-Reymond, Emil, 88
Duchenne de Boulogne, Guillaume-Benjamin, 152; collaboration with Darwin, 161–70; experimental methods from, 163, 168; expressions and, 164, 166; photography and, 164–69 *passim*
Dufour, Guillaume-Henri, 100f, 105f, 107–10, 113, 116; criticism of, 111, 115; instruments and, 102, 109; work of, 92, 111, 115, 117
Dufour map, 113–18 *passim*
Dufriche-Desgenettes, A., 261, 421n68
Duguid, Paul, 12
Durée, 361f, 440n29
Dürer, Albrecht, 31, 44, 49, 150, 375n11; alphabet by, 42ff
Durkheim, Emile, 282
Duruy, Victor, 253f, 256
Dyson, Freeman, 432n32

Easter Islands, 135
Edgerton, Samuel, 207, 228
Edinburgh Review, The: on photography, 142–43, 223f
Edison, Thomas, 81, 250, 267; business phonograph by, 269; graphophone by, 269, 274; kinetoscope by, 268; phonograph of, 274
Education sentimentale (Flaubert), 358
Edwards, Ernest, 156, 231–39 *passim*, 247
Edwards, Jonathan, 192
Egerton, Admiral, 331
Electromagnetic stylus, 262
Electron micrographs, 299

Electron microscopy, 298ff
Electrophysiological research, 163f
Eliot, George, 137
Ellis, Alexander J., 82
Elson, A. W., 240
Emergence, metaphor of, 315
Emotions, expression of, 140, 148–49
Empirical factuality, 62
Encyclopédie ou dictionnaire raisonne des arts et des métiers (Diderot and d'Alembert), 48, 355
Encyclopédistes, 48
Enfant sauvage, 258
Engravers, astronomers and, 235, 245ff
Engravings, 184, 231, 388n52; improvements/alterations and, 233, 241; for maps, 110
Enzensberger, Hans Magnus, 73
Ephemeridum medico-physicarum germanicarum, 37
Epistemic practices, 87, 286f, 297–98, 356
Epistemology, 353, 362, 363–64
Eschmann, Johann, 104ff, 111
Essay for Recording of Illustrious Providences, An (Mather), 27
Essay on Population (Malthus), 393n8
Essays on Physiognomy (Lavater), 149–51, 155
Etchings, 230, 233–34
Evans, P. O., 341f, 436n38
Evelyn, John, 21
Evolutionary theory, 119, 132, 143–49 *passim*, 306, 393n8
Experience: empiricism of, 22–26; experiencedness and, 291; perception and, 14, 352, 353
Experimental reasoning, 286
Experimental systems, 6, 286–89 *passim*, 426n13; differential reproduction of, 291
Explanations (Chambers), 214
Expression of the Emotions in Man and Animals, The (Darwin), 18, 141, 146–47, 148f, 150–55, 181, 399n60; Duchenne and, 165–66, 169; illustrations in, 155–61, 169, 179–80; photographs in, 144ff, 155–62 *passim*, 176, 178, 180; Rejlander and, 170–75 *passim*, 179

Expressions, 141, 148–49, 158, 165, 160, 354; evolutionary significance of, 146–49; facial, 156, 164, 177; illustrating, 152 (table); nature of, 150–51; photographing, 156, 158

Facts: empiricism of, 22–26; legitimacy of, 61–62; literary inscriptions and, 8
Falkland Islands, 133, 137, 345
Falklands War, 345–46, 348
Fanti, Sigismondo, 375n11
Federal Charter of the Swiss Confederation (1832), 99, 388n54
Feer, Johannes, 105
Feliciano, Felice, 41, 45
Fellmann, Ferdinand, 352, 360
Feyerabend, Paul, 2, 4
Fichte, Johann Gottlieb, 80, 85f
Fick, Adolf, 421n71
Film, 10, 360–61
Finsler, Hans Konrad, 98, 100f, 106, 388n57, 389n62
Fishacre, Richard, 72f
Fitzroy, Robert, 121, 129–33 *passim*, 151
Flaubert, Gustave, 358
Fleck, Ludwik, 221, 288
Fletcher, Harvey, 90
Fontenelle, Bernard de, 26
Formalism, 64, 67
Foucault, Michel, 3f, 59, 249, 273, 305, 307f, 325–26, 355f
Fournier, Pierre Simon, 48–50, 377nn25, 26, 28
Fractal Geometry of Nature, The (Mandelbrot), 120
Fractionation, 293f
Franklin, John, 334ff
Fraunhofer, Josef, 190, 208
Freud, Sigmund, 8–10 *passim*, 161, 362, 431n23, 440nn25, 26
"Freud and the Scene of Writing" (Derrida), 9
Frost, Edwin, 239
Fuegians, 129
Fuller, Sarah, 81
Fuseli, Henry, 151f
Future generating machines, 14, 287–90
Future of the Book, The (Nunberg), 12

Gadamer, Hans-Georg, 70
Galapagos Islands, 121, 131f, 137; Darwin
    and, 122–23, 124, 138
Galen, on emotions, 150
Galileo, 1, 20f, 23, 60, 68; language and,
    55–58 passim
Galison, Peter, 167, 224, 248, 285f
Garden, George, 36
Gau, Carl Friedrich, 76
Gauging power, 196
Gelatine, 230f
Gellert, 80
General character, 307–8
Geological Observations on South America
    (Darwin), 128
Geological Observations on the Volcanic Islands
    Visited During the Voyage of the HMS
    Beagle (Darwin), 128
George III, 198
Georgian star, 194
Gerando, J. M. de, 258
Gilfillan, George, 202
Gillieron, Jules, 270
"Ginx's Baby" (Rejlander), 173, 174–75
Glance away, 324, 326
Glass positives, 236–41 passim
Glossophonographic museum/archives,
    277–81 passim
Gnostics, 321
God trick, 315
Godzich, Wlad, 439n21
Goerres, Johann Joseph von, 254
Goethe, Johann Wolfgang von, 70, 90, 134
Gorse, Abbé, 424n110
Gosse, Philip Henry, 161
Grammatology, Of, 4, 18f
Grammè, 9
Graphemes, 6, 15, 287, 295–302
Graphophone, 269
Grégoire, Abbé, 256, 270, 272
Gregory XVI, Pope, 389n63
Grew, Nehemiah, 30
Griffith, Trevor, 345
Gross, Paul, 2f
Grosvenor, Gilbert, 333f
Groundlines (surveying), 102–6 passim
Gruber, Howard, 139, 393n8
Gulf War, 347–50

Gutenberg, Johann, 75
Gutzmann, Hermann, 88–89
Guy, Richard, 316

Habermas, Jurgen, 78ff
Hacking, Ian, 285f, 298, 416n11
Hale, George Ellery, 237
Halftones, 229–38 passim, 248
Haller, Albrecht von, 94ff, 104, 116,
    384n8, 386n31
Haller, Karl Ludwig von, 98
Halley, Edmund, 29–36 passim
Halley's Comet, 235, 242
Hamilton, William Rowan, 211–15 passim
Handeln, Handlung and, 440n29
Haraway, Donna, 315, 432n36
Harris, Roy, 416n14
Hart, Robert, 279
Hassauer, Friederike, 80
Havet, Louis, 260–64 passim, 421n68
Hegel, G. W. F., 5
Heidegger, Martin, 56, 84, 363
Heliogravures, 238, 244
Heliotypes, 156–57, 173, 231
Helmholtz, Hermann, 82, 88f, 251, 261,
    264, 267, 421n71
Helmreich, Stefan, 431n26
Helvetia, 97, 115, 386n26
Helvetic Society, 99
Hemming, John, 344–45
Henzi Conspiracy, 93, 385n12
Hermann, Ludimar, 88, 267
Hermeneutics, 9, 249, 352, 353–56, 360
Herodotus, 24
Herschel, Caroline, 190
Herschel, John, 141f, 144, 186, 189f, 200,
    202, 205f, 218, 396n3; nebular astron-
    omy and, 199–204 passim
Herschel, William, 186, 192, 198f, 201,
    213f, 218f; cosmology of, 193–200
    passim; nebulae and, 196–200 passim,
    212, 216; technology and, 194, 195–
    96; telescopes and, 187, 191–92, 196,
    213
Higgins, Henry, 83f, 381n19
Higher Superstition (Gross and Levitt), 2
Histoire et Mémoires de l'Académie Royale des
    Sciences de Paris, 20f, 26f

Histoires prodigieuses, 25
Historical-constructivism, 18
History of Animals (Aristotle), 22
History of Dogma (Landgraf), 72
History of England (Hume), 127
History of England (Macaulay), 135
Hobbes, Thomas, 22, 183
Hoberman, J., 347
Holden, Edward, 223, 241, 244, 247
Homberg, Wilhelm, 35
Hooke, Robert, 32–36 *passim*
Hooker, J. D., 129, 138
Hour and the Man, The (L'Ouverture), 127
Hubbard, Gardiner Green, 81
Hubbard, Mabel, 81–85 *passim*
Huggins, William, 219–20, 221
"Human material," 363, 441n32
Humboldt, Alexander von, 127, 217
Hume, David, 127
Humphries, Martin, 345–46
Huntford, Roland, 330, 334–35, 336f,
    341f, 344–45, 436n37
Huntington Laboratories, ultracentrifuge
    at, 298
Huntington Memorial Hospital, 428n40
Hussein, Saddam, 349
Husserl, Theodor, 5, 56, 440nn25, 29
Huxley, Leonard, 329
Huxley, Thomas Henry, 129, 177
Huygens, 75, 199

Icarus myth, 313f
Idéologues, 258
Illustration: scientific, 18, 228; strategies
    for, 155–61
Image acoustique, 252, 262
Images: perspectivist, and cartography,
    113–18; reflection and, 351
Image vocale, 252, 262
Imaginary, 10, 13
Imaginer, imago and, 69
Improvements, alterations and, 240–48
Incorporation, fractioned, 292
Incorporeal subject, experience of, 357
Information, 311; biological, 304–5; tech-
    nology and, 12, 310
Instinct, behavior and, 149, 154
Institutionism, 67

"Instruction for Recording in 1/25,000"
    (Dufour), 107–8
International Phonetic Alphabet, *Visible
    Speech* and, 83
Interpretation, 296, 351, 354, 360
Intervention, 18, 229–36
Introduction to the Study of Natural Philosophy
    (Darwin), 142
Intruders, predations of, 132–33
Investigation, local context of, 2
"Investigations of the Limits of Linguistic
    Perception" (Gutzmann), 88
In vitro protein synthesis, 298
Ireland, Darwin on, 132
Iselin, Isaak, 386n26
Island Life (Wallace), 129
Islands, 119f
Isomorphism, 60, 355
Italics, 45
Itard, Jean-Marc-Gaspard, 258

Jacob, François, 288, 306, 310f
Jakobson, Roman, 64
Jammes, André, 46f
Janssen, Pierre, 277
Jaugeon, Jacques, 43
Jespersen, Otto, 269
Journal of Researches, The, 121, 137
Joyce, James, 362
Judgments, 307
July Revolution (1830), 99

Kafka, Franz, 10, 90
Kant, Immanuel, 352
Keeler, James Edward, 225, 237f, 240, 244f
Kempelen, Wolfgang Ritter von, 82, 88
Kindermann, Adolph Diedrich, 398n35
Kirwan, L. P., 342
Kittler, Friedrich, 9, 10, 15–16, 250
Knowledge, 2; mathematical, 56; observa-
    tional, 220; reality and, 356; sociology
    of, 362f
Knuth, Donald, 39, 50–54 *passim*
Koenig, Rudolph, 83, 261, 264, 266–67,
    274, 422n82
Kohler, Robert, 427n13
Koppel, Moshe, 309f, 322f
Koschwitz, Eduard, 270, 277

Kubler, George, 427n16
Kuhn, Thomas, 61, 285

Laboratory Life (Woolgar and Latour), 1
Lacan, Jacques, 4, 16, 74
La forme intellectuelle et morale de France (Renan), 256
La France profonde, vocalism of, 276
Lamarck, Jean-Baptiste, 147
Lambert, Jean-Henri, 385n22
La méthode graphique en sciences expérimentales (Marey), 251f
Landgraf, A. M., 72
Langton, Christopher, 306, 312, 321; A-life and, 305–7, 313, 315–17, 323
Language, 1–2, 5, 21, 66, 68, 70–74 *passim*, 84, 90, 203, 254ff, 257–58, 419n42; autonomy of, 249, 255–56; mathematics and, 57–58; politics of, 252–57, 280; science of, 249–53 *passim*; subjectivity and, 64, 71
Langue, 276, 281
Laplace, Pierre Simon, 201
La revue des patois galloromans, 270
Larynx explorer, 273
Lassell, William, 217, 244
Last Place on Earth, The (television series), 344, 346, 436n44
Latour, Bruno, 17, 91, 109, 285, 296, 368n32, 429n53
Lavater, Johann Caspar, 149–55 *passim*, 164
Lavery, David, 314
Least action, principle of, 261
Le Brun, Charles, 151–55 *passim*, 164
Le Duc, Stephane, 430n1
Leeuwenhoek, Anton von, 32, 35
Leibniz, Gottfried Wilhelm, 21, 75f
Leroi-Gourhan, André, 10–11
Les Passions de l'âme (Descartes), 151
Lesson, Monsieur, 135
Letterforms, 51, 54
Lettre sur les Sourd et Muets (Diderot), 258
L'étude géographique sur la limite de la langue d'oc et de la langue d'oïl (De Tourtoulon), 271
Leviathan and the Air Pump (Shapin and Schaffer), 1

Leviathan of Parsonstown, 17, 185, 187f, 205, 207–17, 221–22
Levitt, Norman, 2, 3
Levy, Max, 408n13
Liberal reform movement, 100–101
Lick Observatory, 225f, 231, 235f, 237f, 244–45
Life, 308, 312, 316f, 324, 357; sovereignty of, 320–27
Life (game), 311f
Life-forms, changes in, 119–20
Lingua Franca, 3
Linguistics, 249–53 *passim*, 259, 260–61, 269–77, 279–80
Linguistic values, 251, 281, 284
Linnaeus, Carolus, 126
Linnean Society, 142
Lipmann, Fritz, 290
Lissajoux, Jules, 261
Lister, Martin, 35
Literary criticism, pictures and, 227
Literary genres, 38, 183
Literary inscriptions, 8, 250
Literary technology, 182ff, 193–202 *passim*, 222
Literature, sociology of, 15
Loftfield, Robert, 289
Logistica, arithmetica and, 68
Lohmann, Johannes, 73
Lombardus, Petrus, 71
London Times, 213, 216, 330
Longuet-Higgins, Christopher, 308, 309
L'Ouverture, Toussaint, 127
Luhmann, Niklas, 14, 356, 361, 438nn5, 8
Lyell, Charles, 121, 136, 157
Lynch, Michael, 207, 228, 293
Lyotard, Jean-François, 3, 352, 438n3

MacArthur, Robert, 120
Macaulay, Thomas, 135
MacCannell, Dean, 436n35
McClintock, Leopold, 311, 366
McClure, Captain, 336
Mach, Ernst, 61
Machines, 77, 312, 431n20
McLuhan, Marshall, 15
Maddox, Brenda, 84
Magellanic Clouds, 200

Maillardoz, Colonel, 111
Mallarmé, Stéphane, 10, 359
Malthus, Robert, 16, 393n8
Mandelbrot, Benoit, 120
Manier, Edward, 143, 144, 166
Mannheim, Karl, 362
Mapmaking, 91f, 117
Maps, 110–18, 385n19; landscape and, 106–13; nations and, 91f, 106–13; politics and, 116
Mardersteig, Giovanni, 375n11
Marey, Etienne-Jules, 180, 249–51, 259–63 passim, 267, 270, 273, 399n60
Marichelle, H., 267f
Markham, Clements, 335–38 passim
Marriage, women and, 86–87
Martin, William, 206
Martineau, Harriet, 127, 185–86
Masculinity, 350; American myths of, 331–34; science and, 328
Maskelyne, Nevil, 190, 195f
Mathematical certainty, 62
Mathematical commands, inclusive/exclusive, 65
Mathematical Theory of Communication, The (Shannon), 311
Mathematics: 56, 60–63 passim, 67f; language and, 55–56, 57–58
Mather, Increase, 27, 30
Maxwell, James Clerk, 219
May, Arthur Dampier, 160
Mead, George Herbert, 362, 440n25
Meaning, 7ff, 303
Mécanisme de la Physiomonie Humaine ou Analyse Électro-physiologique de l'Expression des Passions Applicable a la Pratique des Arts Plastiques (Duchenne), 163, 164–68
Mechanical reproduction, 226, 248
Media, 10, 16, 78f, 352; British/American, 344–47
Medieval symbolic realism, 354
Meillet, Antoine, 419n46
Meissonier, Jean-Louis-Ernst, 399n60
Meister, Ulrich, 117f
Memoirs (Royal Astronomical Society), 200
"Mental Distress" (Rejlander), 173ff

Mentally ill, photographing of, 162–63
Menzel, Adolph, 358
MetaCode, 63–67 passim, 314, 326
Metafont, 50–52, 54, 378n36
METAFONTbook, The (Knuth), 54
Metalanguage, language and, 66
Metamorphosis (Kafka), 10
Metasigns, 13–14, 59
Meta-Subject, 14
Metcalf, Andy, 345–46
Method to Learn to Design the Passions (Le Brun), 151
Metrology, 91, 282
Meyer, Conrad Ferdinand, 75
Micheli du Crest, Jacques Barthelemi, 97, 101, 108, 386nn26, 33, 387n36, 389n66; cartography and, 93–97
Michell, John, 192
Micrographia (Hooke), 32
Microscopes, 298f
Microsomes, 299f
Milky Way, 197, 219, 235, 240, 242
Mill, John Stuart, 203, 206, 214
Milton, John, 16
Miscellaneous Observations, Concerning Things Rare and Wonderful, 27
Mitchell, Timothy, 250–51
Models, 289, 297f, 303
Modernity, American myths of, 331–34
Molecular biology, 14, 304, 427n13
Monnaie mécanique, 425n138
Monod, Jacques, 306, 310f
Monsters, 27f
Monster Telescopes Erected by the Earl of Rosse, The, 211
Montaigne, Michel Eyquem de, 182
Montessori, Maria, 258
Montrose Lunatic Asylum, Browne and, 162
Moon, Holden prints of, 244
Moore, James, 133, 139, 370n42
Moravec, Hans, 325
Moreau de Maupertuis, Pierre Louis, 384n8
Morus, Iwan, 186–87
Motion, perception of, 360–64
Moureau, Frédéric, 358
Movable type, first use of, 39–40

Moyllus (De Moyle), 375n11
Müller, Johannes, 87
Multiculturalism, 3
Muret, Jean-Louis, 97, 386n34
Muybridge, Eadweard, 180, 231, 399n60

Nadar, Félix, 163
Napoleon Bonaparte, 98
Nares, George, 337
Narrative form, gender and, 343–44
Narrative of the Surveying Voyages of His Majesty's Ships Adventure and Beagle Between the Years 1826 and 1836 (Darwin), 121
Nasmyth, James, 207, 217
National Exhibition (1883), mapmaking and, 117
National Gallery of Art, 142
National Gallery of Practical Science, 187
National Geographic, 347, 350
National Geographic Society, 346
Natural history, 22f
Natural History (Pliny), 22
Natural History of New England, 27
Naturalism, 229, 359
Natural philosophy, 20ff, 24, 25–27, 29, 120, 143–44, 183, 371n15; natural selection, 125, 138, 147, 149
Natural theology, 143, 370n42
Natural Theology: or, Evidences of the Existence and Attributes of the Deity (Paley), 154
Nebulae, 196–200, 212, 216–18, 220, 235, 237, 241f, 244f. See also various nebulae by name
Nebular astronomy, 196–97, 198, 203f, 221
Nebular Hypothesis, 201, 219
Nelson, Admiral, 335, 337
Nemethy, Carl, 234
Neue Zürcher Zeitung, 99
Neve, Michael, 130
New historicism, 15
Newton, Isaac, 20, 182
Nichol, John Pringle, 201–2, 206f, 213, 216, 404n61; cosmology of, 200–201, 203, 214–15; Orion nebula and, 201, 202–3, 208, 214, 215

Nietzsche, Friedrich, 56, 71, 304, 307, 315, 317, 327, 359, 362
Nimble and Handsome Calculations for Businessmen (Widmann), 74
North Pole, The (Peary), 332
Northwest Passage, Franklin expedition to, 334
"Note on the Wunderblock (Mystic Writing Pad)" (Freud), 8
"Nova Helvetiae Tabula Geographica," 96
Novalis, 74, 134
Nutt, Emma, 85f

Oates, 341f, 436n37
Objectivity, 18, 229, 369n36
Objects: subjects and, 356; theory and, 6
On Divine Proportions (Pacioli), 43
On Marvellous Things Heard, 22
"On the Seat of the Soul" (Longuet-Higgins), 308f
On the Sensations of Tone as a Physiological Basis for the Theory of Music (Helmholtz), 82
Opticks (Smith), 199
Order of Things, The (Foucault), 307
Oresme, Nicole, 22
Orientalism (Said), 128
Origin of Species, The (Darwin), 16–17, 125, 133, 138, 143f
Orion nebula, 199ff, 202–3, 205f, 208, 213, 214–19 passim, 220, 244–47 passim
Ostwald, Wilhelm, 269

Pacioli, Luca, 43f, 49
Paget, Francis, 336
Painting, 142, 184
Paley, William, 154
Pamela (Richardson), 80
Panuncillo, English and, 394n22
Parallax measurement, 194
Paris, Gaston, 255–57 passim, 270
Paris Academy of Sciences, 20, 24, 43, 51
Paris Observatory, lunar atlas by, 244
Parole, 276
Pascal, Blaise, 62, 385n18
Passions, Cartesian conception of, 151
Patois, 269–72 passim, 276f

Pattee, Howard, 312, 432n37
Peace and War (Rubens), 170
Peary, Robert, 328–29, 332–34, 338, 342–50 *passim*
Peirce, Charles Sanders, 14, 58–59, 61, 64ff
Penalty copy, 39
Perception: experience and, 14, 352f; motion and, 362
Père, Sébastien. *See* Truchet, Father Jean
Perrine, Charles, 240, 246
Person, 64, 68; myth of, 79
Personal Narrative (Humboldt), 127
Petermann, Mary, 300
Phenomenology, 356–57, 362
Philology, 250–60 *passim*, 264, 270, 276, 280
Philosophical Transactions of the Royal Society of London, 20f, 26f, 34
Philosophische und patriotische Träume eines Menschenfreundes (Iselin), 386n26
Phonautograph, 82, 264f
Phonemes, 252, 261, 263, 421n68
Phonetics, 249, 260, 270f, 277, 280, 416n8
Phoniatry, 88–89
Phonic substance, 282
Phonography, 264, 270
Photochemical technology, advancements in, 156
Photoengraving, 229–36, 240–48
Photographic Society of London, 162–63
Photographs, 18, 226, 227–28, 241; altering, 176–77, 229, 236; astronomical, 225–26, 231
Photography, 141, 142–46 *passim*, 168–70, 184, 223–24, 227, 248, 357f
Photogravure, 231ff, 236, 238f, 248
Photogravure & Color Company, 225, 231, 237f, 240, 246f, 409n42
Photophone, 268
Phrenology, 149
Physical Symbol System Theory, 432n37
Physics, linguistics and, 270
Physiognomy, 149ff
Physiology, 249, 267, 270
Pictograms, 108
Pictorial Times, 206
Place, sense of, 275

Plateau phenakistoscope, 268
Plates, 233–34, 248; collodion, 231; copper, 45, 110, 113, 230ff; etching, 230; "star," 409n42; steelfacing, 232
Plato, 4f, 14
Platonism, 67f
Plinian Society, 140, 162
Pliny, 25
Podbielski, Victor von, 86
Point system (typography), 48–50
Polar exploration, 328, 333–34, 336–37
Politics, language and, 257
Polygraph, vocal, 262f
Positives, glass, 236–41 *passim*
Postal services, 79f
Postcolonialism, 120
Powell, Colin, 349
Practical approach, 289
Pratisakhya, 260f, 421n66
Pratt, Mary, 435n35
Prediction, scientific, 66–67
Prescription, description and, 421n66
Presence, metaphysics of, 5
Presentation, absentation and, 297
Principles of Geology (Lyell), 157
Printers, astronomers and, 226, 240f, 245
Printing, 48, 50, 142, 184, 236–40, 248, 355
Process, logic of, 290–95
Prodger, Phillip, 17f
Program music, 359
Proofs, mathematical, 66
Proportion, 39–43 *passim*, 48
Protein synthesis, 288ff, 294
Psychic machinery, 9
Psychoanalysis, media and, 16
Public sphere, 78
Publicus, 78
Pygmalion (Shaw), 83

Quadrumana, facial expressions of, 158
Quine, W. V. O., 64
Quiriquina Island, 135

"Raising of Lazarus, The" (del Piombo), 396n5
Ramage, John, 189, 192
Raphael, 170

Ratiocination, 62
Ray, Tom, 322–27 *passim*
Rayleigh, John William Strutt, 88
Real, transformation of, 10
Realism, 68, 352, 357ff
Reality, knowledge and, 356
"Recent Astronomy and the Nebular Hypothesis" (Spencer), 219
Reflection, image and, 351
Reflectors, 186, 189
Regional accents/dialects, 257, 277
Regis, Ed, 313
Rejlander, Mary, 176
Rejlander, Oscar Gustave, 152, 160f, 170–79
"Remarks on the History of Things" (Kubler), 427n16
Renan, Ernest, 252, 256f, 418n25
Representation, 12, 293–97 *passim*, 318–19
Representative Men in Literature, Science, and Art (Walford), 156
Representing and Intervening (Hacking), 285
Resolvability, 221
Reynolds, Joshua, 142
Rhythm, 361
Richard II (Shakespeare), 126
Richardson, Samuel, 80
Richmond, Herbert, 337
Rimbaud, Arthur, 359
Riviere, Briton, 152, 160f
RNA, 298, 301
Robinson, Thomas Romney, 207–14 *passim*, 404n61
Romains du roi, 48
Roman alphabet, 41–45 *passim*
Romance of Natural History (Gosse), 161
Romano, Giulio, 152
Rosapelly, Charles, 260–63 *passim*, 270, 273
Rose, Gillian, 274
Rosse, Earl of, 192, 197, 203, 205, 208–9, 210f, 213–17 *passim*, 218–22 *passim*, 244; telescopes and, 185ff, 207, 212, 216, 220
Rotman, Brian, 13–14, 303, 314
Rousselot, Abbé Jean-Pierre, 270–76 *passim*, 279, 423n101
Rousselot, Madame, 276

Royal Astronomical Society, 199, 201, 209
Royal Engraving, 235
Royal Geographical Society, 335, 344
Royal Irish Academy, 209, 213
Royal Navy, polar exploration by, 335, 337
Royal Society of London, 24, 29f, 194f, 209, 211–12, 222
RPTV, 86
Rubens, Peter Paul, 170
Rules of Sociological Method (Saussure), 282

Sacrifice: male, 334–38; self-, 336, 342f
Said, Edward, 128
St. Helena, 131
Saint-Hilaire, Geoffroy, 201
St. Jago, 130, 157
St. Paul's Island, 127
San Francisco Chronicle, on Gulf War, 349
Sanskrit, 253, 254–55, 260
Sartor Resartus (Carlyle), 127
Saussure, Ferdinand de, 5, 58, 256, 262, 280–84, 419n42; linguistics and, 6–7, 251f, 281, 419n46, 425n130
Schaffer, Simon, 17f, 61f
Scheler, Max, 362
Scheuchzer, Johann Jakob, 96
Schlegel, Friedrich, 254
Schleicher, August, 253, 255, 419n40
Schrödinger, Erwin, 306, 311, 324
Schutz, Alfred, 362, 440n29
Schwartzkopf, Norman, 349
Science, 183, 184–85, 207, 227, 285, 316, 328, 389n63, 404n58; language of, 57f, 305, 427n13; as practice and culture, 2ff, 285; society and, 183, 285
Sciences de l'homme, 356
Scientific knowledge, 11, 24, 38, 56, 60, 141, 182f
Scientific real, 286, 296
Scientific representations, 183, 226–29, 297
Scoresby, William, 212, 216–17
Scott, Kathleen, 329, 330–31
Scott, Léon, 82
Scott, Robert Falcon, 328ff, 332, 335, 337f, 340ff, 346, 348, 435nn18, 30; death of, 329–30, 340f, 345, 434n10; myth of, 334, 338–43 *passim*

Scott and Amundsen (Huntford), 330, 341, 344f
Scott de Martinville, Leon, 263f
Scott's Last Expedition, 329
Second-order observer, 356
Second Villmerger War (1712), 96
Secular Ark, The (Browne), 127
Sedimentation pattern/coefficient, 300
Selection, 147f
Self-sacrifice, 336, 342f
Semiotics, 1, 58–61 *passim*, 283f
Sentence Commentaries (Fishacre), 72
Sentence Commentaries (Lombardus), 71
Serres, Michel, 326
Shakespeare, William, 16, 126f
Shannon, Claude, 79, 311
Shapin, Stevin, 61f
Shaw, George Bernard, 83
Sholes, Lilian, 86
Siemens, Werner, 89
Sigmund, Karl, 312
Signal points, trigonometric, 108–9
Signans, 5
Signatum, 4–5
Sign-deregulation, 359
Significant difference, 290
Signifier, 7ff, 74, 305, 354, 383
Signifying Nothing (Rotman), 13, 59, 303
Sign language, 84
Signs, 7, 13, 58, 63f, 68f, 91, 252
Sign-writing, 64
Simulation, 319f
Simulations (Baudrillard), 318
Slough, 187–88, 191
Smith, Adam, 59
Smith, Bernard, 228
Smith, Patti, 71
Smith, Robert, 199
Social change, fears of, 136
Social constructivism, 5, 286
Social Text, 2
Société de l'anthropologie, 277, 279
Société de l'Enseignment supérieur, 256
Société de linguistique de Paris, 249–58 *passim*, 261
Sokal, Alan, 2f
Sonderbund War (1847–48), 100
Soul, 151

South, James, 212ff
Spaces, transparency of, 93–97
Species, indigenous/interchanged, 138
Spectrograms, criticism of, 242
Speech, 5, 7, 90, 249, 263f, 272, 280, 283; visible, 257–69
Speech inscriptor, 273
Speech machine, 82
Spencer, Herbert, 219
Spirit–body unity, 353
Spiritual depth, material surface and, 354
Stabilization, 291
Staffordshire, 123f
Stallone, Sylvester, 349
Standards, 50, 284
Stanhope, Earl of, 189
Star clusters, 242, 245
Stars, models of, 195
Steen, Jan, 170
Stefansson, Vilhjalmur, 328–32 *passim*, 344, 434n10
Stenography, 264
Stephan, Heinrich von, 85–89 *passim*
Stevin, Simon, 59
Stokes, Terry, 427n13
Stoney, Bindon, 205, 220
Strange facts, 21, 26–32, 35–38
Strange phenomena, 25–28, 30–31, 32, 35–38
Structuralism, 16, 249
"Structuralism in Modern Linguistics" (Cassirer), 249
Structural Transformation of the Public Sphere (Habermas), 79
Structure and Distribution of Coral Reefs, The (Darwin), 128
Struggle, 144
Struve, Otto, 218–21 *passim*
Stumpf, Carl, 89
SubCode, 67
Subject, 63f, 65–69 *passim*; 70f, 356
Subject-experience, 355
Subjecthood, mathematical reasoning and, 55
Subjectivity: language and, 64; self-manifesting, 90
Subject/object paradigm, 353, 355, 360, 362

Sulloway, Frank, 121
Summa aurea (William of Auxerre), 72
Suppositio formalis, suppositio materialis and, 72
Surrealism, 362
Surveying, 92, 99, 101–6, 110, 113
Svevo, Italo, 362
Sweet, Henry, 83
Swift, Jonathan, 54
Swiss Federal State, cartography in, 92
Swiss Gymnastics Club, 99
Swiss National Exhibition, 116
Swiss Naturalist Society, 98
Symbols, 10, 74f
Symonds, Eliza Grace, 83
Systems of difference, 125

Tables of Proportions (Fournier), 49
Talbot, William Henry Fox, 141f, 144
Tarachow, telephone and, 88
Tasmania, 132
Tâtonnement, 291
Tattiriya-Pratisakhya, 421n66
Taurus, 32f
Technology, 12, 156, 185, 193, 206, 210, 222, 226, 310, 328, 353, 416n11 434n13; Gulf War and, 347–50; literary, 182ff, 195–202 *passim*, 222; media, 10, 352
Technoscience, 12, 304, 315
Technostructure, language of, 314, 432n36
Telegraphs, 76, 80f, 252, 284
Telephones, 76, 80–81, 86ff
Telescopes, 186, 204, 210–11, 212, 222
Tempest, The (Shakespeare), 127
Tennyson, Alfred Lord, 329, 334–35
TEX (Tau Epsilon Chi), 39
Theoretical Biology, 313
Theory, object and, 6
Theory of Island Biogeography (MacArthur and Wilson), 120
Things, 275; history of, 287
Thought, conservation of, 11
Thought experiment, 60–61
Tierra del Fuego, 137f
Tierra program, 322–27 *passim*
Timbre, 88, 266–67
Titian, 170

Toepler, 88
"Topographical Map of Switzerland," 107, 110, 114–15, 117
Topography, 388nn49, 55
Torniello, 375n11, 376n19
Tory, Geoffrey, 45, 375n11, 376n19
Tournachon, Adrien, 152, 163
Towards a Theoretical Biology (Longuet-Higgins), 308
Trace-articulations, 295
Transcendentalism, 314, 316
"Transgressing the Boundaries" (Sokal), 2
Transpositions: rules of, 106–13
Triangulation, 98f, 113, 299
"Triangulation primordiale de la Suisse," 112f
Trifid nebula, 237, 243
Truchet, Father Jean (Père Sébastien), 43–48 *passim*
Turing, Alan, 76f, 309, 430n14
Turnbull, David, 427n13
Twilight of the Idols (Nietzsche), 304, 307
"Two Ways of Life, The" (Rejlander), 171f
Typography, 44, 48–54 *passim*, 76

Über die Sprache und Weisheit der Indier (Schlegel), 254
Ulysses (Tennyson), 329, 334–35
Uniformitarianism, theory of, 136
Universal Exposition (1900), 279

Vaïsse, Leon, 258f, 420n58
Value judgments, 307
Van Diemen's Land, 132, 138
Vaucanson, 309, 431n22
Vedas, 260
Venus and Adonis (Titian), 170
Verdin, Charles, 273f, 423n101
Vergleichende Grammatik (Bopp), 253
Verini, Giovanni Baptista, 375n11
Vestiges of the Natural History of Creation (Chambers), 203, 206ff, 214ff
Vienna Academy of Sciences, 279
Viete, 74
Vietnam War, 346–47, 349
Vinson, Julien, 277, 279
Virilio, Paul, 3

Virtual witnessing, 62, 184
Visible Speech (Bell), 81, 83
Visual representations, 184, 228
Vitruvian ratio, 41
Vocalization, image of, 268
Voice, 88; psychogenetic, 90; women and,
    87, 90
Von Neumann, John, 76f
Von Orel, Eduard, 118
Vowel sounds, 263–64
Voyage of the Beagle, The (Darwin), 121,
    130–35 *passim*
Voyage of the Discovery, The (Scott), 338

Waddington, C. H., 89, 307, 311, 317,
    320, 321
Wagner, Karl Willy, 89
Wagner, Richard, 359
Walford, Edward, 156
Wallace, Alfred Russel, 120, 129
Wallich, George Charles, 398n35
Waser, Johann Heinrich, 97, 386n34
Watson, James, 304, 309
Weber, Eugen, 251
Weber, H. F., 88
Weber, Samuel, 369n36
Wedgwoods, 124, 131, 396n2
Werther (Goethe), 90
West, Benjamin, 151f
Western Electric, 89f
Westminster Review, 201, 219
What Is Life? (Schrödinger), 306
Wheatstone, Charles, 82
Whewell, William, 200–201, 215

White, Lynn, 226
Whitley, Charles Thomas, 142
"Why I Studied My Patois and How I
    Studied It" (Rousselot), 271
Widmann, Johann, 74
Wien, Max, 88
Wiener, Norbert, 309
Wild Child of Aveyron, 258
Wilhelm Meisters Wanderjahre (Goethe),
    134
Wilson, E. A., 436n37
Wilson, Edward, 339
Wilson, Edward O., 120
Winthrop, John, 34f
Wittgenstein, Ludwig, 56, 57–58, 286
Wolf, Joseph, 152, 160–61
Woman and Socialism (Bebel), 79
Wood, Thomas W., 152, 160
Woolgar, Steve, 17, 293, 296
World-experience/world-perception,
    356
Wright, C. L., 238; and Company, 235
Writing, 1, 5–6, 7ff, 15, 419n42
Wundt, Wilhelm, 421n71
Wurstemberger, Ludwig, 98, 100, 388n57

Xenotext, 302–3

Yamas, 260, 263, 422n75
Yerkes Observatory, 235ff
Young Grammarians, 276

Zamecnik, Paul, 288, 297, 300
Zero, 13, 59f, 67, 74f

Library of Congress Cataloging-in-Publication Data

Inscribing science : scientific texts and the materiality of
    communication / edited by Timothy Lenoir.
        p.   cm. — (Writing science)
    Includes bibliographical references and index.
    ISBN 0-8047-2776-7 (cloth). —
ISBN 0-8047-2777-5   (pbk.)
        1. Communication in science.
    2. Communication in science—Philosophy.
    I. Lenoir, Timothy.   II. Series.
    Q223.I497   1998
    501'.4—dc21   97-29112
                                                        CIP

⊛ This book is printed on acid-free, recycled paper.

Original printing 1998
Last figure below indicates year of this printing:
07   06   05   04   03   02   01   00   99   98